TABLE OF ATOMIC WEIGHTS

(Compared to the relative atomic mass of carbon-12 equaling exactly 12)

The atomic weight of many elements can change slightly, depending on the origin and treatment of the material. The footnotes to this table explain the types of variations to be expected for individual elements. The values of the atomic weight given here apply to elements as they exist naturally on earth and to certain artificial elements. With the footnotes taken into consideration, the atomic weights are considered reliable to ±3 in the last digit when followed by an asterisk (*). Values in parentheses are used for certain radioactive elements whose atomic weights cannot be stated precisely without knowing where the element came from; the value given is the mass number of the isotope of that element with the longest known half-life.

ALPHABETICAL ORDER OF ELEMENTS		ATOMIC NUMBER	ATOMIC WEIGHT
NAME	SYMBOL		
Actinium[a]	Ac	89	227.0278
Aluminum	Al	13	26.98154
Americium	Am	95	(243)
Antimony	Sb	51	121.75*
Argon[b,c]	Ar	18	39.948*
Arsenic	As	33	74.9216
Astatine	At	85	(210)
Barium[c]	Ba	56	137.33
Berkelium	Bk	97	(247)
Beryllium	Be	4	9.01218
Bismuth	Bi	83	208.9804
Boron[b,d]	B	5	10.81
Bromine	Br	35	79.904
Cadmium[c]	Cd	48	112.41
Calcium[c]	Ca	20	40.08
Californium	Cf	98	(251)
Carbon[b]	C	6	12.011
Cerium[c]	Ce	58	140.12
Cesium	Cs	55	132.9054
Chlorine	Cl	17	35.453
Chromium	Cr	24	51.996
Cobalt	Co	27	58.9332
Copper[b]	Cu	29	63.546*
Curium	Cm	96	(247)
Dysprosium	Dy	66	162.50*
Einsteinium	Es	99	(254)
Erbium	Er	68	167.26*
Europium[c]	Eu	63	151.96
Fermium	Fm	100	(257)
Fluorine	F	9	18.998403
Francium	Fr	87	(223)
Gadolinium[c]	Gd	64	157.25*
Gallium	Ga	31	69.72
Germanium	Ge	32	72.59*
Gold	Au	79	196.9665

[a]The atomic weight of these elements is that of the radioactive isotope of longest half-life.

[b]There are natural variations in the isotopic composition of these elements that prevent a more precise atomic weight from being given.

[c]Some samples of this element have an isotopic composition that varies a great deal from most of the samples found. The atomic weight of those few samples may be different than that given in the table.

[d]Commercially available samples of this element may have an atomic weight that differs from that given in the table. The difference may be due to purification methods or other industrial processes.

ALPHABETICAL ORDER OF ELEMENTS		ATOMIC NUMBER	ATOMIC WEIGHT	ALPHABETICAL ORDER OF ELEMENTS		ATOMIC NUMBER	ATOMIC WEIGHT
NAME	SYMBOL			NAME	SYMBOL		
Hafnium	Hf	72	178.49*	Promethium	Pm	61	(145)
Helium[c]	He	2	4.00260	Protactinium[a]	Pa	91	231.0359
Holmium	Ho	67	164.9304	Radium[a,c]	Ra	88	226.0254
Hydrogen[b]	H	1	1.0079	Radon	Rn	86	(222)
Indium[c]	In	49	114.82	Rhenium	Re	75	186.207
Iodine	I	53	126.9045	Rhodium	Rh	45	102.9055
Iridium	Ir	77	192.22*	Rubidium[c]	Rb	37	85.4678*
Iron	Fe	26	55.847*	Ruthenium[c]	Ru	44	101.07*
Krypton[c,d]	Kr	36	83.80	Samarium[c]	Sm	62	150.4
Lanthanum[c]	La	57	138.9055*	Scandium	Sc	21	44.9559
Lawrencium	Lr	103	(260)	Selenium	Se	34	78.96*
Lead[b,c]	Pb	82	207.2	Silicon	Si	14	28.0855*
Lithium[b-d]	Li	3	6.941*	Silver[c]	Ag	47	107.868
Lutetium	Lu	71	174.97	Sodium	Na	11	22.98977
Magnesium[c]	Mg	12	24.305	Strontium[c]	Sr	38	87.62
Manganese	Mn	25	54.9380	Sulfur[b]	S	16	32.06
Mendelevium	Md	101	(258)	Tantalum	Ta	73	180.9479*
Mercury	Hg	80	200.59*	Technetium	Tc	43	(97)
Molybdenum	Mo	42	95.94	Tellurium[c]	Te	52	127.60*
Neodymium[c]	Nd	60	144.24*	Terbium	Tb	65	158.9254
Neon[d]	Ne	10	20.179*	Thallium	Tl	81	204.37*
Neptunium[a]	Np	93	237.0482	Thorium[a,c]	Th	90	232.0381
Nickel	Ni	28	58.70	Thulium	Tm	69	168.9342
Niobium	Nb	41	92.9064	Tin	Sn	50	118.69*
Nitrogen	N	7	14.0067	Titanium	Ti	22	47.90*
Nobelium	No	102	(259)	Tungsten (Wolfram)	W	74	183.85*
Osmium[c]	Os	76	190.2	Uranium[c,d]	U	92	238.029
Oxygen[b]	O	8	15.9994*	Vanadium	V	23	50.9414*
Palladium[c]	Pd	46	106.4	Xenon[c,d]	Xe	54	131.30
Phosphorus	P	15	30.97376	Ytterbium	Yb	70	173.04*
Platinum	Pt	78	195.09*	Yttrium	Y	39	88.9059
Plutonium	Pu	94	(244)	Zinc	Zn	30	65.38
Polonium	Po	84	(209)	Zirconium[c]	Zr	40	91.22
Potassium	K	19	39.0983*				
Praseodymium	Pr	59	140.9077				

source: *Pure and Applied Chemistry*, Vol. 47, pp. 80–81, 1976. Headings and footnotes have been paraphrased from those in the original table.

PHYSICAL CONSTANTS

Avogadro's number	6.0221367×10^{23} particles/mol
Ideal gas constant	0.08206 atm·L/mol·K
Mass of electron	0.00054858 u
Mass of neutron	1.0086650 u
Mass of proton	1.0072765 u
Speed of light	2.997925×10^8 m/s

PREVIEW OF GENERAL CHEMISTRY

PREVIEW OF GENERAL CHEMISTRY

Harvey F. Carroll
Kingsborough Community College
The City University of New York

WILEY

JOHN WILEY & SONS
New York   Chichester   Brisbane
Toronto   Singapore

TO
LINDA
AND
ADAM

Cover: A computer generated model of the crystal structure of a superconductor. This material is superconducting up to about 90 K and contains
barium, yttrium, copper, and oxygen atoms. *IBM Photo.*

Library of Congress Cataloging in Publication Data:

Carroll, Harvey F.
Preview of general chemistry.

Includes index.
1. Chemistry.   I. Title.

QD31.2.C35 1989   540   88-5735
ISBN 0-471-86002-6 (pbk.)

Printed in the United States of America

10 9 8 7 6 5 4 3 2

Printed and bound by the Arcata Graphics Company

PREFACE

Preview of General Chemistry is designed to help students acquire the skills needed to succeed in general chemistry. It is assumed that they intend to take such a main sequence chemistry course.

Students will benefit most from this book if they have been exposed to arithmetic and some elementary algebra. Necessary topics from these subjects are reviewed in some detail as needed. No previous background in chemistry is assumed.

Many students find general chemistry difficult, to a large extent because of the abstraction and problem solving involved. I feel that if students come into a general chemistry course with good problem-solving techniques and some feeling for the abstract nature of chemistry, they can then concentrate on getting a deeper understanding of the subject. They won't get bogged down puzzling over scientific notation, significant figures, unit conversions, stoichiometry, and the simpler aspects of atomic theory.

This book purposely covers a limited number of topics in great detail, and the order of topics is intentionally somewhat different from that in most books. For instance, I believe that since our students are in a chemistry course, they should start learning about chemistry, not mathematics. Computation is important, but it is not an end in itself. After a student has painfully learned how to do, say, long division, are there exciting applications? More long division problems hardly seem exciting. Contrast this with learning to read. When a student learns to read, there are wonderful books to discover as the payoff. Therefore, I have tried to treat computation as a skill that can be applied to chemistry rather than as an end product. Doing chemistry is the reward for learning computation.

In keeping with this philosophy, the first chapter discusses atoms and isotopes. This material is not difficult, and mastery of it builds confidence.

The second chapter discusses atomic weight and shows how a weighted average is used. The calculations require a calculator. Calculator instructions are provided in Appendix 2-2 and are kept separate from the discussion so as not to distract from it. The derivation of the weighted average presents a gold mine of mathematical techniques involving fractions, percent, and decimals and will serve as a good review.

The third and fourth chapters cover balancing equations and elements, compounds, and mixtures. The fifth chapter begins a three-chapter sequence on mathematics—scientific notation, significant figures, and unit conversions. In the chapter on scientific notation, the calculator instructions are kept separate from the main discussion in the beginning sections. This arrangement will encourage the student to learn, without a calculator, how to manipulate numbers and keep track of the decimal point in scientific notation. There are no appendices covering mathematics at the back of the book—in a course of this kind, the math should be taught as an integral part of learning chemistry.

In Chapters 8 through 11, this mathematics is applied to stoichiometry. Chapters 12 through 15 cover gases, atomic structure, and chemical bonding. The book concludes with chemical nomenclature, oxidation numbers, redox equations, logarithms, an extensive glossary, and solutions to all problems.

If the instructor prefers to cover mathematics first, a suggested sequence of chapters is 5, 6, 7, 1, 2, 3, 4, 8, 9, ...

Nomenclature is deliberately covered before oxidation numbers. If time is limited, this allows nomenclature to be taught without covering oxidation numbers. Most general chemistry textbooks cover nomenclature in an early chapter, long before oxidation numbers are discussed.

Two unique features of this text are the detail with which topics are discussed and the extensive calculator instructions provided in chapters that use more than simple arithmetic. Students who use this book will benefit most if they own a scientific calculator, preferably a solar model. These are available for under $20 and should serve the students throughout their careers.

Other unique features of this book are

1. The discussion of binding energy in Chapter 2.
2. The discussion, in Chapter 14, of the reason why orbital degeneracy is removed in multielectron atoms.
3. The discussion, in Chapter 17, of the oxidation numbers of atoms in organic molecules using structural formulas.

This material is not usually covered at the level of this book. However, I feel that my discussion of binding energy may help the student understand why isotopic masses are noninteger. The discussion of removing orbital degeneracy helps explain why the orbitals arrange themselves in energy levels the way they do. The discussion of the oxidation numbers in organic molecules is fairly easy and introduces students to structural formulas in an interesting way. This material could be omitted without loss of continuity.

As far as possible, I have written the section headings as declarative sentences. Reading the section headings should give the student a better idea of the chapter's contents than would simple phrases.

The writing style is conversational and as simple and clear as I could make it. One of the principal reasons why my students like the book is that they find it very easy to read.

Depending on the length of the course and the needs of the students, it is likely that only part of the book can be covered in one semester. It is my opinion that the first nine chapters are crucial and should be taught if at all possible.

This book contains no formal sections on descriptive and applied chemistry. Although these topics are vital to the education of a chemist, I feel that in a preparatory course it is more important to concentrate on the skills needed for general chemistry. I have, however, presented numerous applications of chemistry in the examples and problems.

There are many worked examples in the text. In Chapters 7 through 9, I use the given (G) and asked for (AF) notation to help the student understand the problem. The problems at the end of each chapter are divided into two types. A *keyed problem* is almost identical to the example of the same number in the chapter. These problems are designed to allow students to mimic a problem-solving technique, thus increasing the probability of a correct solution. The *supplemental problems* cover all types of problems from the chapter. I feel students should work out all the keyed problems and a selection of supplemental problems.

Rockaway Park, New York   HARVEY F. CARROLL
June 1988

PHOTO CREDITS

ACKNOWLEDGMENTS

I wish to thank the hundreds of students who have used earlier versions of this book and made many useful suggestions.

Sir Rudolph E. Peierls read Chapter 13 and agreed that my treatment of the uncertainty principle was reasonable at this level. In addition, Professor Roald Hoffmann of Cornell University was kind enough to read Appendix 14-1 on orbital spacing. Since I had not seen an elementary discussion of this topic, his acknowledgment that my treatment is reasonable was very encouraging. I offer my sincere thanks to both distinguished scientists.

Professor Sidney Emerman of Kingsborough Community College has been my most severe critic while remaining a good friend. He has used earlier versions of this text, and his advice has always been of the highest quality.

I typed the manuscript using Peregrine Falcon's technical word processor, The EGG. This IBM-compatible word processor is easy to learn and use, and it displays equations and chemical symbols on screen in WYSIWYG. The index was prepared using Kensa Software's INSORT. Both programs are exceptional.

The staff at John Wiley are a group of highly competent professionals with whom it has been a pleasure to work. My editor, Dennis Sawicki, believed in this project and saw it through. Supervising copy editor Priscilla Todd, photo researchers Anne Manning and Stella Kupferberg, designer Dawn Stanley, production supervisor Dawn Reitz, and copy editor Robert Golden all made major contributions to the accuracy, readability, layout, and design of the book.

Finally, thanks to my colleagues who read the manuscript and made numerous useful suggestions. The book is much better because of the efforts of Stanley Asbaugh, Orange Coast College; Rufus Cox, Community College of Philadelphia; John Egger, Erie Community College; William Fateley, Kansas State University; Lindsay Foote, Western Michigan University; Hans Gunderson, Northern Arizona University; John Healy, Chabot College; William Huggins, Pensacola Junior College; James Petrich, San Antonio College; Howard Powell, Eastern Kentucky University; Donald Roach, Miami Dade Community College; C. V. Senoff, University of Guelph (Ontario); Donald Slavin, Community College of Philadelphia; and Mary Wadman, Northern Essex Community College.

H. F. C.

BRIEF CONTENTS

CONTENTS

TO THE READER

The only way to learn how to solve problems in chemistry is to solve a great number of problems. You cannot learn problem solving by reading a book or watching an instructor lecturing. Just as in learning how to drive, ski, ice skate, play chess, play a musical instrument, play baseball, or any other complex human activity, you acquire skill only through repeated practice, not by reading about an activity or watching other people do it.

Therefore, to get the most benefit from this book, you should work as many of the problems as possible. To help you in this endeavor, the problems at the end of each chapter (except Chapter 13) are divided into two parts. The *Keyed Problems* are "keyed" to the almost identical *Example* of the same number in the body of the chapter. The *Supplemental Problems* cover all the topics in the chapter in a random order.

You should try to solve all the *Keyed Problems* and at least a selection of the *Supplemental Problems*. You will probably learn most efficiently if you solve a *Keyed Problem* right after you read the corresponding *Example* in the chapter. When you've finished the chapter, try solving a selection of supplemental problems as assigned by your instructor.

One of the best ways to learn is to teach. So try to explain the solutions to the problems to others in the class when you are studying. But whatever your study techniques, solve a great number of problems.

As a study aid, solutions to all the end-of-chapter problems can be found at the back of the book. These are preceded by a glossary in which you can look up the definitions of terms.

1

ATOMS

Suppose that you took a small piece of iron and chopped it up as finely as you could. Then, somehow, you found someone who could chop it even more finely. The pieces became finer and finer and finer, finer than it is really possible to imagine. What would eventually happen?

At some point in the chopping, you would find that you couldn't chop any more and still have iron. This is because you already had the smallest particles of iron possible. These particles are called **atoms.** If you chopped up the iron atoms, they would no longer be iron atoms.

In this book we will discuss some of the properties and reactions of atoms. We will also look at the structure of atoms in some detail. Since atoms are very important in the study of chemistry, let's start our discussion with the structure of atoms.

1-1
ATOMS CONSIST OF A POSITIVE NUCLEUS
AND NEGATIVE ELECTRONS

One of the basic particles of matter is called the **atom.** The atom consists of two parts. In the center is a **nucleus,** which is small and has a positive electrical charge. The nucleus also contains almost all of the mass of the atom.

Mass is the inherent amount of material of an object. When the force of gravity pulls an object down, we say that the object has weight. An object always has the same mass, but its weight can vary. The weight depends on the gravity force. Thus the astronauts who went to the moon had the same mass as they had on earth, but they weighed only one-sixth of their weight

(Opposite) Lightning striking the Washington Monument. Lightning occurs when electrons are separated from atoms. Eventually, because the air cannot support this charge separation, there is a giant "spark."

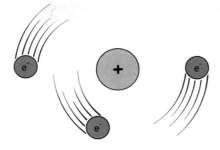

FIGURE 1-1 A very rough picture of the atom.

on earth. The gravity force on the moon is only one-sixth the gravity force on the earth. That's why they could jump around so easily.

The second part of the atom surrounds the nucleus and consists of one or more **electrons.** The way the electrons surround the nucleus will be discussed in detail in Chapter 13.

The electrons have a negative electrical charge and have very little mass. Compared to the mass of the nucleus, the electron is a real lightweight. Figure 1-1 shows a very rough picture of an atom. It will be modified later as you learn more.

In Figure 1-1 the circle with the plus sign stands for the nucleus with its positive charge. Each e⁻ stands for an electron. The minus sign on the e shows that the electron has a negative charge. This negative charge turns out to be the smallest unit of negative charge that can exist.

1-2
THE ATOM STAYS TOGETHER BECAUSE OF ELECTRICAL ATTRACTION

A very important rule of nature is that *opposite electrical charges attract each other* and *like electrical charges repel each other.* Therefore, the positive charge of the nucleus attracts the negative charge of the electron. It is this electrical attraction that holds the atom together.

You might ask why the nucleus doesn't attract the electron with so much force that the electron falls into the nucleus. One reason is that the electron is in constant motion "around" the nucleus. This constant motion "around" the nucleus *creates* a force called **centrifugal force.** The electrical attraction and the centrifugal force exactly balance each other, as is shown in Figure 1-2. Thus the electron doesn't fall into the nucleus.

You are probably familiar with centrifugal force. As pictured in Figure 1-3, if you tie a rock to a string and whirl it around your head, the centrifugal force on the rock keeps the string pulled tight.

If you drive around a corner too fast, the centrifugal force can become greater than the force holding the tires on the road, and the car skids or even

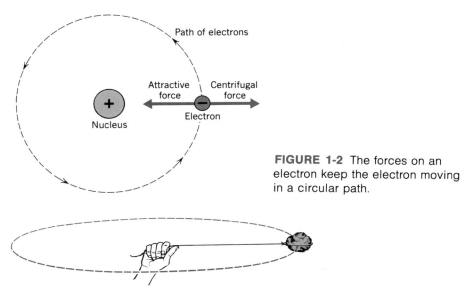

FIGURE 1-2 The forces on an electron keep the electron moving in a circular path.

FIGURE 1-3 The rock moves in a circular path because of the forces on it.

turns over. A centrifugal force is generated *only* when you change the direction of the motion of an object.

1-3
THE PROPERTIES OF ELECTRONS, PROTONS, AND NEUTRONS ARE RELATED TO THEIR MASS AND CHARGE

Although each isolated electron is small, when electrons surround the nucleus they take up a lot of space. For instance, if the nucleus were the size of a basketball, the electrons would surround the nucleus up to 10 miles from it. This is just an example to give you a feeling for the relative size; but remember, the atom is *very, very* small.

Now we will go back and look at what makes up the nucleus. The nucleus of an atom contains **protons** and **neutrons.** The exception is the simplest nucleus, which is just a proton. The proton has one unit of positive electrical charge. This positive electrical charge is equal to the negative charge on the electron outside the nucleus but has the opposite sign. The proton and the neutron are very heavy compared to the electron. However, the neutron has *no* electrical charge. It is neutral, which is why it is called a neutron.

The proton and the neutron have very nearly the same mass. Since these particles are so basic to the structure of the atom, they each are assigned a mass of almost exactly **one atomic mass unit.** (The reason that their mass is not exactly one atomic mass unit will have to wait until we define the atomic mass unit in Chapter 2.) The abbreviation for atomic mass unit is u. The electron, on the other hand, is very light and has a mass of only $1/1823$ u. Thus a proton or a neutron is almost 2000 times heavier than an electron.

Since the space that the atom takes up is mostly filled with very light electrons, it is interesting to think what matter would be like if we could collect a sample of just the nuclei alone. If we could do this, and the nuclei were touching each other like marbles in a jar, we would have a material that is very dense. In fact, a piece of it the size of the ball of a fine-point ball point pen would weigh around as much as the ocean liner Queen Elizabeth 2. There are actually some stars where this compression of matter is thought to have taken place. It is believed that the electrons of the atoms in these "neutron" stars are forced into the protons of the nuclei to make neutrons; this gets around the problem of the positive nuclei repelling each other (like charges repel each other). The forces needed to compress matter like this are tremendous and could occur only in large stars as a result of the fantastic gravitational force that they exert. If a star is big enough, it may become a neutron star as a stage in its death process. Our star, the sun, will not become a neutron star when it dies in five or six billion years; it is too small.

Table 1-1 gives a summary of the properties of the proton, neutron, and electron.

Most of the time matter is electrically neutral. The reason for this is that atoms have the same number of protons *in* the nucleus as they have electrons *surrounding* the nucleus. Since the positive charge on the proton is the same size as the negative charge on the electron, the two charges cancel each other and the atom is neutral. When an atom isn't neutral, it is because some of the electrons have been removed, leaving more protons than electrons. Thus there are more positive charges than negative ones, and the atom as a whole is positive.

TABLE 1-1

A SUMMARY OF THE PROPERTIES OF THE PROTON, NEUTRON, AND ELECTRON

PARTICLE	APPROXIMATE ATOMIC MASS UNITS (EXACT VALUES IN PARENTHESES)	CHARGE	LOCATION IN ATOM	AMOUNT OF SPACE THEY OCCUPY IN THE ATOM
Proton	1 (1.0072765)	+1	Nucleus	A little
Neutron	1 (1.0086650)	0	Nucleus	A little
Electron	$1/1823$ (0.00054858)	−1	Space around the nucleus	A lot

The torn-off electrons can, among other other things, make a spark of electricity or an electric current. This separation of charge occurs only under special conditions and is obvious when you turn on a television set, rub your feet along a carpet and then touch a metal object, or see a flash of lightning.

To complete the story, we should mention that extra electrons can also add to some atoms, making them negative. We will discuss positively and negatively charged atoms in Chapter 15.

The appendix at the end of this chapter gives a very brief survey of some of the problems scientists have had in figuring out the structure of the atom. Chapters 13 and 14 discuss atomic structure in more detail.

1-4
THERE ARE THREE KINDS OF HYDROGEN ATOMS: PROTIUM, DEUTERIUM, AND TRITIUM

We can now start to describe some real atoms. The simplest possible arrangement is an atom made of one proton in the nucleus and one electron around the nucleus. A rough diagram of this atom is shown in Figure 1-4. The $1p^+$ represents the one proton (which has a single positive charge) in the nucleus. The $1e^-$ shows that the atom has one electron. The half-circle in the diagram separating the 1 and the e^- represents the electron surrounding the nucleus. The reason that the electron is not shown going around the nucleus in a full circle is that scientists do not believe that it goes around in a circle. Just think of it as surrounding the nucleus in some manner.

This simple atom, consisting of one proton and one electron, is called a **hydrogen atom.** It has a mass of about 1 u, which comes almost entirely from the proton. The mass of the electron is so small that we can ignore it for now.

Each of the different kinds of atoms that exist is given a name, a symbol that stands for the name, and a number. For example, the name of the atom described above is hydrogen; the symbol for a hydrogen atom is H. The number is called the **atomic number** and represents the number of protons in the nucleus. Since hydrogen has one proton in the nucleus, its atomic number is 1. Atomic numbers for all the different kinds of atoms are listed in Tables 1-2 and 2-2.

Please note that the atom with one proton and one electron is also called **protium,** to distinguish it from the atoms of deuterium and tritium, which are discussed below.

FIGURE 1-4 A diagram of the simplest atom.

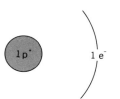

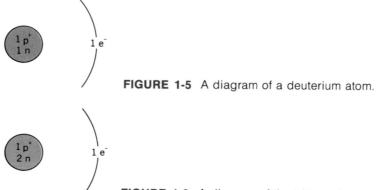

FIGURE 1-5 A diagram of a deuterium atom.

FIGURE 1-6 A diagram of the tritium atom.

The next simplest atom has one proton and one neutron in the nucleus and one electron surrounding the nucleus. (Remember that for neutral atoms, the number of protons equals the number of electrons.) It has a mass of about 2 u, one each from the proton and the neutron. It is called **deuterium** and is usually given the symbol D. The atomic number of deuterium is also 1, since there is one proton in the nucleus. To sketch a picture of the deuterium nucleus (Figure 1-5), put a 1n under $1p^+$. This 1n stands for the one neutron.

With one proton and two neutrons in the nucleus and one electron outside, we have **tritium,** which is usually given the symbol T. The tritium atom has a mass of about 3 u (1 proton + 2 neutrons). The atomic number of tritium is also 1. A diagram of tritium is shown in Figure 1-6.

The process of adding another neutron to the nucleus has not been successful: If we try to make a nucleus with one proton and three neutrons, we cannot; they do not stick together. So after tritium there must be two protons in the nucleus as a minimum. But before we continue, let's go back and look at protium, deuterium, and tritium a little more closely.

1-5
THE THREE ISOTOPES OF HYDROGEN
ARE THE SIMPLEST ELEMENTS

Protium, deuterium, and tritium atoms all have the same number of protons and electrons but a different number of neutrons in the nucleus. The atomic number of all three atoms is the same, namely, 1. Atoms like this are called isotopes. **Isotopes** are atoms with the same number of protons (and, of course, electrons) but different numbers of neutrons.

Chemical properties are determined primarily by the electrons that surround the nucleus and depend only a little on the mass of the nucleus. Since the number of electrons is the same in each atom of protium, deuterium, and tritium, they have very similar chemical properties. The atoms

protium, deuterium, and tritium are spoken of as *isotopes of the element that has one proton in its nucleus.*

An **element** is all atoms with a given atomic number, or number of protons in the nucleus. If there is only one proton in the nucleus of an atom, the element is called **hydrogen.** The neutrons only change the mass of the atom but do not change which element it is. The atoms of an element are the smallest particles in nature that have a "chemical" identity. If you chop up these atoms, you have protons, neutrons, and electrons. These particles do not have a chemical identity.

Of the hundred or so known elements, only hydrogen (with one proton in its nucleus) has isotopes that are given special names. The special names are used because these isotopes are so useful in atomic research. When a chemist says "hydrogen," the reference is *either* to the naturally occurring mixture of isotopes of the element with atomic number 1 *or* to the lightest isotope of that element (namely, protium). You should also know that some people call deuterium "heavy hydrogen" and tritium "very heavy hydrogen." Historically, the element hydrogen was named long before anybody knew about isotopes or, for that matter, protons, neutrons, and electrons.

The word "isotope" is used in two subtly different ways. For instance, you may have heard the expression "radioactive isotopes can be dangerous." In this use, no particular element is referred to, just various nuclei that happen to be radioactive. Nuclei that are unstable and fly apart are said to be **radioactive.** Second, as we have been discussing with hydrogen, the word isotope can refer to different nuclei of the same element. We have been speaking of the three isotopes of the element hydrogen. Later we will discuss the isotopes of other elements.

Naturally occurring hydrogen consists mostly of the light isotope protium, the one we usually call "hydrogen." On earth, for about every 5000 protium atoms, there is one deuterium atom. There is also an extremely small amount of tritium in nature, although large quantities of tritium have been made in nuclear reactors and are very useful in research.

The reason that hardly any tritium exists in nature is that its nucleus is unstable and flies apart. Tritium is radioactive and is the lightest radioactive isotope that exists. Since half the tritium atoms in any sample would decay

The Browns Ferry Nuclear Plant near Athens, Alabama. It is capable of generating 3.5 million kilowatts of electric power.

(**decay** is the technical word for radioactive decomposition) in about 12 years and the world is billions of years old, you might wonder why there is *any* tritium at all in nature. The reason is that a very little bit is always being made in the upper atmosphere by cosmic rays. Scientists can make tritium for experimental use by putting lithium metal in a nuclear reactor. The neutrons that are produced by the reactor react with the lithium to produce tritium.

1-6
THE NOTATION FOR WRITING ISOTOPES TELLS US THE NUMBER OF PROTONS AND NEUTRONS IN THE NUCLEUS

There is a notation for atoms that makes it very easy to tell what element and what isotope we are dealing with. Figure 1-7 shows diagrams of the three isotopes of hydrogen. Consider protium, H.

If we write the number of protons in the nucleus in the bottom left corner of the H symbol (i.e., $_1$H) and write the total number of protons *plus* neutrons (for this one isotope there are no neutrons) in the top left corner (i.e., ^{1}H) and combine both, we have $_1^1$H. This is the **isotopic symbol** for the

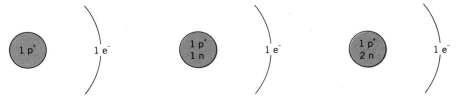

FIGURE 1-7 Diagrams of the three isotopes of hydrogen—protium, deuterium, and tritium.

lightest isotope of hydrogen. Since the number of electrons in the neutral atom equals the number of protons, the bottom number also tells us the number of electrons.

Now let's look at deuterium in Figure 1-7. The isotopic symbol would be $_1^2D$, where the 2 represents the sum of 1 proton + 1 neutron.

A diagram of tritium is also shown in Figure 1-7. Tritium has the isotopic symbol $_1^3T$, where the 3 represents the sum of 1 proton + 2 neutrons.

Sometimes you may see $_1^2D$ and $_1^3T$ written as $_1^2H$ and $_1^3H$. This is because some authors do not use the symbols D and T for deuterium and tritium.

Let's summarize the notation for isotopes. If E stands for the symbol of any element, then the isotopic symbol for the element E would written in the following way:

$$_{\text{number of protons}}^{\text{number of protons + number of neutrons}}E$$

EXAMPLE 1   The most common isotope of the element oxygen has 8 protons and 8 neutrons in the nucleus. The symbol for oxygen is O. Write the isotopic symbol for this isotope.

Solution: The isotopic symbol is $_8^{16}O$, where 8 protons + 8 neutrons = 16. ■

1-7
THE ATOMIC NUMBER AND THE MASS NUMBER ARE USED IN WRITING ISOTOPIC SYMBOLS

There are special names given to the numbers that we wrote in front of the letter symbols for the isotopes. As you already know, the bottom number is called the **atomic number** and is the number of protons in the nucleus. The top number is called the **mass number** of the isotope and is the sum of the protons and neutrons. Since the mass of each proton and neutron is very close to 1 u, the mass number gives us an approximate total mass of the atom (again ignoring those light electrons). Referring to our symbol representing any element, E, we can write

$$\begin{matrix} \text{mass number} \\ \text{atomic number} \end{matrix} E$$

A reminder: *The actual mass of an atom in atomic mass units (u) is not the same as the mass number.*

EXAMPLE 2   What is the atomic number and the mass number of the oxygen isotope $^{16}_{8}O$?

Solution: The atomic number is 8, and the mass number is 16. ■

EXAMPLE 3   If an atom has a mass number of 25 and an atomic number of 12, how many neutrons are in the nucleus? Also, how many electrons surround the nucleus?

Solution: Since the mass number is equal to the protons + neutrons, we can write the simple equation

mass number = number of protons + number of neutrons

or

number of neutrons = mass number − number of protons

Substituting numbers from this example, we can write

number of neutrons = 25 − 12 = 13

Since the number of electrons equals the number of protons, the atom has 12 electrons. ■

1-8
ATOMS OF ELEMENTS HEAVIER THAN HYDROGEN HAVE MORE PROTONS, NEUTRONS, AND ELECTRONS

Now we can examine atoms with more than one proton in the nucleus. A nucleus with just two protons cannot exist. Both protons have a positive

charge, but positive charges repel each other. A neutron is needed to stabilize two protons so that they stick together. Neutrons act as a sort of nuclear
"glue" to hold the nucleus together. However, the story is much more
complicated than just calling neutrons nuclear "glue"; too many neutrons
can actually make a nucleus unstable.

The element that contains two protons in the nucleus is helium (symbol
He). Two isotopes of helium exist in nature. The first has two protons and
one neutron; it has a mass number of 3. This isotope is very rare in nature.

EXAMPLE 4 Sketch a diagram and write down the isotopic symbol for the
helium isotope with a mass number of 3.

Solution: The diagram and symbol are shown in Figure 1-8.   ■

FIGURE 1-8 The diagrams and symbols
of the two isotopes of helium.

The second isotope of helium has two protons and two neutrons in the
nucleus; its mass number is 4. This is by far the most abundant isotope of He
in nature. The diagram and isotope symbol of the atom is also shown in
Figure 1-8.

If more neutrons are added to a $^4_2$He nucleus, it becomes unstable and
flies apart.

For light nuclei, a combination of about half neutrons and half protons
results in a stable nucleus. For heavier nuclei, the number of neutrons is
larger than the number of protons.

After helium, the next stable nucleus contains three protons. The element with three protons in the nucleus is lithium (symbol Li). In nature,
lithium exists in two isotopic forms: One form has 3 neutrons in the nucleus,
and the other has 4 neutrons in the nucleus.

EXAMPLE 5 Sketch both isotopes of lithium and write down their isotopic
symbols.

Solution: The solution is shown in Figure 1-9.   ■

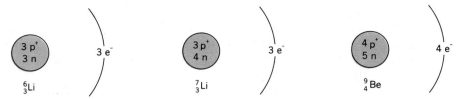

FIGURE 1-9 The diagrams and symbols of the naturally occurring isotopes of lithium and beryllium.

Isotopes are sometimes referred to with the mass number following the name, such as lithium-6 and lithium-7. When naming isotopes by this method, the element name is not abbreviated. Thus writing Li-6 for lithium-6 is not correct. $^3_2$He would be called helium-3, $^4_2$He would be called helium-4, and $^2_1$H would be called hydrogen-2 (most people would call it deuterium).

After lithium comes beryllium, which has the symbol Be. Beryllium has only one naturally occurring isotope, which has a mass number of 9. The symbol and diagram are shown in Figure 1-9.

More and more protons and neutrons can be added (along with the appropriate number of electrons), giving many more elements and their isotopes. More than 105 different elements are now known. Of this number, about 300 isotopes of the various elements occur in nature. Some of these isotopes are radioactive, but most are not. In addition to the naturally occurring isotopes, nuclear scientists have made over 1300 artificial isotopes. These are all radioactive and do not occur in nature.

Table 1-2 lists the elements, along with their naturally occurring isotopes. A few of the elements have no stable isotopes; these elements do not exist in nature (at least not on the earth) but have been made by scientists. Such elements are indicated in the table by a dash in the column called "Natural isotopes (mass numbers)."

Usually, isotopes that occur in nature are not radioactive, but if they are radioactive they decay very slowly. The reason is that the elements (i.e., the ones we know) were all formed in some manner long before our star, the sun, and its planets were made. These elements then became part of our sun and its planets. Since our earth is about 4.5 billion years old, any elements that existed at the beginning of the earth must also be at least 4.5 billion years old. Any isotope that decayed in a much shorter time than 4.5 billion years would simply have disappeared by now.

A few isotopes *are* younger than the earth. This is because they have been formed since the earth was made. There are two ways that this can happen. The first way is that these young isotopes are formed from the radioactive decay of other isotopes. The second way is that isotopes are made by cosmic rays hitting the earth's atmosphere. Examples of isotopes that formed by means of the first way are radioactive radium-226 and a stable isotope of lead, namely, lead-206. Both of these elements come from the

TABLE 1-2

THE NATURAL ISOTOPES OF THE ELEMENTS[a]

ATOMIC NUMBER	SYMBOL	NATURAL ISOTOPES (MASS NUMBERS)
1	H	1, 2
2	He	4, 3
3	Li	7, 6
4	Be	9
5	B	11, 10
6	C	12, 13
7	N	14, 15
8	O	16, 18, 17
9	F	19
10	Ne	20, 22, 21
11	Na	23
12	Mg	24, 25, 26
13	Al	27
14	Si	28, 29, 30
15	P	31
16	S	32, 34, 33, 36
17	Cl	35, 37
18	Ar	40, 36, 38
19	K	39, 41, 40
20	Ca	40, 44, 42, 48, 43, 46
21	Sc	45
22	Ti	48, 46, 47, 49, 50
23	V	51, 50
24	Cr	52, 53, 50, 54
25	Mn	55
26	Fe	56, 54, 57, 58
27	Co	59
28	Ni	58, 60, 62, 61, 64
29	Cu	63, 65
30	Zn	64, 66, 68, 67, 70
31	Ga	69, 71
32	Ge	74, 72, 70, 73, 76
33	As	75
34	Se	80, 78, 76, 82, 77, 74

(*Table 1-2 continues on next page*)

TABLE 1-2 *(continued)*

THE NATURAL ISOTOPES OF THE ELEMENTS[a]

ATOMIC NUMBER	SYMBOL	NATURAL ISOTOPES (MASS NUMBERS)
35	Br	79, 81
36	Kr	84, 86, 82, 83, 80, 78
37	Rb	85, 87
38	Sr	88, 86, 87, 84
39	Y	89
40	Zr	90, 94, 92, 91, 96
41	Nb	93
42	Mo	98, 96, 95, 92, 94, 97, 100
43	Tc	—
44	Ru	102, 104, 101, 99, 100, 96, 98
45	Rh	103
46	Pd	106, 108, 105, 110, 104, 102
47	Ag	107, 109
48	Cd	114, 112, 111, 110, 113, 116, 106, 108
49	In	115, 113
50	Sn	120, 118, 116, 119, 117, 124, 122, 112, 114, 115
51	Sb	121, 123
52	Te	130, 128, 126, 125, 124, 122, 123, 120
53	I	127
54	Xe	132, 129, 131, 134, 136, 130, 128, 124, 126
55	Cs	133
56	Ba	138, 137, 136, 135, 134, 130, 132
57	La	139, 138
58	Ce	140, 142, 138, 136
59	Pr	141
60	Nd	142, 144, 146, 143, 145, 148, 150
61	Pm	—
62	Sm	152, 154, 147, 149, 148, 150, 144
63	Eu	153, 151
64	Gd	158, 160, 156, 157, 155, 154, 152
65	Tb	159
66	Dy	164, 162, 163, 161, 158, 160
67	Ho	165
68	Er	166, 168, 167, 170, 164, 162

TABLE 1-2 *(continued)*

THE NATURAL ISOTOPES OF THE ELEMENTS[a]

ATOMIC NUMBER	SYMBOL	NATURAL ISOTOPES (MASS NUMBERS)
69	Tm	169
70	Yb	174, 172, 173, 171, 176, 170, 168
71	Lu	175, 176
72	Hf	180, 178, 177, 179, 176, 174
73	Ta	181
74	W	184, 186, 182, 180
75	Re	187, 185
76	Os	192, 190, 189, 188, 187, 186, 184
77	Ir	193, 191
78	Pt	195, 194, 196, 198, 192, 190
79	Au	197
80	Hg	202, 200, 199, 201, 198, 204, 196
81	Tl	205, 203, 206, 207, 208, 210
82	Pb	208, 206, 207, 204, 210, 211, 212, 214
83	Bi	209, 210, 211, 212, 214
84	Po	210, 211, 212, 214, 215, 216, 218
85	At	218, 215
86	Rn	222, 220, 219
87	Fr	223
88	Ra	223, 224, 226, 228
89	Ac	227, 228
90	Th	227, 228, 230, 231, 232, 234
91	Pa	231, 234
92	U	238, 235, 234

[a]The isotopes of each element are listed in the order of abundance (the most abundant being given first). A dash (—) signifies that the element has no natural isotopes.

radioactive decay of uranium-238. Examples of isotopes that formed by means of the second way are tritium (which has already been mentioned in Section 1-5) and a radioactive isotope of carbon, namely, carbon-14.

EXAMPLE 6   Referring to Table 1-2, write the diagrams and the symbols for the three isotopes of uranium, element number 92.

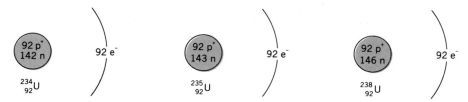

FIGURE 1-10 The diagrams and symbols for the naturally occurring isotopes of uranium.

Solution: The solution is shown in Figure 1-10. ■

APPENDIX 1-1
A BRIEF HISTORY OF THE STRUCTURE OF THE ATOM

The electron was discovered in 1897 by the British physicist J. J. Thomson (1856–1940). Some positively charged particles associated with atoms were also discovered at about the same time. These discoveries led to the following question: How are the negative electrons and the positive particles arranged to form an atom?

We now believe that the electrons surround a small, heavy, positively charged nucleus. The positive charge of the nucleus attracts the negative charge of the electrons; this attraction is what holds the atom together. But around 1900 the situation wasn't that clear. Scientists of that era knew that when electrons traveled in a curved path, such as circles, they would radiate (give off) energy. This was a basic rule of the physics at that time. This energy loss would quickly cause the electrons to "spiral" into the nucleus, as an earth satellite will do when it loses energy in the earth's atmosphere. Scientists estimated that if electrons were really outside the nucleus, all the atoms in the universe would collapse in about one second. Therefore, the scientists of 1900 felt that it was impossible for electrons to be outside the nucleus.

To try to answer these problems, J. J. Thomson proposed a model of the atom that had the electrons embedded in a large blob of nuclear "goo." At the time, no one really thought that this was a very good model, but nobody could think of anything better.

Then, in 1911, after a long series of experiments, Ernest Rutherford (1871–1937), a New Zealander working in England, found that the atom really did have a small, heavy positive nucleus with the electrons outside the nucleus. But this model, called the **planetary model,** couldn't be explained by the laws of physics of 1911.

In 1913, the Danish physicist, Niels Bohr (1885–1962), proposed a way out of the dilemma. He said that the electrons didn't spiral into the nucleus because they *couldn't.* He assumed that *new* laws of physics had to be used to explain the structure of the atom. His revolutionary ideas agreed with certain experiments and were the start of a new theory of the atom, called **quantum theory.** This theory was developed in its present form in the 1920s by many people. Our present description of the atom is based on the results of quantum theory. We will briefly discuss some of its ideas and applications to chemistry in Chapters 13 through 15.

PROBLEMS[1]

KEYED PROBLEMS

1. An isotope of rubidium has 37 protons and 48 neutrons in the nucleus. The symbol for rubidium is Rb. Write the isotopic symbol for rubidium.

2. What is the atomic number and the mass number of the gold isotope $^{197}_{79}Au$?

3. If an atom has a mass number of 41 and an atomic number of 19, how many neutrons are in the nucleus. Also, how many electrons surround the nucleus?

4. Sketch a diagram and write down the isotopic symbol for the fluorine isotope with a mass number of 19. The atomic number of fluorine is 9.

5. The atomic number of nitrogen is 7. Nitrogen has two naturally occurring isotopes with mass numbers 14 and 15. Sketch both isotopes of nitrogen and write down their isotopic symbols.

6. Referring to Table 1-2, write the diagrams and the symbols for the three isotopes of radon, element number 86.

SUPPLEMENTAL PROBLEMS

7. Define the following terms.
   a. atomic number   d. isotope   g. tritium
   b. mass number   e. radioactive
   c. element   f. deuterium

8. Sketch the diagram and write the isotopic symbol for all the isotopes of the following elements (refer to Table 1-2).
   a. Ne   d. Cu   g. I   j. Bi
   b. S   e. Se   h. Eu
   c. Ti   f. Y   i. Ir

For Problems 9 and 10, refer to Table 2-2 for the information you will need.

9. Using very fast chemical separation techniques, a team of chemists at Lawrence Berkeley Laboratories has observed the chemical properties of ele-

---

[1]Solutions to all problems can be found at the end of the book.

ment 105, namely, hahnium-262. Dr. Darleane Hoffman, a team leader, bombarded berkelium-249 with oxygen-18 atoms to produce hahnium, which decays in a few minutes. Sketch the diagram and write the isotopic symbol for each of the three isotopes used in the experiment. (The chemical symbol for hahnium is Ha.)

10. Workers at Lawrence Livermore National Laboratory have recently made some long-lived isotopes of lawrencium. The new isotopes, namely, lawrencium-261 and lawrencium-262, were made by bombarding einsteinium-254 with neon-22 in a cyclotron. Sketch the diagram and write the isotopic symbol for these isotopes.

11. Fill in the blank spaces in the following chart.

| ISOTOPIC SYMBOL | NUMBER OF | | | ATOMIC NUMBER | MASS NUMBER |
	PROTONS	NEUTRONS	ELECTRONS		
$^{24}_{12}\text{Mg}$					
	15	16			
				20	40
			17		37
$_{18}\text{Ar}$		22			
		40		32	

2

RELATIVE ATOMIC WEIGHT

In our discussion so far, we have referred to the mass number of an atom. As you know, the mass number is the sum of the number of protons and neutrons in the nucleus. But the mass number is *not* the exact mass of an atom. The following discussion will explain why this is true and then introduce the concept of the relative atomic weight.

2-1
THE ATOMIC MASS UNIT IS A UNIT USED
TO MEASURE THE MASS OF AN ATOM

Scientists find it necessary to know the exact mass of an atom. But because they cannot actually weigh an individual atom, they have had to come up with the concept of relative masses of atoms. Of course, to talk about relative masses we have to have a standard with which to compare things. The standard of comparison that scientists use for atoms is one of the isotopes of carbon, namely $^{12}_{6}C$. An atom of carbon-12 is assigned a mass of *exactly* 12 atomic mass units (u). When we say that an atom of carbon-12 is assigned a mass of 12 u, we are including both the nucleus and the electrons.

In our discussion in Chapter 1, we said that the mass of a proton or a neutron is *about,* but not exactly, 1 u. The exact definition of an atomic mass unit is as follows. **One atomic mass unit** is exactly one-twelfth the mass of an atom of one of the isotopes of carbon, namely, carbon-12. On the basis of this definition, the mass of any proton is found, by experiment, to be 1.0072765 u. The mass of any neutron is found to be 1.0086650 u. The reason for the masses not equaling exactly 1 u will be discussed in Section 2-5, when we discuss binding energy.

Now let's discuss the experimentally determined masses of two well-known atoms. Consider $^{16}_{8}O$, the most common isotope of oxygen. The mass

(Opposite) A nuclear test in the atmosphere. The energy released from nuclear fission, fusion, or both is the power source for such a device. We can hope that in the future this enormous amount of energy will be used for peaceful purposes, such as power generation.

of oxygen-16 is experimentally found to be 15.99491 u. Consider sodium (symbol Na), which has only one stable isotope, namely, $^{23}_{11}$Na. The mass of sodium-23 is found, by experiment, to be 22.98977 u. The reason that you cannot determine an atom's exact mass by simply adding up the masses of its protons, neutrons, and electrons will also be discussed in Section 2-5, when we talk about binding energy.

2-2
THE RELATIVE ATOMIC WEIGHT TAKES INTO ACCOUNT ALL
THE NATURALLY OCCURRING ISOTOPES OF AN ELEMENT

From our discussion in Chapter 1, we have seen that most elements have more than one naturally occurring isotope. If you look at Table 1-2, you will see that the element with the most naturally occurring isotopes is tin (symbol Sn, atomic number 50); tin has *ten* naturally occurring isotopes.

The naturally occurring isotopes of an element do not usually occur in equal amounts. The average mass of all the naturally occurring isotopes that make up an element is called the **relative atomic weight** of that element. The word "relative" is used because all the masses are compared to (or related to) carbon-12. The unit of relative atomic weight is the atomic mass unit (u).

The next two sections will show you how to compute the relative atomic weight of an element.

2-3
THE WEIGHTED AVERAGE GIVES MORE WEIGHT
TO NUMBERS THAT APPEAR MORE OFTEN

In the previous section, the relative atomic weight was defined as the average mass of all the naturally occurring isotopes that make up an element. The way that this average mass is calculated is the subject of this and the following section.

First, let's review how to calculate the usual average with which you are familiar. As an example, consider how you would average the weight of two people who weigh 120 pounds and 130 pounds. (The abbreviation for pound is lb.) The average weight is the sum of the two weights divided by 2:

$$\begin{array}{r} 120 \text{ lb} \\ + 130 \text{ lb} \\ \hline 250 \text{ lb} \end{array} \qquad \frac{250 \text{ lb}}{2} = 125 \text{ lb}$$

As one equation, this becomes

$$\frac{120 \text{ lb} + 130 \text{ lb}}{2} = \frac{250 \text{ lb}}{2} = 125 \text{ lb}$$

Another example would be to average the weights of six people. Assume the weights are 120, 120, 130, 130, 130, and 150 lb. The average is

$$
\begin{array}{r}
120 \text{ lb} \\
120 \text{ lb} \\
130 \text{ lb} \\
130 \text{ lb} \\
130 \text{ lb} \\
+150 \text{ lb} \\
\hline
780 \text{ lb}
\end{array}
\qquad
\frac{780 \text{ lb}}{6} = 130 \text{ lb}
$$

Or, in one equation, we have

$$
\frac{120 \text{ lb} + 120 \text{ lb} + 130 \text{ lb} + 130 \text{ lb} + 130 \text{ lb} + 150 \text{ lb}}{6} = \frac{780 \text{ lb}}{6} = 130 \text{ lb}
$$

There is another way of writing

$$
120 + 120 + 130 + 130 + 130 + 150.
$$

That way is

$$
(2)(120) + (3)(130) + (1)(150)
$$

where (2)(120) is the same as 2×120, (3)(130) is the same as 3×130, and (1)(150) is the same as 1×150. Thus our expression for the average weight can be written as

$$
\frac{(2)(120) + (3)(130) + (1)(150)}{6}
$$

Applying the distributive law of algebra (see Appendix 2-1 at the end of this chapter) to this expression, we get

$$
\frac{(2)(120) + (3)(130) + (1)(150)}{6} = \frac{(2)(120)}{6} + \frac{(3)(130)}{6} + \frac{(1)(150)}{6}
$$

This can be rearranged slightly to give

$$
(\tfrac{2}{6})(120) + (\tfrac{3}{6})(130) + (\tfrac{1}{6})(150)
$$

Let's now do the division of each fraction to get a decimal.

$$
\tfrac{2}{6} = \tfrac{1}{3} = 0.333 \qquad \tfrac{3}{6} = \tfrac{1}{2} = 0.500 \qquad \tfrac{1}{6} = 0.167
$$

Our expression for the average weight thus becomes

$$
(0.333)(120) + (0.500)(130) + (0.167)(150)
$$

Thus we have shown that

$$\frac{120 + 120 + 130 + 130 + 130 + 150}{6}$$

$$= (0.333)(120) + (0.500)(130) + (0.167)(150)$$

Since the left side of this equation gives the average weight of the six people, the right side must also give the average weight of the six people, which is 130 lb.

Let's go back to the original group of six people. Their weights were 120 lb, 120 lb, 130 lb, 130 lb, 130 lb, and 150 lb. Note that 2 of the 6 people weighed 120 lb, 3 of the 6 weighed 130 lb, and 1 of the 6 weighed 150 lb. The *percentage* of the people who had each weight is given in the following table.

WEIGHT	FRACTION	DECIMAL	PERCENTAGE
120 lb	$\frac{2}{6}$	0.333	33.3%
130 lb	$\frac{3}{6}$	0.500	50.0%
150 lb	$\frac{1}{6}$	0.167	16.7%

The percentage is formed from the decimal by moving the decimal point two places to the right.

We have discovered an interesting thing. All you have to do to take an average weight is multiply each weight by the decimal equivalent of the people who had that weight and add all the terms. This way of taking an average is called a **"weighted" average** because it gives more "weight" to the numbers that appear more often. (Don't confuse the word "weight" in *weighted average* with weight in pounds. The term *weighted average* is used for *all* kinds of averaging, not just those involving pounds.)

Example 1 gives an example of the use of the weighted average for averaging money. Try to work it out yourself before looking at the solution. Appendix 2-2 at the end of this chapter will show you how to work out the examples in this chapter by using a calculator.

EXAMPLE 1   Using the method of the weighted average, find the average annual salary of Ph.D. chemistry professors in the United States in 1986 from the following data.

SALARY	THOSE MAKING THIS SALARY, percent
$50,000	50%
$38,000	30%
$30,000	10%
$25,000	10%

Solution: First we must calculate the decimal equivalent of the percents. This is done by moving the decimal point two places to the *left*. If you do not see a decimal point in a number, assume it is after the last digit. For example, 10% could be written as 10.%, 40% as 40.%, and so on.

PERCENT	DECIMAL EQUIVALENT
50%	0.50
30%	0.30
10%	0.10
10%	0.10

The average salary is calculated as follows:

$$(0.50)(50,000) + (0.30)(38,000) + (0.10)(30,000) + (0.10)(25,000)$$
$$= 25,000 + 11,400 + 3,000 + 2,500 = \$41,900$$

The average annual salary is thus $41,900. ■

The nice thing about taking the weighted average in Example 1 is that we didn't have to know how many professors we were talking about. All we had to know were the salaries and the percent of the professors making that salary. As you shall soon see, this is the kind of information we usually have available about atoms. We know the isotopic masses and the percentage of the atoms with a particular mass. We do not know how many atoms we have. The weighted average is thus an ideal method for calculating the relative atomic weights of the elements.

2-4
CALCULATING RELATIVE ATOMIC WEIGHTS USES THE WEIGHTED AVERAGE

We can now define relative atomic weights in terms of weighted averages. The **relative atomic weight** is the weighted average mass of all the naturally occurring isotopes of an element.

Table 2-1 lists the percent of each isotope of some elements and the corresponding isotopic masses. We can use the information in Table 2-1 to calculate some relative atomic weights. Table 2-2 lists the relative atomic weights for 103 elements. The relative atomic weights calculated from Table 2-1 should agree closely with those listed in Table 2-2.

TABLE 2-1

PERCENT ABUNDANCES AND MASSES OF SOME ISOTOPES

ISOTOPE	ABUNDANCE IN NATURE, %	MASS, u
Hydrogen-1	99.985	1.007825
Hydrogen-2	0.015	2.01410
Boron-10	20.0	10.01294
Boron-11	80.0	11.00931
Carbon-12	98.89	12.00000
Carbon-13	1.11	13.00335
Nitrogen-14	99.64	14.00307
Nitrogen-15	0.36	15.00011
Oxygen-16	99.76	15.99491
Oxygen-17	0.04	16.99913
Oxygen-18	0.2	17.99916
Fluorine-19	100.000	18.9984
Neon-20	90.51	19.99244
Neon-21	0.27	20.99385
Neon-22	9.22	21.99138
Chlorine-35	75.77	34.96885
Chlorine-37	24.23	36.96590
Iron-54	5.8	53.9396
Iron-56	91.8	55.9349
Iron-57	2.1	56.9394
Iron-58	0.3	57.9333
Tin-112	1.0	111.9040
Tin-114	0.7	113.9030
Tin-115	0.4	114.9035
Tin-116	14.7	115.9021
Tin-117	7.7	116.9031
Tin-118	24.3	117.9018
Tin-119	8.6	118.9034
Tin-120	32.4	119.9021
Tin-122	4.6	121.9034
Tin-124	5.6	123.9052
Uranium-234	0.005	234.0409
Uranium-235	0.720	235.0439
Uranium-238	99.275	238.0508

TABLE 2-2

TABLE OF ATOMIC WEIGHTS

(Compared to the relative atomic mass of carbon-12 equaling exactly 12)
    The atomic weight of many elements can change slightly, depending on
the origin and treatment of the material. The footnotes to this table explain the
types of variations to be expected for individual elements. The values of the
atomic weight given here apply to elements as they exist naturally on earth and
to certain artificial elements. With the footnotes taken into consideration, the
atomic weights are considered reliable to ±3 in the last digit when followed by
an asterisk (*). Values in parentheses are used for certain radioactive elements
whose atomic weights cannot be stated precisely without knowing where the
element came from; the value given is the mass number of the isotope of that
element with the longest known half-life.

ALPHABETICAL ORDER OF ELEMENTS		ATOMIC	ATOMIC
NAME	SYMBOL	NUMBER	WEIGHT
Actinium[a]	Ac	89	227.0278
Aluminum	Al	13	26.98154
Americium	Am	95	(243)
Antimony	Sb	51	121.75*
Argon[b,c]	Ar	18	39.948*
Arsenic	As	33	74.9216
Astatine	At	85	(210)
Barium[c]	Ba	56	137.33
Berkelium	Bk	97	(247)
Beryllium	Be	4	9.01218
Bismuth	Bi	83	208.9804
Boron[b,d]	B	5	10.81
Bromine	Br	35	79.904
Cadmium[c]	Cd	48	112.41
Calcium[c]	Ca	20	40.08
Californium	Cf	98	(251)
Carbon[b]	C	6	12.011
Cerium[c]	Ce	58	140.12
Cesium	Cs	55	132.9054
Chlorine	Cl	17	35.453
Chromium	Cr	24	51.996
Cobalt	Co	27	58.9332
Copper[b]	Cu	29	63.546*
Curium	Cm	96	(247)
Dysprosium	Dy	66	162.50*
Einsteinium	Es	99	(254)

(*Table 2-2 continues on next page*)

TABLE 2-2 *(continued)*

ALPHABETICAL ORDER OF ELEMENTS		ATOMIC	ATOMIC
NAME	SYMBOL	NUMBER	WEIGHT
Erbium	Er	68	167.26*
Europium[c]	Eu	63	151.96
Fermium	Fm	100	(257)
Fluorine	F	9	18.998403
Francium	Fr	87	(223)
Gadolinium[c]	Gd	64	157.25*
Gallium	Ga	31	69.72
Germanium	Ge	32	72.59*
Gold	Au	79	196.9665
Hafnium	Hf	72	178.49*
Helium[c]	He	2	4.00260
Holmium	Ho	67	164.9304
Hydrogen[b]	H	1	1.0079
Indium[c]	In	49	114.82
Iodine	I	53	126.9045
Iridium	Ir	77	192.22*
Iron	Fe	26	55.847*
Krypton[c,d]	Kr	36	83.80
Lanthanum[c]	La	57	138.9055*
Lawrencium	Lr	103	(260)
Lead[b,c]	Pb	82	207.2
Lithium[b–d]	Li	3	6.941*
Lutetium	Lu	71	174.97
Magnesium[c]	Mg	12	24.305
Manganese	Mn	25	54.9380
Mendelevium	Md	101	(258)
Mercury	Hg	80	200.59*
Molybdenum	Mo	42	95.94
Neodymium[c]	Nd	60	144.24*
Neon[d]	Ne	10	20.179*
Neptunium[a]	Np	93	237.0482
Nickel	Ni	28	58.70
Niobium	Nb	41	92.9064
Nitrogen	N	7	14.0067

TABLE 2-2 *(continued)*

ALPHABETICAL ORDER OF ELEMENTS		ATOMIC	ATOMIC
NAME	SYMBOL	NUMBER	WEIGHT
Nobelium	No	102	(259)
Osmium[c]	Os	76	190.2
Oxygen[b]	O	8	15.9994*
Palladium[c]	Pd	46	106.4
Phosphorus	P	15	30.97376
Platinum	Pt	78	195.09*
Plutonium	Pu	94	(244)
Polonium	Po	84	(209)
Potassium	K	19	39.0983*
Praseodymium	Pr	59	140.9077
Promethium	Pm	61	(145)
Protactinium[a]	Pa	91	231.0359
Radium[a,c]	Ra	88	226.0254
Radon	Rn	86	(222)
Rhenium	Re	75	186.207
Rhodium	Rh	45	102.9055
Rubidium[c]	Rb	37	85.4678*
Ruthenium[c]	Ru	44	101.07*
Samarium[c]	Sm	62	150.4
Scandium	Sc	21	44.9559
Selenium	Se	34	78.96*
Silicon	Si	14	28.0855*
Silver[c]	Ag	47	107.868
Sodium	Na	11	22.98977
Strontium[c]	Sr	38	87.62
Sulfur[b]	S	16	32.06
Tantalum	Ta	73	180.9479*
Technetium	Tc	43	(97)
Tellurium[c]	Te	52	127.60*
Terbium	Tb	65	158.9254
Thallium	Tl	81	204.37*
Thorium[a,c]	Th	90	232.0381
Thulium	Tm	69	168.9342

(Table 2-2 continues on next page)

TABLE 2-2 *(continued)*

ALPHABETICAL ORDER OF ELEMENTS		ATOMIC	ATOMIC
NAME	SYMBOL	NUMBER	WEIGHT
Tin	Sn	50	118.69*
Titanium	Ti	22	47.90*
Tungsten (Wolfram)	W	74	183.85*
Uranium[c,d]	U	92	238.029
Vanadium	V	23	50.9414*
Xenon[c,d]	Xe	54	131.30
Ytterbium	Yb	70	173.04*
Yttrium	Y	39	88.9059
Zinc	Zn	30	65.38
Zirconium[c]	Zr	40	91.22

[a]The atomic weight of these elements is that of the radioactive isotope of longest half-life.

[b]There are natural variations in the isotopic composition of these elements that prevent a more precise atomic weight from being given.

[c]Some samples of this element have an isotopic composition that varies a great deal from most of the samples found. The atomic weight of those few samples may be different than that given in the table.

[d]Commercially available samples of this element may have an atomic weight that differs from that given in the table. The difference may be due to purification methods or other industrial processes.

source: *Pure and Applied Chemistry,* Vol. 47, pp. 80–81, 1976. Headings and footnotes have been paraphrased from those in the original table.

A word about nomenclature. The percent, by atoms, of each isotope of an element is called the **percent abundance.** The decimal equivalent of the percent abundance is called the **fractional abundance.**

EXAMPLE 2   From Table 2-1 we see that nitrogen has two naturally occurring isotopes, namely, nitrogen-14 and nitrogen-15. Calculate the relative atomic weight of nitrogen.

Solution: From Table 2-1 we have the following information.

ISOTOPE	MASS, u	ABUNDANCE, %	FRACTIONAL ABUNDANCE
nitrogen-14	14.00307	99.64	0.9964
nitrogen-15	15.00011	0.36	0.0036

The weighted average is thus

$(0.9964)(14.00307) + (0.0036)(15.00011)$

$$= 13.952659 + 0.0540004 = 14.0067 \text{ u}$$

The value for the relative atomic weight listed in Table 2-2 is 14.0067 u, which is the same as our calculated value, rounded to six figures. ■

EXAMPLE 3   Oxygen has three naturally occurring isotopes: oxygen-16, oxygen-17, and oxygen-18. Calculate the relative atomic weight of oxygen using the values listed in Table 2-1.

Solution: From Table 2-1 we have the following information.

ISOTOPE	MASS, u	ABUNDANCE, %	FRACTIONAL ABUNDANCE
oxygen-16	15.99491	99.76	0.9976
oxygen-17	16.99913	0.04	0.0004
oxygen-18	17.99916	0.2	0.002

The weighted average is

$(0.9976)(15.99491) + (0.0004)(16.99913) + (0.002)(17.99916)$

$$= 15.956522 + 0.0067997 + 0.0359983 = 15.99932 \text{ u}$$

This is within the error limits of the value found in Table 2-2. ■

You might wonder why some of the relative atomic weights in Table 2-2 are listed to four or five decimal places, whereas some are listed to only two or three decimal places. The reason is that the isotopic abundances of some elements are not known to the same accuracy as are others. The footnotes to Table 2-2 describe the problems encountered with these elements.

For those of you without a calculator, you might want to try calculating an approximate relative atomic weight. You can get a reasonably close answer by using the mass number instead of the isotopic mass.

EXAMPLE 4   Calculate the approximate relative atomic weight of nitrogen by using the mass number instead of the isotopic mass.

Solution: Using the information from Example 2, we have the following.

ISOTOPE	MASS NUMBER	FRACTIONAL ABUNDANCE
nitrogen-14	14	0.9964
nitrogen-15	15	0.0036

The approximate relative atomic weight is

$$(0.9964)(14) + (0.0036)(15) = 13.9496 + 0.054 = 14.0036 \text{ u}$$

Compare this to the actual relative atomic weight of 14.0067 u. ∎

The next example is just a bit of practice in reading Table 2-2.

EXAMPLE 5   From Table 2-2, what are the relative atomic weights of the following elements: hydrogen, copper, gold, tantalum, and zirconium?

Solution:

hydrogen	1.0079 u
copper	63.546 u
gold	196.9665 u
tantalum	180.9479 u
zirconium	91.22 u ∎

2-5
BINDING ENERGY HOLDS THE NUCLEUS TOGETHER

This section will help explain why the mass of a proton and a neutron is not exactly 1 u. It will also explain why the mass of an atom is not exactly the mass number of that atom.

To begin, there is a natural tendency for things to become more stable by releasing energy if they can. One example is the explosion of TNT. The energy released makes the products of the explosion more stable than the TNT was. This tendency to become more stable by releasing energy is central to the discussion of binding energy.

Binding energy is responsible for the fact that a nucleus made up of more than one nucleon (a **nucleon** is a nuclear particle, i.e., a proton or a neutron) weighs less than the sum of the weights of its uncombined protons and neutrons. This is because some energy is released to stabilize the nucleus when it is formed from its nucleons. This energy comes from the mass of the nucleons. A little bit of the mass has been converted into energy. This energy is the binding energy of the nucleus.

The amount of energy can be calculated by using the Einstein mass–energy equation, which is $E = mc^2$. In this equation, E is the energy, m is the mass, and c is the speed of light. Since $c = 3.00 \times 10^8$ meters/second or 186,000 miles/second, the speed of light is a large number. The formula shows that a small amount of mass is equivalent to a very large amount of energy.

Remember that an atom of carbon-12 has a mass of exactly 12.00000 u. It is instructive to compare some other atoms with carbon-12 as far as binding energy is concerned. Atoms that weigh less than carbon-12 should have less binding energy than carbon-12. Atoms that weigh more than carbon-12 should have more binding energy than carbon-12.

If you look up fluorine in Table 1-2, you see that it only has one naturally occurring isotope, $^{19}_{9}F$. The mass number of this isotope is 19. If you look up fluorine's atomic weight in Table 2-2, you find that it is 18.998403 u. Thus, even with the mass of the nine electrons counted in, the atomic weight is a bit less than 19. Fluorine-19 has more nucleons and more binding energy than carbon-12.

Now look at Table 1-1. Notice that because carbon-12 is used as a mass reference, each separate proton and neutron has a mass slightly greater than 1 u. The exact mass of a proton is 1.0072765 u, and the exact mass of a neutron is 1.0086650 u. This is to be expected, since protons and neutrons do not have a nuclear binding energy when they are not in a nucleus. (There is no need for you to memorize the exact masses of the proton and neutron. But in the mid-1950s a 16-year-old girl who was an expert on atomic energy lost $64,000 on a TV quiz show because she could not remember the exact mass of a neutron.)

Looking again at Table 1-2, you can see that beryllium has one naturally occurring isotope, $^{9}_{4}Be$. Beryllium's relative atomic weight is 9.01218 u. This is larger than 9 and is expected, since beryllium-9 has three fewer nucleons than carbon-12. Thus beryllium-9 has less binding energy than carbon-12.

In Table 2-1, you will find that, in general, the isotopic masses for elements heavier than carbon-12 are *less* than the atomic mass numbers. Elements lighter than carbon-12 have isotopic masses *greater* than the mass number. This is consistent with what we have said about binding energy.

You might notice that the isotopes carbon-13, nitrogen-14, and nitrogen-15 seem not to be consistent with what we have said in the previous paragraph. In fact, these isotopes *do* have more binding energy than carbon-12. However, it is not enough additional binding energy to reduce the isotopic mass below the mass number.

Although binding energy involves only very small mass changes, the energy that this mass becomes is very large. You can see and feel this energy every day. In fact, without this energy, you would not be alive. This is because the binding energy that is released when four $^{1}_{1}H$ nuclei form a $^{4}_{2}He$

The sixteen-foot-diameter target chamber of the Nova Laser Facility. Ten powerful laser beams will converge on a small pellet contianing solid deuterium and tritium. The laser beams will heat the pellet to 100 million degrees Celsius, causing the isotopes to fuse, form helium, and release energy.

nucleus is the energy that powers our sun and many other stars. This combining of light nuclei to make heavier nuclei is called **nuclear fusion**. On earth, scientists have used a similar reaction (deuterium plus tritium to give helium) in the hydrogen bomb. And this same fusion reaction, when controlled, may lead to almost unlimited power generation on earth.

It turns out that as elements get heavier than iron, their binding energy per nucleon[1] *decreases* slightly with respect to lighter elements. Around mass number 215, this decrease causes the isotopic mass to be *greater* than the mass number, as can be seen for the uranium isotopes listed in Table 2-1. This slight decrease in binding energy per nucleon makes uranium-235 unstable with respect to certain lighter nuclei. The splitting of uranium-235, called **nuclear fission**, supplies the energy for the atomic bomb and nuclear reactors that can be used for power generation. Other isotopes that can undergo nuclear fission are uranium-233 and plutonium-239.

APPENDIX 2-1
THE DISTRIBUTIVE LAW

The simplest definition of the distributive law of algebra, for any variables a, b, and c, is

$$a(b + c) = ab + ac$$

---

[1] The binding energy per nucleon will not be defined in this book. Nuclear fission can be appreciated without a definition.

where $ab = (a)(b) = a \times b$, and so on. To show that the distributive law is reasonable, let's test it with real numbers, say 3, 4, and 5: Does $3(4 + 5) = (3)(4) + (3)(5)$? Well, $3(4 + 5) = 3 \times 9 = 27$ and $(3)(4) + (3)(5) = 12 + 15 = 27$. So the answer is yes. It is true that $3(4 + 5) = (3)(4) + (3)(5)$. You can try any numbers you wish—the distributive law always works!

If there were more numbers (or symbols), such as a, b, c, d, and e, the distributive law would look like

$$a(b + c + d + e) = ab + ac + ad + ae$$

Suppose that we have a division of the form $(x + y)/z$ and we want to apply the distributive law. (We have changed variables to x, y, and z for variety.) The first thing to recognize is that division by a number is the same as multiplication by 1 over the number ("1 over a number" is called the **reciprocal** of the number). In our case, z is the number and $1/z$ is the reciprocal of the number. We can now write

$$\frac{x + y}{z} = \frac{1}{z}(x + y)$$

Let's test this out with some actual numbers. Does

$$\frac{7 + 8}{5} = \frac{1}{5}(7 + 8)?$$

Yes, since $\dfrac{7 + 8}{5} = \dfrac{15}{5} = 3$ and $\dfrac{1}{5}(7 + 8) = \dfrac{1}{5}(15) = 3$

Now let's apply the distributive law to

$$\frac{1}{z}(x + y)$$

We get

$$\frac{1}{z}(x + y) = \frac{x}{z} + \frac{y}{z}$$

So it must be true that applying the distributive law to $(x + y)/z$ gives

$$\frac{x + y}{z} = \frac{x}{z} + \frac{y}{z}$$

since

$$\frac{x + y}{z} = \frac{1}{z}(x + y) = \frac{x}{z} + \frac{y}{z}$$

Therefore, in our discussion of the weighted average, we used the distributive law correctly, and it is true that

$$\frac{(2)(120) + (3)(130) + (1)(150)}{6} = \frac{(2)(120)}{6} + \frac{(3)(130)}{6} + \frac{(1)(150)}{6}$$

APPENDIX 2-2
USING A CALCULATOR

We will assume that you have a full scientific calculator that uses algebraic logic. If you have a Hewlett–Packard calculator that uses Reverse Polish Notation (RPN), please consult the instruction manual that comes with your calculator.

Some students may have an older calculator that doesn't have algebraic logic. Since the instructions in this appendix assume algebraic logic, you should test your calculator according to the following procedure.

Compute $(2)(3) + (4)(5)$ by pushing the following buttons on your keyboard: $2, \times, 3, +, 4, \times, 5, =$

You should read the following in the display.

BUTTON PUSHED	DISPLAY READS
2	2.
×	2.
3	3.
+	6.
4	4.
×	4.
5	5.
=	26.

The correct answer is 26. Doing the calculation by hand we get

$$(2)(3) + (4)(5) = 6 + 20 = 26$$

If your calculator doesn't have algebraic logic, your answer will probably be 50. The calculator doesn't follow the distributive law correctly. It is doing the problem this way:

$$(2)(3) + (4)(5) = (6 + 4)(5) = (10)(5) = 50$$

If you have one of these calculators, you must write down intermediate steps or store them in memory.

Let's see how to work out the examples in this chapter using a calculator. We won't list the display readings—just the buttons you're supposed to push. For clarity, we have written numbers less than 1 with a 0 before the decimal point, but you don't have to enter the 0 into the calculator. For example, .3 is the same as 0.3.

EXAMPLE 1   Calculator entries are (we have left out the commas in the numbers for clarity)

0.5, ×, 50 000, +, 0.3, ×, 38 000, +, 0.1, ×, 30 000, +, 0.1, ×, 25 000, =

The answer is 41,900.   ■

EXAMPLE 2   Calculator entries are

0.9964, ×, 14.00307, +, 0.0036, ×, 15.00011, =

The answer is 14.006659 u.   ■

EXAMPLE 3   Calculator entries are

0.9976, ×, 15.99491, +, 0.0004, ×, 16.99913, +, 0.002, ×, 17.99916, =

The answer is 15.99932 u.   ■

EXAMPLE 4   Calculator entries are

0.9964, ×, 14, +, 0.0036, ×, 15, =

The answer is 14.0036 u.   ■

There are additional calculator instructions throughout the text.

PROBLEMS

KEYED PROBLEMS

1. Using the method of the weighted average, find the average annual salary of Ph.D. chemists in the United States in 1986 from the following data.

JOB	SALARY	THOSE MAKING THIS SALARY, %
College/University	$41,900	26
Industry	$49,900	74

2. From Table 2-1, we see that chlorine has two naturally occurring isotopes: chlorine-35 and chlorine-37. Calculate the relative atomic weight of chlorine. Compare your answer with the actual value in Table 2-2.

3. From Table 2-1, we see that neon has three naturally occurring isotopes: neon-20, neon-21, and neon-22. Calculate the relative atomic weight of neon. Compare your answer with the actual value listed in Table 2-2.

4. Calculate the approximate relative atomic weight of chlorine by using the mass numbers instead of the isotopic masses. Compare your answer with the actual value listed in Table 2-2.

5. From Table 2-2, what are the relative atomic weights of the following elements: bromine, helium, cobalt, silver, antimony.

SUPPLEMENTAL PROBLEMS

6. Define the following terms.
   a. atomic mass unit
   b. relative atomic weight
   c. percent abundance of an isotope
   d. fractional abundance of an isotope
   e. nucleon
   f. binding energy
   g. nuclear fusion
   h. nuclear fission
   i. distributive law of algebra

7. Calculate the relative atomic weight of the following elements listed in Table 2-1.
   a. hydrogen   d. iron
   b. boron   e. tin
   c. carbon   f. uranium
   In each case, compare your answer with the value listed in Table 2-2.

8. Calculate the approximate relative atomic weights of the elements listed in Problem 7 by using the mass numbers instead of the isotopic masses.

9. An $^{16}_{8}O$ atom has 8 protons, 8 neutrons, and 8 electrons. Using the exact weights for these particles listed in Table 1-1, calculate what an $^{16}_{8}O$ atom would weigh if there were no such thing as binding energy. The difference

between this value and the isotopic mass of 15.99491 u for $^{16}_{8}O$ is the mass that has been converted into binding energy. What is the size of the mass (in atomic mass units) that has been converted into binding energy?

SPECIAL PERCENTAGE PROBLEMS

10. Convert the following decimals to percents.
   a. 0.75   d. 0.031
   b. 0.42   e. 0.101
   c. 1.61   f. 0.002

11. Convert the following percents into decimals.
   a. 38%   d. 0.24%
   b. 7%   e. 1.46%
   c. 135%   f. 11.77%

12. Find 25% of 70.   16. What percent of 12 is 7?

13. Find 7.2% of 250.   17. What percent of 180 is 30?

14. Find 0.02% of 150.   18. What percent of 7 is 0.03?

15. Find 150% of 30.

3

MOLECULES AND THE BALANCED CHEMICAL EQUATION

So far we have discussed atoms and their composition. But that's just the beginning of the study of chemistry. Most of the things that you see around you are made up of different combinations of atoms. All these different combinations give us wood, plastics, water, alloys, plants, animals, people, air, and so on. Thus, in order to really get into chemistry, we must look at combinations of atoms.

3-1
MOLECULES CONSIST OF TWO OR MORE ATOMS

There are over 105 different known elements and thus over 105 different kinds of atoms (not counting isotopes of the different elements).

It is rather common for some of these atoms to combine with one another to form molecules. A **molecule** is two or more atoms that are bound together.

As an example of a molecule, let's look first at two hydrogen atoms. Remember that the symbol for a hydrogen atom is H. Two of these atoms can combine to form a molecule called a **hydrogen molecule**. Three representations of the hydrogen molecule are shown in Figure 3-1.

In these diagrams, each H stands for a hydrogen atom. Each of the diagrams is a way of *representing* a hydrogen molecule. The straight line between the two H atoms represents a chemical bond. A **chemical bond** is the attractive force that holds atoms together. The spring between them also represents the chemical bond but reminds us that the bond acts something like a spring and that the H atoms vibrate back and forth. The hazy cloud indicates that the electrons of the hydrogen atoms form the chemical bond.

(Opposite) A forest fire illustrates a vigorous chemical reaction that releases heat.

FIGURE 3-1 Three representations of the hydrogen molecule.

Of the three, the most common way of representing the hydrogen molecule is with the straight line connecting the two atoms, because it is so easy to do. You just have to remember that the line stands for a chemical bond.

Writing a hydrogen molecule as H—H can be abbreviated even further. Chemists frequently write a hydrogen molecule as H_2. The 2 following the H means that there are two H atoms combined in a molecule. The 2 is placed half a space down and called a subscript. H_2 is pronounced "H two" and is called the **chemical formula** of the hydrogen molecule.

3-2
BALANCING CHEMICAL EQUATIONS CONSERVES ATOMS

Chemists have developed a notation to describe what happens when two hydrogen atoms combine to form a hydrogen molecule. This notation is called the **chemical equation** and is useful for describing *any* chemical reaction. For the hydrogen reaction the chemical equation can be written as

$$H + H \rightarrow H—H$$

and is read "One hydrogen atom plus one hydrogen atom forms (or yields) one hydrogen molecule." Another way to read it is "One hydrogen atom reacts with another hydrogen atom to form a hydrogen molecule." The hydrogen atoms on the left side of the arrow are called the reactants, and the hydrogen molecule on the right side of the arrow is called the product. In general, in a chemical equation, **reactants** are substances written to the left of the arrow, and **products** are substances written to the right of the arrow.

The arrow ($\rightarrow$) separating the reactants from the products can be read: "reacts to give" or "yields" or "forms." Chemists also use other symbols to separate the reactants and the products. We won't use them in this book, but we will mention them here because you may run across them in other books and articles. The other symbols commonly used to separate the reactants and products are the double arrow ($\leftrightarrows$) and the equals sign ($=$).

A shorter way of writing the chemical equation for making a hydrogen molecule is

$$H + H \rightarrow H_2$$

This is sometimes written in another, even shorter, way,

$$2H \rightarrow H_2$$

where the 2H stands for two hydrogen atoms.

There is a great difference between 2H and H_2. The 2H are separate hydrogen atoms that really have nothing to do with one another except that they happen to be about to combine to form a hydrogen molecule. The H_2 means that the atoms have already been joined by a chemical bond to form a hydrogen molecule.

H atoms "want" to combine to make H_2 in a very vigorous way. The hottest welding torch uses this reaction to generate its heat. The torch is called the **atomic hydrogen torch**.

EXAMPLE 1   In the equation $H_2 \rightarrow 2H$, what are the reactants and what are the products?

Solution: H_2 is the reactant and 2H are the products. Note that we have reversed our original equation, which was $2H \rightarrow H_2$; accordingly, the reactants and the products are also reversed.   ■

The vast majority of atoms can combine to form molecules, and a molecule can be made up of many atoms. When there are two atoms in molecule, such as the hydrogen molecule, we say that it is a **diatomic molecule** ("di" means two).

EXAMPLE 2   Write the chemical equation for the formation of the diatomic oxygen molecule from two oxygen atoms.

Solution: $O + O \rightarrow O_2$ or $2O \rightarrow O_2$. The O atoms are the reactants, and the O_2 molecule is the product.   ■

Let's look at some more molecules and equations. A common reaction is the burning of hydrogen molecules and oxygen molecules to make water molecules. This reaction can be written as

$$H_2 + O_2 \rightarrow H_2O$$

where H_2 and O_2 are the reactants, and H_2O is the product. (What we have written here is not the final version; please read on. If you are also wondering how O_2 becomes O in H_2O, read on.) As we said before, hydrogen exists as a diatomic molecule under normal conditions, so in the equation above we have written it as H_2.

We write O_2 for the same reason. Oxygen atoms "want" to combine to form molecules. All the oxygen in the air (about 21% of the air is oxygen) exists as diatomic molecules of O_2.

When chemists talk about hydrogen and oxygen gases, they actually mean H_2 and O_2, the diatomic molecules. When they mean atoms, they say so: "Speaking of hydrogen atoms and oxygen atoms . . . ," or "Speaking of monatomic hydrogen and monatomic oxygen . . ." (**monatomic** means one atom) is what you will hear when they are referring to H atoms and O atoms. However, chemists say that water, H_2O, consists of hydrogen and oxygen. This could be misleading because here they mean hydrogen and oxygen atoms.

The chemical formula H_2O represents a water molecule. This formula tells us that the water molecule is made up of two hydrogen atoms and one oxygen atom. It is usually drawn

$$H \diagup \overset{\displaystyle O}{} \diagdown H$$

because this is a way to show how it looks. Remember, the lines represent chemical bonds. The letter symbols H and O represent the H atom and the O atom, respectively.

From the diagram, you might think that the formula for water should be written as HOH. Many times it is, but H_2O is more popular. Water is never written as OH_2.

Chemical formulas don't usually tell you how the atoms are connected to each other. For that information, a diagram of the molecule must be drawn as was done above for the water molecule. This diagram is called the **structural formula** of the molecule because it shows how the atoms of a molecule are connected to one another.

Now let's go back to the equation

$$H_2 + O_2 \rightarrow H_2O$$

There is something wrong with it. If you count the number of atoms of oxygen on the left side of the arrow, you will find two oxygen atoms in the oxygen molecule. But on the right side of the arrow there is only one oxygen atom in the water molecule. What happened to the other oxygen atom?

The other oxygen atom must have been used to make another water molecule. We can draw what happens, as follows:

$$\begin{matrix} H-H \\ H-H \end{matrix} + O-O \rightarrow \begin{matrix} H \diagup \overset{\displaystyle O}{} \diagdown H \\ H \diagup \overset{\displaystyle O}{} \diagdown H \end{matrix}$$

Now we have used up all the oxygen on the left of the arrow to make two water molecules. However, the second water molecule needed two hydrogen atoms, so we had to use another H_2 on the left side of the arrow.

We can write this structural formula equation much more easily as the chemical equation

$$2H_2 + O_2 \rightarrow 2H_2O$$

The notation $2H_2$ means two separate H_2 molecules. The notation $2H_2O$ means two separate H_2O molecules. In words, you can read the equation as "Two hydrogen molecules react with one oxygen molecule to form two water molecules."

The numbers put in front of the molecules are called **coefficients**.

The equation for making water from hydrogen and oxygen illustrates something that is very important, namely: **The same number of atoms of each element must appear on each side of the arrow in a chemical equation**, whether they are combined in molecules or appear as free atoms. We say that in a chemical reaction the atoms of each element are **conserved**, which simply means that atoms aren't created or destroyed in a chemical reaction; they must all be accounted for.

Note that molecules are *not* usually conserved. This is reasonable, and in fact necessary, since chemical reactions make and break up molecules.

Chemical equations that have the same number of atoms of each element on each side of the arrow are said to be **balanced**. The process of figuring out the correct coefficients is called "balancing the equation."

There is one further point about chemical equations you should be aware of at this time. Chemical equations tell us what the reactants and products are. However, they don't usually tell us *how* the reacting substances collided to form the products. The how of chemical reactions is usually very complicated and not very easy to discover. You will appreciate this more when you have studied more chemistry.

EXAMPLE 3   The symbol for the element carbon is C. Carbon can combine with hydrogen to make, among other things, a molecule called *methane*. Methane is the main component of natural (cooking) gas and has the formula CH_4. How many atoms of each element are there in a CH_4 molecule? In two CH_4 molecules?

Solution: There are one C atom and four H atoms in each methane molecule. In two methane molecules, $2CH_4$, there are two C atoms and 8 H atoms. If we draw the structural formula of methane twice, you can just count the atoms.

$$\begin{array}{ccc}
& \text{H} & & \text{H} \\
& | & & | \\
\text{H}-&\text{C}&-\text{H} \quad \text{H}-&\text{C}&-\text{H} \quad \blacksquare \\
& | & & | \\
& \text{H} & & \text{H}
\end{array}$$

EXAMPLE 4   The symbol for the element nitrogen is N. Nitrogen gas, which makes up about 78% of the air, contains diatomic molecules N_2. In the hot flame inside an automobile engine (or any other hot flame), N_2 and O_2 combine to form nitric oxide, NO, which is one of the causes of air pollution. Write the balanced equation for the reaction.

Solution: $N_2 + O_2 \rightarrow 2NO$. The structural formulas are

$$N{\equiv}N + O{=}O \rightarrow \begin{matrix} N{=}O \\ N{=}O \end{matrix}$$

The equation can be read "One molecule of nitrogen reacts with one molecule of oxygen to form two molecules of nitric oxide."   ∎

As you can see from the structural formulas in Example 4, some molecules have more than one bond joining the atoms together. The atoms in N_2 are joined by a triple bond, represented by three lines; the atoms in O_2 and NO are joined by a double bond, represented by two lines. We will discuss bonding in Chapter 15.

NOTE: In our discussion of the water equation above, we drew the oxygen molecule with one bond. That was done so we wouldn't have to stop the discussion to explain that the oxygen molecule really has a double bond connecting the atoms.

A few more words about writing balanced equations are in order. We can use the water equation
$$2H_2 + O_2 \rightarrow 2H_2O$$

as an example.

We could have balanced the equation by writing, for instance,

$$4H_2 + 2O_2 \rightarrow 4H_2O$$

Atoms are conserved (there are as many of each element on the left side of the arrow as on the right side), but there are twice as many as the minimum needed to balance the equation. This is not done: **The coefficients should be the smallest possible whole numbers**.

As another possibility, you might ask if it is acceptable to write

$$H_2 + \tfrac{1}{2}O_2 \rightarrow H_2O$$

One-half of O_2 is just one O atom, and the equation is balanced. This method of using fractions to balance chemical equations is not incorrect, but it is not used as commonly as whole numbers.

Actually, any chemical equation usually represents a large number of atoms and molecules. So taking one-half of O_2 is not really dividing the molecule into atoms; we're just taking one-half of a large number of O_2 molecules. For example, if we have 1000 oxygen molecules, the fraction $\tfrac{1}{2}$ denotes 500 of them.

If you use fractions as coefficients to balance a chemical equation, they must be "reasonable" fractions. What are "unreasonable" fractions? If you tried to balance the water equation in the following way you would be "unreasonable":

$$\tfrac{1}{2}H_2 + \tfrac{1}{4}O_2 \rightarrow \tfrac{1}{2}H_2O \quad \textbf{unreasonable}$$

It is unreasonable because it is more complicated. Chemists like to keep things clear and simple.

There is another incorrect way that people sometimes try to balance chemical equations. For example, the water equation is sometimes incorrectly written as

$$H_2 + O_2 \rightarrow H_2O_2 \quad \textbf{!! wrong !!}$$

This equation is balanced but is wrong. The H_2O has been incorrectly changed to H_2O_2, which is the formula for hydrogen peroxide. If you have ever put hydrogen peroxide on your hair to bleach it, you *know* that it is not water. So when you balance chemical equations, please, you must *never* change the chemical formulas of the atoms or molecules.

The following is a summary of the rules for balancing chemical equations.

1. The same number of atoms of each element must appear on each side of the equation.
2. **Never** change the chemical formula of any of the atoms or molecules appearing in the equation. This means that you should not change subscripts in a chemical formula when you balance a chemical equation.
3. You *may* take multiples or fractions of the atoms and molecules by changing the coefficients.
4. The coefficients must be the smallest possible whole numbers or reasonable fractions.
5. The object is to keep things as simple and as clear as possible.

The following examples of how to balance equations may help you to really understand how to use these rules.

EXAMPLE 5   Methane, CH_4, can burn in oxygen to give carbon dioxide (CO_2) and water. Write the balanced equation for this reaction.

Solution: The unbalanced equation is

$$CH_4 + O_2 \rightarrow CO_2 + H_2O$$

Since there is one C atom on each side of the arrow, let's start by counting the H atoms. There are 4 on the left and 2 on the right. Thus we need to take 2 H_2O molecules to give 4 H atoms on the right side of the equation.

$$CH_4 + O_2 \rightarrow CO_2 + 2H_2O$$

Counting the O atoms, we find 2 on the left and 4 on the right. We need to take 2 O_2 molecules to get 4 O atoms on the left:

$$CH_4 + 2O_2 \rightarrow CO_2 + 2H_2O$$

The equation is now balanced. Writing structural formulas we have

Here the CO_2 has been drawn with two bonds to each oxygen atom to make the picture more realistic. ■

EXAMPLE 6   The metal aluminum, Al, reacts with oxygen to form aluminum oxide, Al_2O_3, which is usually a white powder. However, Al_2O_3 can be "grown" into large crystals that are clear and colorless like glass. When certain metal atoms are included in the crystal, it becomes colored. Rubies and sapphires are two colored Al_2O_3 crystals. Write the balanced equation for the reaction between Al and O_2 to form Al_2O_3.

Solution: The unbalanced equation is

$$Al + O_2 \rightarrow Al_2O_3$$

To begin balancing the equation, look at the number of oxygen atoms first. There are 2 O atoms on the left and 3 on the right. If we take 3 O_2 molecules and 2 Al_2O_3 molecules, there are 6 O atoms on each side:

$$Al + 3O_2 \rightarrow 2Al_2O_3$$

To finish the balancing, we need 4 Al atoms on the left side to equal the 4 on the right:

$$4Al + 3O_2 \rightarrow 2Al_2O_3 \quad ■$$

EXAMPLE 7   Octane, C_8H_{18}, is a major component of gasoline. It burns in oxygen to give CO_2 and water. Write the balanced equation for the combustion of octane.

Solution: The unbalanced equation is

$$C_8H_{18} + O_2 \rightarrow CO_2 + H_2O$$

The best way to start is to balance the C atoms:

$$C_8H_{18} + O_2 \rightarrow 8CO_2 + H_2O$$

Then balance the H atoms:

$$C_8H_{18} + O_2 \rightarrow 8CO_2 + 9H_2O$$

This gives us 16 O atoms from the 8 CO_2 molecules and gives us 9 O atoms from the 9 H_2O molecules; altogether we have 25 O atoms on the right side of the equation. The best way to put 25 O atoms on the left side is to write $\frac{25}{2}O_2$. Remember that $\frac{1}{2}O_2$ is one O atom. Thus 25 times that is 25 O atoms. The balanced equation is

$$C_8H_{18} + \tfrac{25}{2}O_2 \rightarrow 8CO_2 + 9H_2O$$

This equation is a good example of the use of fractional coefficients. If you want only whole number coefficients, multiply all the coefficients by 2:

$$2C_8H_{18} + 25O_2 \rightarrow 16CO_2 + 18H_2O \quad \blacksquare$$

You might sometimes wonder just how to get started in balancing an equation. For instance, what atoms do you balance first, second, and so on. The following approach is helpful.

1. *First, balance the atoms for elements that occur in only one substance on each side of the equation.* For example, in the equation discussed in Example 7, namely,

$$C_8H_{18} + O_2 \rightarrow CO_2 + H_2O$$

we balanced the C and the H first, since they occur in only one substance on each side of the equation.

2. *Second, balance free atoms (such as Fe or Al) or elemental molecules (such as O_2 or Cl_2) last.* This is because you can put coefficients in front of them without changing the number of atoms of any other elements in the equation. This is why we balanced the oxygen last in Example 7.

As is usual with most things that are unfamiliar, the best way to learn how to balance equations is to balance a great number of them.

3-3
BALANCE GROUPS AS A WHOLE IN EQUATIONS
WITH MORE COMPLEX MOLECULAR FORMULAS

So far, the substances we have discussed have had rather simple formulas: H_2, O_2, H_2O, CH_4, C_8H_{18}, Al_2O_3, and so on. There are a large number of substances that have more complicated formulas. For example, the substance "papermaker's alum," or aluminum sulfate, has the formula

$$Al_2(SO_4)_3$$

The group of atoms represented by the SO_4 is called the *sulfate group*.[1] Three sulfate groups combine with two aluminum atoms to make aluminum sulfate. The parentheses allow us to take three sulfate groups:

$Al_2(SO_4)_3$ ← This 3 indicates three sulfate groups.

As far as counting atoms in the formula of aluminum sulfate, we have

2 Al atoms   3 S atoms   12 O atoms

The following two examples will give you some practice in working with chemical formulas that contain parentheses. To read the formula $Al_2(SO_4)_3$, you would say "ay el two ess oh four taken three times."

EXAMPLE 8   How many atoms of each element are in the formula $Ca_3(PO_4)_2$?

Solution: There are 3 Ca atoms, 2 P atoms, and 8 O atoms. The substance is called *calcium phosphate* and PO_4 is the *phosphate group*. The PO_4 group has a −3 charge and is another example of a polyatomic ion.   ■

EXAMPLE 9   Balance the following equation,

$$Ca(OH)_2 + H_3PO_4 \rightarrow Ca_3(PO_4)_2 + HOH$$

where we have written water as HOH. This will make balancing the equation easier.

Solution: One way to simplify the balancing is to count the groups, some of which may be in parentheses, as units, if possible. This would mean counting the OH and PO_4 groups as units in this equation.

---

[1] The SO_4 group is not considered a molecule because it always has a −2 charge. It is called a **polyatomic ion**. In Chapter 16, you will learn more about polyatomic ions.

First balance the Ca, since it appears in only one substance on each side of the arrow:

$$3Ca(OH)_2 + H_3PO_4 \rightarrow Ca_3(PO_4)_2 + HOH$$

Next balance the PO_4 group, since it also appears in only one substance on each side of the arrow:

$$3Ca(OH)_2 + 2H_3PO_4 \rightarrow Ca_3(PO_4)_2 + HOH$$

Then balance the H and the OH. There are 6 OH groups and 6 H atoms on the left; we need 6 HOH molecules on the right:

$$3Ca(OH)_2 + 2H_3PO_4 \rightarrow Ca_3(PO_4)_2 + 6HOH$$

The equation is now balanced.   ∎

To see how easy balancing can be when you keep the groups in parentheses together, try balancing the equation in Example 9 without keeping the groups together. Just balance it using individual atoms.

PROBLEMS

KEYED PROBLEMS

1. In the equation $N_2 \rightarrow 2N$, what is the reactant and what are the products?

2. Write the chemical equation for the formation of the diatomic chlorine molecule from two chlorine atoms. The symbol for chlorine is Cl.

3. The symbol for the element nitrogen is N. Nitrogen can combine with hydrogen to make a molecule called ammonia, NH_3. How many atoms of each element are there in an NH_3 molecule? In two NH_3 molecules?

4. Nitric oxide, NO, reacts slowly with the oxygen, O_2, in air to form nitrogen dioxide, NO_2. NO_2 is the primary cause of the reddish-brown haze you can see in polluted air. Write the balanced equation for the reaction between NO and O_2 to form NO_2.

5. Ethane, C_2H_6, can burn in oxygen to give carbon dioxide, CO_2, and water. Write the balanced equation for this reaction.

6. The metal iron, Fe, reacts with oxygen, O_2, to give the reddish rust called ferric oxide, Fe_2O_3. Write the balanced equation for this reaction.

7. $C_{12}H_{26}$ is a component of kerosene. It burns in oxygen to give CO_2 and water. Write the balanced equation for the combustion of $C_{12}H_{26}$.

8. How many atoms of each element are in the formula $Fe_2(SO_3)_3$.

9. Balance the following equation:

$$Al(OH)_3 + H_2SO_4 \rightarrow Al_2(SO_4)_3 + HOH$$

SUPPLEMENTAL PROBLEMS

10. Define the following terms.
    a. molecule   d. diatomic molecule
    b. chemical formula   e. structural formula
    c. balanced chemical equation   f. coefficient

11. How many atoms of each element are in the following compounds?
    a. KCl   e. C_3H_7COOH i. $Mg(CN)_2$
    b. $KClO_4$ f. H_2S j. $Ca(HCO_3)_2$
    c. $AgNO_3$ g. $(NH_4)_2CO_3$
    d. C_6H_{14} h. $Zn(NO_3)_2$

12. Balance the following equations with the smallest possible whole numbers or reasonable fractions.
    a. $SO_2 + O_2 \rightarrow SO_3$
    b. $PCl_5 \rightarrow PCl_3 + Cl_2$
    c. $CaH_2 + H_2O \rightarrow Ca(OH)_2 + H_2$
    d. $(NH_4)_2Cr_2O_7 \rightarrow Cr_2O_3 + N_2 + H_2O$
    e. $Na + O_2 \rightarrow Na_2O$
    f. $H_2 + Cl_2 \rightarrow HCl$
    g. $P + O_2 \rightarrow P_2O_3$
    h. $NH_3 + H_2SO_4 \rightarrow (NH_4)_2SO_4$
    i. $Zn + Pb(NO_3)_2 \rightarrow Zn(NO_3)_2 + Pb$
    j. $Cu + S \rightarrow Cu_2S$
    k. $Al + H_3PO_4 \rightarrow H_2 + AlPO_4$
    l. $NaNO_3 \rightarrow NaNO_2 + O_2$
    m. $H_2O_2 \rightarrow H_2O + O_2$
    n. $BaO_2 \rightarrow BaO + O_2$
    o. $Al + Cl_2 \rightarrow AlCl_3$
    p. $P_4 + O_2 \rightarrow P_4O_{10}$
    q. $H_2 + N_2 \rightarrow NH_3$
    r. $BaCl_2 + (NH_4)_2CO_3 \rightarrow BaCO_3 + NH_4Cl$
    s. $PbO_2 \rightarrow PbO + O_2$
    t. $Al + HCl \rightarrow AlCl_3 + H_2$
    u. $Fe_2(SO_4)_3 + Ba(OH)_2 \rightarrow BaSO_4 + Fe(OH)_3$
    v. $KClO_3 \rightarrow KCl + O_2$
    w. $Mg + N_2 \rightarrow Mg_3N_2$
    x. $C_3H_7CHO + O_2 \rightarrow CO_2 + H_2O$
    y. $NaHCO_3 + HCl \rightarrow NaCl + H_2O + CO_2$
    z. $Zn(OH)_2 + H_2SO_4 \rightarrow ZnSO_4 + HOH$

aa. $C_4H_9OH + O_2 \rightarrow CO_2 + H_2O$
bb. $CaC_2 + H_2O \rightarrow C_2H_2 + Ca(OH)_2$
cc. $CaCO_3 + H_3PO_4 \rightarrow Ca_3(PO_4)_2 + CO_2 + H_2O$
dd. $C_3H_7COOH + O_2 \rightarrow CO_2 + H_2O$

13. A rocket propellant consisting of hydrazine, N_2H_4, and hydrogen peroxide, H_2O_2, reacts to form nitrogen, N_2, and water. Write the balanced equation for the reaction.

14. Yeast can ferment glucose, $C_6H_{12}O_6$, to give ethanol, C_2H_5OH, and carbon dioxide. Write the balanced equation for this reaction.

15. Carbon dioxide can be removed from the air of a spacecraft by reacting it with lithium hydroxide, LiOH, to form lithium carbonate, Li_2CO_3, and water. Write the balanced equation for this reaction.

16. "Quicklime," CaO, can react with water to form "slaked lime," $Ca(OH)_2$. Write the balanced equation for this reaction.

17. In a blast furnace, iron ore, Fe_2O_3, is burned with carbon monoxide, CO, to give iron metal, Fe, and carbon dioxide, CO_2. Write the balanced equation for this reaction.

ELEMENTS, COMPOUNDS, AND MIXTURES

Different elements exist as solids, liquids, or gases. The arrangement of the atoms in different elements also varies. In this chapter we will briefly discuss such differences.

4-1
EACH ELEMENT CONSISTS OF A SINGLE KIND OF ATOM

The following is a list of some of the elements and their formulas that we have mentioned so far.

hydrogen	H_2	aluminum	Al	argon	Ar
oxygen	O_2	radium	Ra	calcium	Ca
nitrogen	N_2	lead	Pb		
carbon	C	helium	He		

Some of the elements can exist as individual atoms at room temperature. Gases such as helium (He) and argon (Ar) exist as individual atoms. They are monatomic.

Some elements exist as individual molecules at room temperature. The ones like this that we have talked about are H_2, O_2, and N_2, which all exist as diatomic gas molecules.

Other elements that we have mentioned (C, Al, U, Ra, Fe, Pb) exist as atoms bonded together in a solid at room temperature. One example of such a solid is a metal; another is a diamond, which is a nonmetal. It is impossible to find individual molecules in these solids, so we write the chemical formulas as single atoms. This is because it's simple to do and nobody has thought of a better way of writing it.

(Opposite) A scanning electron microscope photograph of lead tin telluride crystals. The magnification is about 50 times.

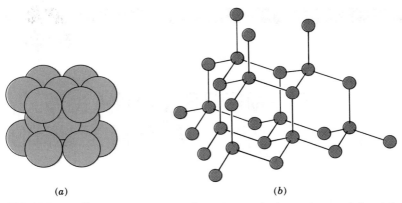

FIGURE 4-1 Two arrangements that atoms of elements can take: (a) metal atoms; (b) carbon atoms in diamond. There are no separate molecules in either.

Figure 4-1 illustrates some of the arrangements that atoms of elements can take. Notice that in a diamond, each carbon atom is connected to several other carbon atoms. It is impossible to find individual molecules. A diamond crystal large enough to see consists of billions and billions of carbon atoms. It can be thought of as a giant molecule. We will refer to such an arrangement as a **giant array** of atoms. Even though there are billions of carbon atoms in the giant array, we still write the chemical formula of diamond as C. This may seem strange because the chemical formula makes it appear as though the diamond is made up of separate carbon atoms. However, in this case the formula simply indicates that diamonds are are made up only of carbon atoms.

Metals can also be considered as consisting of a giant array of atoms.

Two elements, mercury (Hg) and bromine (Br_2), are liquid at room temperature.

4-2
COMPOUNDS CONSIST OF DIFFERENT KINDS OF ATOMS

Suppose that a substance contains two or more atoms of *different* atomic number connected with a chemical bond. We call this type of substance a **compound**. A quick way to spot a compound is to look at its chemical formula. If the chemical formula is made up of two or more *different* symbols (elements), the substance is a compound. Examples of compounds we have already mentioned are water, H_2O; calcium phosphate, $Ca_3(PO_4)_2$; carbon dioxide, CO_2; aluminum sulfate, $Al_2(SO_4)_3$; methane, CH_4; phosphoric acid, H_3PO_4; aluminum oxide, Al_2O_3; calcium hydroxide, $Ca(OH)_2$; nitric oxide, NO; octane, C_8H_{18}; hydrogen peroxide, H_2O_2.

Another way to tell whether the formula represents an element or a compound is as follows. If there is only *one* capital letter in the formula, it's an element. If there are *two or more* capital letters, it's a compound.

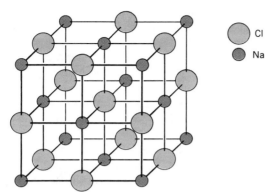

Cl

Na

FIGURE 4-2 The sodium chloride structure.

Again, as was true of elements, some of the compounds just mentioned, like water, exist as separate molecules at room temperature. Water is, of course, a liquid at room temperature. The separate molecules exist in the liquid.

In other compounds like sodium chloride (common table salt, NaCl), there are no individual molecules because each atom is connected to several other atoms. We discussed a similar situation with the diamond. In a piece of NaCl large enough to see, there are billions and billions of atoms. This type of arrangement, illustrated in Figure 4-2, is also called a giant array of atoms. However, we still write the chemical formula of sodium chloride as NaCl.

Crystals of sodium chloride.

Although this way of writing the chemical formula makes it appear that sodium chloride is made up of separate molecules, the formula only indicates that there is one Na atom for each Cl atom. Chemists generally can determine whether the formula of a substance represents separate atoms or molecules or a giant array of atoms. As you learn more chemistry, you will also be able to determine this.

In addition to liquids and solids, many compounds exist as gases at room temperature. Examples are CO_2, CH_4, and NO. The molecules in the gas are separate.

Please do not get the idea, from what we have said, that the type of bonding is the same in substances that are as different as NaCl, H_2O, diamond, and metals such as aluminum. The bonding is very different; for the full story, you will have to study more chemistry than is in this book. However, some details are given in Chapter 15.

EXAMPLE 1 Which of the following formulas represent elements, and which represent compounds? The formulas are Cl_2, N_2O_4, C_2H_6, S_8, P_4O_{10}, K, SO_2, Fe, CO, and I_2.

Solution: It doesn't matter whether you know the names of the substances to answer this one. However, for the curious, we will include the names in the answer.

ELEMENTS		COMPOUNDS	
Cl_2	chlorine	N_2O_4	dinitrogen tetraoxide
S_8	sulfur	C_2H_6	ethane
K	potassium	P_4O_{10}	tetraphosphorus decaoxide
Fe	iron	SO_2	sulfur dioxide
I_2	iodine	CO	carbon monoxide

4-3
MIXTURES CONSIST OF TWO OR MORE SUBSTANCES

A **mixture** has two or more substances (compounds, elements, or both) mixed up together. These different substances are not connected to each other by chemical bonds and can be mixed in *any* proportions.

For example, the compound sodium chloride (NaCl) and the element iron (as a powder) mixed together form a mixture. Another mixture, usually called a **solution**, is the compound sugar dissolved in the compound water. The key here is that you can separate the substances in the mixture without breaking chemical bonds. You could get the iron powder out of the salt with a magnet, and you could get the sugar out of the water by evaporating the water.

Some chemists might argue that solutions are not true mixtures because of the weak "interactions" (most chemists do not think of them as real chemical bonds) between the water molecules and the molecules of the dissolved substance. Nor can you usually dissolve as much of something in water as you might want to, since most substances have a limited solubility. However, up to the solubility limit, proportions can vary. Many chemists do think of solutions as mixtures, and in this book we shall consider solutions as mixtures.

EXAMPLE 2   Which of the following sets of substances are mixtures, which are compounds, and which are elements?

salt and sugar	potassium metal
sulfur and aluminum powder	salt
	S_8
air (separate molecules of O_2, N_2, Ar, CO_2, H_2O, etc.)	PCl_3
	pure ice
liquid oxygen	ginger ale

Solution:

MIXTURES	COMPOUNDS	ELEMENTS
salt and sugar	salt	liquid oxygen
sulfur and aluminum powder	PCl_3	potassium metal
air	pure ice	S_8
ginger ale		

PROBLEMS

KEYED PROBLEMS

1. Which of the following formulas represent elements, and which represent compounds? The formulas are F_2, N_2O_5, C_3H_8, P_4, PCl_5, Na, SO_2, Cr, CO_2, and Br_2.

2. Which of the following sets of substances are mixtures, which are compounds, and which are elements?

salt and pepper	sugar
iron powder and charcoal	P_4
polluted air	SF_6
liquid nitrogen	cherry soda
rubidium metal	steam

SUPPLEMENTAL PROBLEMS

3. Define the following terms.
   a. element
   b. compound
   c. "giant array" of atoms
   d. mixture
   e. monatomic substance

4. List two examples of substances that fit into the following categories.
   a. elements that are liquid at room temperature
   b. monatomic elements
   c. diatomic elements
   d. compounds that are liquid at room temperature
   e. compounds that are gaseous at room temperature
   f. compounds that are a giant array of atoms
   g. elements that are a giant array of atoms

5

SCIENTIFIC NOTATION

We said in Chapter 1 that atoms are very, very small. Because of this extreme smallness, a tremendous number of atoms are needed to make up an amount of matter that you could see or weigh. In the tip of your little finger, there are about this many atoms:

$$10,000,000,000,000,000,000,000$$

This number is ten sextillion.

If you took this number of atoms and laid them side by side, you could make a line of them that would reach out to the sun and back about five times. If you took this number of water molecules and mixed them with all the water in the oceans of the world, any glass of water you took would probably contain one or two of the original water molecules. For you to become as small as a water molecule, you would have to shrink your length down over ten billion times.

You can see that when we are dealing with atoms and molecules, we need to be able to use very big and very small numbers. For this reason, scientists have developed a very handy notation for large and small numbers. This notation is called **scientific notation** and is the subject of this chapter.

5-1
AN EXPONENT TELLS HOW MANY TIMES
TO MULTIPLY A NUMBER BY ITSELF

The key to using scientific notation is to understand numbers that have exponents attached to them. Look at the equation

$$3 \times 3 = 3^2 = 9$$

(Opposite) The great galaxy in Andromeda is about 10^{22} miles from earth. It contains about 10^{11} stars and is about 10^{20} miles in diameter.

The "2" that is a half-space above the number 3 is called the **exponent** or **power** of the number 3. You would read 3^2 as "three to the second power" or, in this case, because exponents equal to 2 are so common, "three squared." Whatever you call it, the exponent is the number of times that you multiply the number by itself.

Now look at the expression 2^{10}. This says that you multiply 2 by itself 10 times. We say that we are "raising 2 to the 10th power," or "taking 2 to the 10th." What you get when you do this is

$$2^{10} = 2 \times 2 \times 2 \times 2 \times 2 \times 2 \times 2 \times 2 \times 2 \times 2 = 1024$$

Now you can see how nice **exponential notation** is. It is much shorter to write 2^{10} than it is to write $2 \times 2 \times 2 \times 2 \times 2 \times 2 \times 2 \times 2 \times 2 \times 2$. But you might ask: Why not simply write 2^{10} as 1024? Isn't that just about as easy? Well, it is for 2^{10}, but what about 2^{100}? This is 100 twos multiplied together. Not only is that difficult to write out, but the actual number it multiplies out to be is about this big:

$$1,000,000,000,000,000,000,000,000,000,000$$

In this case there is no contest; 2^{100} beats them all as the shortest notation.

5-2
A TEN WRITTEN WITH EXPONENTS ALLOWS US TO WRITE NUMBERS IN SCIENTIFIC NOTATION

Since our number system is based on the number 10, it is especially useful to write the number 10 with exponents. Some of the powers of 10 are the following:

$$10^1 = 10$$

$$10^2 = 10 \times 10 = 100$$

$$10^3 = 10 \times 10 \times 10 = 1000$$

$$10^4 = 10 \times 10 \times 10 \times 10 = 10,000$$

This list can be written in a slightly different way. We will multiply each number that has an exponent by "1":

$$10 = 1 \times 10^1$$

$$100 = 1 \times 10^2$$

$$1000 = 1 \times 10^3$$

$$10,000 = 1 \times 10^4$$

Let's try writing a few other numbers in a similar manner:

$$200 = 2 \times 100 = 2 \times 10^2$$

$$3000 = 3 \times 1000 = 3 \times 10^3$$

$$5,000,000 = 5 \times 1,000,000 = 5 \times 10^6$$

For the last number, multiply 10 by itself six times to get 1,000,000. To read 5×10^6, say "five times ten to the sixth."

Another example that really shows the great space saving of exponential notation is

$$6,000,000,000,000,000,000,000,000,000,000,000 = 6 \times 10^{33}$$

Read it "six times ten to the thirty-three" or "six times ten to the thirty-third." Notice that there are 33 zeros after the 6 and that the exponent of the 10 is 33. Thus, to figure out the exponent for this number, just count the zeros.

EXAMPLE 1   Write the number 90,000,000,000,000,000 as a 9 multiplied by a 10 with some exponent.

Solution: Since there are 16 zeros after the 9, the answer is 9×10^{16}, which can be read "nine times ten to the sixteenth."   ∎

This way of writing exponential notation is called **scientific notation,** probably because scientists use it a lot. Writing numbers in scientific notation has the following layout. You first write a number between 1 and 10, then a "times sign," and finally a 10 with some exponent. Let's show this as follows:

number between 1 and 10 $\times$ 10^{exponent}

Actually, a "one" can also be used as the number. For instance, 1×10^6 is a perfectly good number in scientific notation. One reason that scientists like this notation is that it makes it very easy to keep track of the decimal point when multiplying and dividing numbers. And anything that makes it easy to keep track of the decimal point is welcome news indeed.

So far, we haven't written any decimal points. So let's begin. Suppose we want to write the number 362 in scientific notation. We would write the following:

$$362 = 3.62 \times 100 = 3.62 \times 10^2$$

Another example is to write 47,625 in scientific notation:

$$47,625 = 4.7625 \times 10,000 = 4.7625 \times 10^4$$

If it isn't clear how we did these conversions to scientific notation, the following discussion on some points of math may help.

5-3
SOME DETAILS OF MULTIPLICATION AND FRACTIONS

A basic rule of our number system is that the **order of addition does not matter.** For example,

$$2 + 3 + 4 = 3 + 2 + 4 = 4 + 3 + 2 = 9$$

There are even more ways of mixing up the order than we have shown. Adding in any order would add up to nine.

Another rule of our number system is that the **order of multiplication does not matter.** For example,

$$4 \times 3 = 3 \times 4 = 12$$

EXAMPLE 2   Write the product $3 \times 4 \times 5$ in all the six possible ways that the numbers can be mixed up. Calculate the answer of each way.

Solution: The six ways and the product of each way are as follows:

$$3 \times 4 \times 5 = 12 \times 5 = 60$$
$$3 \times 5 \times 4 = 15 \times 4 = 60$$
$$4 \times 3 \times 5 = 12 \times 5 = 60$$
$$4 \times 5 \times 3 = 20 \times 3 = 60$$
$$5 \times 4 \times 3 = 20 \times 3 = 60$$
$$5 \times 3 \times 4 = 15 \times 4 = 60 \quad \blacksquare$$

There is a subrule of the rule just mentioned. The subrule says that **the grouping of numbers in multiplication does not matter.** This subrule refers to using parentheses around the numbers. For example, $3 \times 4 \times 5$ can be written as

$$(3 \times 4) \times 5 \quad \text{or as} \quad 3 \times (4 \times 5)$$

The first notation says that you are supposed to multiply the 3×4 first:

$$(3 \times 4) \times 5 = 12 \times 5 = 60$$

The second notation says that you multiply the 4×5 first:

$$3 \times (4 \times 5) = 3 \times 20 = 60$$

Any way you do it, you always get the same answer. Parentheses are very useful because they break mathematical expressions into smaller pieces that are easier to work with.

The multiplication sign is often left out, and the parentheses serve just as well. For instance,

$$3 \times (4 \times 5) = 3(4 \times 5)$$

EXAMPLE 3   In the product $8 \times 10 \times 2$, insert parentheses in reasonable places and multiply.

Solution: You will learn from experience how to decide what is "reasonable." For instance, writing $(8 \times 10 \times 2)$ is not very reasonable, because it doesn't change anything. Here are some reasonable possibilities:

$$8(10 \times 2) = 8 \times 20 = 160$$

and

$$(8 \times 10)2 = 80 \times 2 = 160$$

You can also replace all the "$\times$" signs by parentheses if you want to:

$$(8)(10)(2) = 160$$

The order in which you multiply the numbers does not matter.   ■

We can treat fractions in the same manner that we treat whole numbers. We can thus write

$$\tfrac{1}{2} \times 10 = 10 \times \tfrac{1}{2} = 5$$

EXAMPLE 4   Write the product $\tfrac{1}{2} \times 10 \times 4$ with parentheses in reasonable places and multiply.

Solution:

$$(\tfrac{1}{2} \times 10)4 = 5 \times 4 = 20$$
$$\tfrac{1}{2}(10 \times 4) = \tfrac{1}{2} \times 40 = 20$$
$$(\tfrac{1}{2})(10)(4) = 20 \quad \blacksquare$$

If you look at the equation

$$\tfrac{1}{2} \times 10 = \tfrac{10}{2} = 5$$

you see that multiplication by a fraction whose numerator (top part) is 1 is the same thing as division by the denominator (bottom part) of the fraction. This, together with the rules of our number system already given, allows us to make some very useful manipulations with fractions.

Consider the multiplication $\tfrac{1}{2} \times 10$. We can easily put a 1 under the 10 so that it looks like a fraction. This is just for our convenience: When any number is divided by 1, the number doesn't change. Doing this we have

$$\tfrac{1}{2} \times 10 = \tfrac{1}{2} \times \tfrac{10}{1}$$

We can then write the two fractions as one fraction because the rule for multiplying fractions says that you "multiply the tops and multiply the bottoms":

$$\frac{1}{2} \times \frac{10}{1} = \frac{1 \times 10}{2 \times 1}$$

Now we have a product in both the numerator (top) and denominator (bottom) of the fraction. Since the order of multiplication doesn't matter, it is all right to interchange the 1 and the 2 in the denominator:

$$\frac{1 \times 10}{2 \times 1} = \frac{1 \times 10}{1 \times 2}$$

Splitting the fraction up again gives

$$\frac{1 \times 10}{1 \times 2} = \frac{1}{1} \times \frac{10}{2}$$

which of course comes out to be

$$1 \times \tfrac{10}{2} = 1 \times 5 = 5$$

Writing the whole thing out in one step gives us

$$\frac{1}{2} \times 10 = \frac{1}{2} \times \frac{10}{1} = \frac{1 \times 10}{2 \times 1} = \frac{1 \times 10}{1 \times 2} = \frac{1}{1} \times \frac{10}{2} = 1 \times 5 = 5$$

The nice thing about math is that it is so flexible. Once you understand the rules, you can make things work for you by simplifying and rearranging.

EXAMPLE 5   Multiply the following after rearranging it as was just demonstrated.

$$30 \times \tfrac{1}{3}$$

Solution: First write it as one big fraction, then rearrange and divide:

$$30 \times \frac{1}{3} = \frac{30}{1} \times \frac{1}{3} = \frac{30 \times 1}{1 \times 3} = \frac{30 \times 1}{3 \times 1} = \frac{30}{3} \times \frac{1}{1} = 10 \times 1 = 10 \quad \blacksquare$$

5-4
SCIENTIFIC NOTATION USING POSITIVE EXPONENTS

Now we can go back and see how we wrote 362 in scientific notation. Certainly we can multiply 362 by 1 and it doesn't change: $362 \times 1 = 362$. We can, however, write 1 in a special way—namely, as $\frac{100}{100}$. Therefore,

$$\frac{100}{100} = 1$$

Then if we remember that any number can be multiplied by 1 in any form and the number doesn't change, we can write

$$362 = 362 \times 1 = 362 \times \frac{100}{100}$$

We can rearrange things a bit and turn 362 into scientific notation:

$$362 \times \frac{100}{100} = \frac{362}{1} \times \frac{100}{100} = \frac{362}{100} \times \frac{100}{1} = 3.62 \times 100 = 3.62 \times 10^2$$

Once you understand the reason for it, it is perfectly all right to just count the decimal places to convert numbers to scientific notation. Since 362 can be written as 362., to get 3.62 you count two decimal places to the left.

Then, since moving the decimal point two places to the left is the same as dividing by 100 (after all, 3.62 is 100 times smaller than 362.), you have to multiply the 3.62 by 100 or 10^2 so you don't change its value.

Moving the decimal point is illustrated below using arrows under the numbers.

$$362. = 3\,6\,2. = 3.62 \times 10^2$$

Two arrows mean we have moved the decimal point two places to the left.

EXAMPLE 6   Convert 47,625. to scientific notation.

Solution: 47625. = 4 7 6 2 5. = 4.7625×10^4. Notice again that the exponent of the 10 is the same number as the number of places (represented by arrows) that you moved the decimal point to the left.   ■

EXAMPLE 7   Convert 94.7 to scientific notation.

Solution: 94.7 = 9 4.7 = 9.47×10^1 ■

EXAMPLE 8   Convert 5,000,764 to scientific notation.

Solution: 5,000,764. = 5 0 0 0 7 6 4. = 5.000764×10^6 ■

To convert numbers from scientific notation to regular notation, all you have to do is to reverse the procedure shown in Examples 6 through 8. **Regular notation** is the usual everyday way of writing numbers without exponents.

EXAMPLE 9   Convert 6.4×10^5 to regular notation.

Solution: 6.4×10^5 = 6.4 0 0 0 0 = 640,000., or 640,000 without the decimal point.   ■

Notice that we have moved the decimal point five places to the right (represented by five arrows) in Example 9. That is because $10^5 = 100,000$, and to multiply a number by 100,000 you have to move the decimal point five places to the right.

EXAMPLE 10   Convert 4.09×10^8 to regular notation.

Solution: $4.09 \times 10^8 = 4.\underrightarrow{0}\,\underrightarrow{9}\,\underrightarrow{0}\,\underrightarrow{0}\,\underrightarrow{0}\,\underrightarrow{0}\,\underrightarrow{0}\,\underrightarrow{0} = 409,000,000.$, or $409,000,000$ without the decimal point. The decimal point has been moved eight places to the right. ■

EXAMPLE 11   Convert 37.57×10^7 to regular notation.

Solution: Notice that 37.57×10^7 is not in scientific notation because 37.57 is not between 1 and 10. This does not matter. Just move the decimal point seven places to the right: $37.57 \times 10^7 = 37.\underrightarrow{5}\,\underrightarrow{7}\,\underrightarrow{0}\,\underrightarrow{0}\,\underrightarrow{0}\,\underrightarrow{0}\,\underrightarrow{0} = 375,700,000.$, or $375,700,000$ without the decimal. ■

We can summarize the rules for converting numbers from regular notation to scientific notation and vice versa.

1. To convert from regular notation to scientific notation, move the decimal point to the *left*.
2. To convert from scientific notation to regular notation, move the decimal point to the *right*.
3. Rules 1 and 2 apply to numbers greater than 1. We will discuss numbers less than 1 later in this chapter.

You can multiply numbers in scientific notation very easily if the following rule is used: **When you multiply numbers with exponents, you add the exponents.** Of course, the numbers that have the exponents, called **base numbers** or **bases,** have to be the same. This will be explained after the following examples. To illustrate the rule just given, multiply $10^2 \times 10^3$. We know that

$$10^2 = 100 \quad \text{and} \quad 10^3 = 1000$$

Since

$$100 \times 1000 = 100,000 = 10^5$$

it must be true that

$$10^2 \times 10^3 = 10^5$$

where we have added the exponents $(2 + 3 = 5)$.

The reason for adding the exponents might be clearer from the following way of looking at this illustration. Let's write out all the tens that we are dealing with:

$$100 = 10 \times 10 \quad \text{and} \quad 1000 = 10 \times 10 \times 10$$

Multiplying $100 \times 1000 = 10 \times 10 \times 10 \times 10 \times 10 = 10^5$, which is the same as we got by adding the exponents. You can see from this that each ten in the product is, in a way, "added," so that the exponent on the 10^5 is the "sum" of the *number of tens* that were multiplied together.

Now you can see why the base numbers have to be the same when you add the exponents. If they were not, you couldn't multiply the base numbers together the way we did. It is this multiplication of base numbers that gives us the rule for adding exponents.

EXAMPLE 12   Multiply $10^8 \times 10^{11}$.

Solution: Adding exponents we get

$$10^8 \times 10^{11} = 10^{8+11} = 10^{19}$$

The 10^8 means that there are eight tens multiplied together. The 10^{11} means that there are 11 tens multiplied together. So all in all there are 19 tens multiplied together to give 10^{19}. ■

We can now proceed to multiply more complicated numbers. For instance, multiply 3×10^5 by 2×10^7. The key to doing this is to (1) group together the "plain" numbers and (2) group together the numbers having exponents:

$$3 \times 10^5 \times 2 \times 10^7 = 3 \times 2 \times 10^5 \times 10^7 = 6 \times 10^{5+7} = 6 \times 10^{12}$$

The rearrangement is allowed because the order of multiplication doesn't matter.

EXAMPLE 13   Multiply 6×10^4 by 9×10^{27}.

Solution: First regroup and then multiply:

$$6 \times 10^4 \times 9 \times 10^{27} = 6 \times 9 \times 10^4 \times 10^{27} = 54 \times 10^{4+27} = 54 \times 10^{31}$$

To get the answer into scientific notation, we must write it as 5.4×10^{32}. To get this form, we have to convert 54 to scientific notation:

$$54 \times 10^{31} = 5.4 \times 10^1 \times 10^{31} = 5.4 \times 10^{32} \quad ■$$

EXAMPLE 14 Multiply the following and put the answer into scientific notation: $(9 \times 10^{15})(8 \times 10^{20})$.

Solution: As before, first regroup and then multiply:

$$(9 \times 10^{15})(8 \times 10^{20}) = 9 \times 10^{15} \times 8 \times 10^{20}$$

$$= 9 \times 8 \times 10^{15} \times 10^{20} = 72 \times 10^{35}$$

Converting 72×10^{35} into scientific notation gives 7.2×10^{36}. ∎

The next thing that we have to look at is the division of numbers expressed in scientific notation. But first, let's look at the division of some numbers expressed in powers of 10. You know that

$$\frac{1000}{100} = 10$$

If we write it in a slightly different way, we have

$$\frac{1000}{100} = \frac{10 \times 10 \times 10}{10 \times 10} = \frac{10}{1} \times \frac{10}{10} \times \frac{10}{10} = \frac{10}{1} \times 1 \times 1 = 10$$

You can see that each 10 in the denominator divides a 10 in the numerator, giving 1.

In other words, each 10 on the bottom "removes" a 10 on the top. At the start there are three 10's on the top and two 10's on the bottom. After division, there is only one 10 on the top.

This method of "removing" the 10's gives us the idea that is used in dividing numbers with exponents. But don't forget that "removing" 10's is really dividing 10's. Since $1000 = 10^3$ and $100 = 10^2$, we can write

$$\frac{1000}{100} = \frac{10^3}{10^2}$$

We know that the division 1000/100 gives 10. so it must be true that

$$\frac{10^3}{10^2} = 10$$

Since we can write 10 as 10^1, we have

$$\frac{10^3}{10^2} = 10^1$$

And if you look carefully, you realize that $3 - 2 = 1$. So, you can **divide numbers with exponents by subtracting the exponents.** Writing our problem with the subtraction clearly indicated, we have

$$\frac{10^3}{10^2} = 10^{3-2} = 10^1 = 10$$

As we said before, we can do this subtraction of the exponents because the two 10's on the bottom "removed" two of the 10's on the top, leaving only one 10.

EXAMPLE 15   Perform the following division:

$$\frac{10^8}{10^3}$$

Solution: By subtracting exponents we get

$$\frac{10^8}{10^3} = 10^{8-3} = 10^5$$

To see it more clearly, let's write out all the 10's:

$$\frac{10^8}{10^3} = \frac{10 \times 10 \times 10 \times \cancel{10} \times \cancel{10} \times \cancel{10} \times 10 \times 10}{\cancel{10} \times \cancel{10} \times \cancel{10}}$$

$$= 10 \times 10 \times 10 \times 1 \times 1 \times 1 \times 10 \times 10$$

$$= 10 \times 10 \times 10 \times 10 \times 10$$

$$= 10^5 \quad \blacksquare$$

If there are numbers multiplying the 10's that have exponents, just rearrange. Then you can divide the "plain" numbers first; after that, you can divide the 10's that have exponents.

EXAMPLE 16   Compute

$$\frac{6 \times 10^7}{4 \times 10^2}$$

Solution: Separating the expression into two parts we have

$$\frac{6}{4} \times \frac{10^7}{10^2} = 1.5 \times 10^5 \quad \blacksquare$$

EXAMPLE 17   Compute

$$\frac{(8 \times 10^3)(3 \times 10^7)}{(6 \times 10^2)(2 \times 10^4)}$$

Solution: Regroup, multiply the numerators and the denominators, and then divide:

$$\frac{8 \times 3 \times 10^3 \times 10^7}{6 \times 2 \times 10^2 \times 10^4} = \frac{24 \times 10^{10}}{12 \times 10^6} = \frac{24}{12} \times \frac{10^{10}}{10^6} = 2 \times 10^4 \quad \blacksquare$$

5-5
SCIENTIFIC NOTATION USING NEGATIVE EXPONENTS

Suppose now, in doing division with exponents, you run into the following expression:

$$\frac{10^2}{10^4}$$

If you follow the usual rule and subtract exponents, you will get

$$\frac{10^2}{10^4} = 10^{2-4} = 10^{-2}$$

What does 10^{-2} mean? Is it reasonable? Let's see. First write out all the 10's:

$$\frac{10^2}{10^4} = \frac{10 \times 10}{10 \times 10 \times 10 \times 10} = \frac{10}{10} \times \frac{10}{10} \times \frac{1}{10} \times \frac{1}{10} = 1 \times 1 \times \frac{1}{10} \times \frac{1}{10} = \frac{1}{100}$$

It is clear, then, that

$$\frac{10^2}{10^4} = \frac{1}{100}$$

The two 10's on the top have "removed" two of the four 10's on the bottom, leaving two 10's on the bottom.

Since $10^2/10^4 = 10^{-2}$ by our rule of subtracting exponents, and since $10^2/10^4 = 1/100$ by division as was shown above, it must be true that

$$10^{-2} = \frac{1}{100}$$

Furthermore, since $100 = 10^2$, we have that $1/100 = 1/10^2$ and thus

$$10^{-2} = \frac{1}{10^2}$$

What we have shown in this case, and what is true for any numbers, is that **a number with a negative exponent really represents 1 over the same number with a positive exponent.**

If we let n stand for *any* number, we can say that

$$10^{-n} = \frac{1}{10^n}$$

EXAMPLE 18   Divide

$$\frac{10^3}{10^4}$$

Solution: By using the subtraction rule we get

$$\frac{10^3}{10^4} = 10^{3-4} = 10^{-1} = \frac{1}{10^1} = \frac{1}{10}$$

EXAMPLE 19   Divide

$$\frac{10^3}{10^3}$$

Solution: By using the subtraction rule we get

$$\frac{10^3}{10^3} = 10^{3-3} = 10^0$$

The expression 10^0 may look a bit strange, but if we work it out by dividing the 10's, we see that

$$\frac{10^3}{10^3} = \frac{10 \times 10 \times 10}{10 \times 10 \times 10} = 1$$

From Example 19, it would seem that

$$10^0 = 1$$

In fact, a basic rule of exponents is that **any number to the zero power equals one.**

Now let's look at some more properties of negative exponents. We show below that we can represent a 10 with a negative exponent as a decimal that is less than 1.

Since

$$\frac{1}{10} = 0.1 \quad \text{and} \quad \frac{1}{10} = 10^{-1}$$

it must be true that

$$10^{-1} = 0.1$$

Similarly,

$$\frac{1}{100} = 0.01 = 10^{-2}$$

We can make a table of negative exponents and their decimal equivalents.

NUMBER	DECIMAL	EXPONENT FORM
$\dfrac{1}{10}$	0.1	10^{-1}
$\dfrac{1}{100}$	0.01	10^{-2}
$\dfrac{1}{1000}$	0.001	10^{-3}
$\dfrac{1}{10,000}$	0.0001	10^{-4}

EXAMPLE 20   Compute

$$\frac{4 \times 10^{-3}}{2 \times 10^4}$$

Solution: Group and divide as we have done before:

$$\frac{4 \times 10^{-3}}{2 \times 10^4} = \frac{4}{2} \times \frac{10^{-3}}{10^4} = 2 \times 10^{-3-4} = 2 \times 10^{-7}$$

Remember that -3 "take away" 4 is written as $-3 - 4$.   ■

EXAMPLE 21   Compute

$$\frac{7 \times 10^2}{3 \times 10^{-4}}$$

Solution: After grouping and dividing we have

$$\frac{7 \times 10^2}{3 \times 10^{-4}} = \frac{7}{3} \times \frac{10^2}{10^{-4}} = 2.33 \times 10^{2-(-4)}$$

$$= 2.33 \times 10^{2+4} = 2.33 \times 10^6 \quad \blacksquare$$

Example 21 illustrates *subtraction of a negative number.* Notice that $2 - (-4) = 2 + 4 = 6$, since a "minus times a minus equals a plus" and the expression $- (-4)$ is really just like -1×-4 or $(-1)(-4)$. The expression $(-1)(-4)$ can be written as $-1(-4)$, where we have left out the first set of parentheses. Now all we have to do is leave off the 1 and we have $-(-4)$, which equals $+4$.

EXAMPLE 22   Compute

$$\frac{9 \times 10^{-2}}{4 \times 10^{-6}}$$

Solution: Again, regroup and then divide:

$$\frac{9 \times 10^{-2}}{4 \times 10^{-6}} = \frac{9}{4} \times \frac{10^{-2}}{10^{-6}} = 2.25 \times 10^{-2-(-6)}$$

$$= 2.25 \times 10^{-2+6} = 2.25 \times 10^4 \quad \blacksquare$$

EXAMPLE 23   Compute

$$\frac{6.4 \times 10^{-4}}{2}$$

Solution: The answer is

$$\frac{6.4 \times 10^{-4}}{2} = \frac{6.4}{2} \times 10^{-4} = 3.2 \times 10^{-4} \quad \blacksquare$$

You might occasionally see someone try to solve Example 23 in the following way:

$$\frac{6.4 \times 10^{-4}}{2} = 3.2 \times 10^{-2} \quad \leftarrow \text{ !! THIS IS WRONG !!}$$

They have forgotten one of the rules of multiplication. They have not grouped and multiplied correctly. (See Examples 3 and 4 in this chapter for a review of grouping.) Since division by 2 is like multiplication by $\frac{1}{2}$, we can write

$$\frac{6.4 \times 10^{-4}}{2} = \tfrac{1}{2} \times 6.4 \times 10^{-4}$$

and using the rules we have already learned,

$$\tfrac{1}{2} \times 6.4 \times 10^{-4} = 3.2 \times 10^{-4}$$

EXAMPLE 24   Compute

$$\frac{9.6 \times 10^{-8}}{10^{-6}}$$

Solution:

$$\frac{9.6 \times 10^{-8}}{10^{-6}} = 9.6 \times \frac{10^{-8}}{10^{-6}} = 9.6 \times 10^{-8-(-6)} = 9.6 \times 10^{-2} \quad \blacksquare$$

EXAMPLE 25   Compute

$$\frac{5}{2 \times 10^{11}}$$

Solution:

$$\frac{5 \times 1}{2 \times 10^{11}} = \frac{5}{2} \times \frac{1}{10^{11}} = 2.5 \times 10^{-11}$$

where we have used the rule that $1/10^n = 10^{-n}$. Another way to look at this is to remember that $1 = 10^0$:

$$\frac{1}{10^{11}} = \frac{10^0}{10^{11}} = 10^{0-11} = 10^{-11} \quad \blacksquare$$

We will now discuss how to change numbers less than 1 from regular notation to scientific notation. We will also discuss the reverse procedure, namely, changing numbers less than 1 from scientific notation to regular notation.

EXAMPLE 26   Convert 0.34 into scientific notation.

Solution: To do this, we will use the fact that $10/10 = 1$ and that multiplication by 1 doesn't change a number:

$$0.34 = 0.34 \times \tfrac{10}{10} = 0.34 \times 10 \times \tfrac{1}{10} = 0.34 \times 10 \times 10^{-1} = 3.4 \times 10^{-1} \quad \blacksquare$$

Once you understand what you are doing, you can simply move the decimal point. In Example 26, you would move the decimal point one place to the right and then multiply by 10^{-1}.

EXAMPLE 27   Convert 0.00076 to scientific notation.

Solution: We will do this problem by moving the decimal point. We need to move the decimal point four places to the right to get 7.6. At the same time we have to multiply by 10^{-4}, since moving the decimal point four places to the right has multiplied the number by 10^4:

$$0.0\,0\,0\,7\,6 = 7.6 \times 10^{-4}$$

Notice that the negative exponent is the same number as the number of places (arrows) that we moved the decimal point to the *right*.   ■

EXAMPLE 28   Convert 6.47×10^{-3} to regular notation.

Solution: We will do this in two ways:

first way: $6.47 \times 10^{-3} = 6.47 \times 0.001 = 0.00647$

second way: $6.47 \times 10^{-3} = 0\,0\,0\,6.4\,7 = 0.00647$

In the second way, we moved the decimal point three places to the *left*.   ■

We can summarize the rules for converting numbers from regular notation to scientific notation and vice versa, when the numbers are *less* than 1.

1. To convert from regular notation to scientific notation, move the decimal point to the *right*.
2. To convert from scientific notation to regular notation, move the decimal point to the *left*.

5-6
USING A CALCULATOR IN SCIENTIFIC NOTATION

Although it is important that you know how to perform calculations without a calculator, using a calculator can certainly save you a great amount of work. To do calculations in this section, you will need a full scientific calculator. We will assume that you have a calculator that uses algebraic logic, not Reverse Polish Notation (RPN, in calculators made by Hewlett–Packard). If you have one of the RPN calculators, see your instruction manual.

To enter the exponent part of a number in scientific notation, calculators have either an **EE** or an **exp** button. The following instructions will use the **EE** notation. Let's look at how you would enter some of the examples from this chapter into a calculator. Refer to previous sections of this chapter for the numerical answers. Each of the following examples refers to the same example number previously discussed in this chapter.

Example 1. 9×10^{16} is entered by punching the buttons **9, EE, 16**. Notice that a "times sign" is *not* punched between the number part and the exponent part, even though there is one when you write the number. The exponent part is shown on the right side of the display, which will look something like this: $\underline{9.\qquad 16}$ The decimal point is entered automatically.

Example 12. $10^8 \times 10^{11}$. Entry: **1, EE, 8, ×, 1, EE, 11, =**. You may have to put a **1** before the **EE**. Some calculators put it in for you, whereas others consider **EE, 8** as if you had entered **0, EE, 8**, which of course equals zero. You might check yours.

Example 13. $6 \times 10^4 \times 9 \times 10^{27}$. Entry: **6, EE, 4, ×, 9, EE, 27, =**.

Example 14. $(9 \times 10^{15})(8 \times 10^{20})$. Entry: **9, EE, 15, ×, 8, EE, 20, =**.

Example 15. $10^8/10^3$. Entry: **1, EE, 8, ÷, 1, EE, 3, =**.

Example 16. $(6 \times 10^7)/(4 \times 10^2)$. Entry: **6, EE, 7, ÷, 4, EE, 2, =**.

NOTE: The "/" in Example 16 means the same thing as the "÷" sign. It is the line separating the numerator and the denominator.

Example 17.

$$\frac{(8 \times 10^3)(3 \times 10^7)}{(6 \times 10^2)(2 \times 10^4)}$$

Entry: **8, EE, 3, ×, 3, EE, 7, ÷, 6, EE, 2, ÷, 2, EE, 4, =**. Notice that in this example you need two ÷ signs. The second term in the denominator *divides* the fraction. Don't make the mistake of punching in a × sign.

Example 18. $10^3/10^4$. Entry: **1, EE, 3, ÷, 1, EE, 4, =**.

Example 19. $10^3/10^3$. Entry: **1, EE, 3, ÷, 1, EE, 3, =**.

Example 20. $(4 \times 10^{-3})/(2 \times 10^4)$. Entry: **4, EE, +/−, 3, ÷, 2, EE, 4, =**.

Notice that the **+/−** button is used to enter a negative exponent. On most calculators, the following entries would both give 4×10^{-3}: **4, EE,**

+/−, **3** or **4**, **EE**, **3**, +/−. On a few calculators, only the second choice works.

Example 21. $(7 \times 10^2)/(3 \times 10^{-4})$. Entry: **7**, **EE**, **2**, ÷, **3**, **EE**, +/−, **4**, =.

Example 22. $(9 \times 10^{-2})/(4 \times 10^{-6})$. Entry: **9**, **EE**, +/−, **2**, ÷, **4**, **EE**, +/−, **6**, =.

Example 23. $(6.4 \times 10^{-4})/2$. Entry: **6.4**, **EE**, +/−, **4**, ÷, **2**, =.

Example 24. $(9.6 \times 10^{-8})/10^{-6}$. Entry: **9.6**, **EE**, +/−, **8**, ÷, **1**, **EE**, +/−, **6**, =.

Example 25. $5/(2 \times 10^{11})$. Entry: **5**, ÷, **2**, **EE**, **11**, =.

Notice that the calculator always displays an answer in scientific notation if the number is too big or too small to fit in the display. On some calculators, an answer will always be displayed in scientific notation if the numbers are entered in scientific notation. On others, the answer will be in regular notation if it will fit in the display.

Most calculators have the ability to convert numbers in the display back and forth between regular notation and scientific notation. The following are two ways they do this (your calculator may do one or the other or neither).

BUTTON(S)	WHAT THEY CONVERT
F ↔ E	Regular notation to scientific notation
F ↔ E	Scientific notation to regular notation
INV,EE,=	Scientific notation to regular notation
EE,=	Regular notation to scientific notation

5-7
POWERS AND ROOTS OF NUMBERS

The **square root** of a number is another number that, when multiplied by itself, gives the original number. Thus 2 is the square root of 4 because $2 \times 2 = 4$.

The expression $\sqrt{4}$ means the square root of 4. Another way of expressing the square root of 4 is to write

$$\sqrt{4} = 4^{1/2}$$

where the $\frac{1}{2}$ is a superscript on the 4. Since $\sqrt{4} = 2$, it must be that $4^{1/2} = 2$. You can convince yourself that $4^{1/2}$ is a sensible way to write the square root of 4 by looking at the following:

$$2 \times 2 = 4^{1/2} \times 4^{1/2} = 4^{1/2+1/2} = 4^1 = 4$$

Since $4^{1/2} \times 4^{1/2} = 4$, $4^{1/2}$ must be the square root of 4 from our definition of the square root. We have just used the rule that you add exponents when you multiply numbers.

How do we simplify an expression like $(9 \times 4)^{1/2}$? Let's multiply and see:

$$(9 \times 4)^{1/2} = 36^{1/2} = 6$$

But a rule of math says that we can also write

$$(9 \times 4)^{1/2} = 9^{1/2} \times 4^{1/2} = 3 \times 2 = 6$$

Since the answer is the same whether we multiply first and then take the square root, or take the square root of each term and then multiply, this rule seems reasonable. In general, for any numbers a and b and exponent $1/n$, we obtain

$$(a \times b)^{1/n} = a^{1/n} \times b^{1/n}$$

If you want to raise $a \times b$ to a power, say m, then

$$(a \times b)^m = a^m \times b^m$$

EXAMPLE 29   Calculate $(25 \times 4)^{1/2}$.

Solution:

$$(25 \times 4)^{1/2} = 25^{1/2} \times 4^{1/2} = 5 \times 2 = 10$$

or

$$(25 \times 4)^{1/2} = (100)^{1/2} = 10 \quad \blacksquare$$

EXAMPLE 30   Calculate $(4 \times 5)^2$.

Solution:

$$(4 \times 5)^2 = 4^2 \times 5^2 = 16 \times 25 = 400$$

or

$$(4 \times 5)^2 = (20)^2 = 20 \times 20 = 400 \quad \blacksquare$$

EXAMPLE 31   Using a calculator, find $7^{1/2}$.

Solution: Press the following keys: **7, $\sqrt{}$** or **7, INV, x^2**. The display will read 2.6457513. Notice that the sequence **INV, x^2** is the same as square root. **INV** stands for "inverse." The reason the button is called "inverse" is that taking a square root and squaring are "opposite" operations. Addition and subtraction are also inverse or "opposite" operations, as are multiplication and division. ∎

Any whole number, fraction, or decimal can be used as an exponent. For instance, what does $27^{1/3}$ mean? The expression $27^{1/3}$ is a number that, when cubed (**cubed** means multiplied by itself three times), gives 27. The expression $27^{1/3}$ is read "27 to the one-third" or "the cube root of 27." The **cube root** of a number is another number that, when multiplied by itself three times, gives the original number. To understand it, look at the following:

$$27^{1/3} \times 27^{1/3} \times 27^{1/3} = 27^{1/3+1/3+1/3} = 27^1 = 27$$

What is the numerical value of $27^{1/3}$? It is 3. This is so because

$$3 \times 3 \times 3 = 27$$

EXAMPLE 32   What is $8^{2/3}$?

Solution:

$$8^{2/3} = 8^{1/3 \times 2} = (8^{1/3})^2 = 8^{1/3} \times 8^{1/3} = 2 \times 2 = 4 \quad ∎$$

Only simple cases, where things "work out," can be done in your head. The more complicated cases require a calculator.

EXAMPLE 33   Compute $5^{2/3}$ on a calculator.

Solution: $2/3 = 0.66666 \ldots$. Press **5, y^x, 0.6666 . . . , =.** The display reads 2.92402. If your calculator has parentheses buttons, you can use them in the following way. Press **5, y^x, (, 2, ÷, 3,), =.** The notation $0.6666 \ldots$ means that the 6's repeat indefinitely. The author entered seven 6's into his calculator for this problem. Since you cannot enter fractions as exponents in a calculator, you must convert the fraction to its decimal equivalent or use parentheses.

NOTE: On some calculators, the y^x key may read x^y, a^x, or a^y. ∎

EXAMPLE 34   Compute $8^{4.67}$ on a calculator.

Solution: Press **8, y^x, 4.67, =.** The display will read 16498.   ∎

Even negative powers and roots can easily be handled on a calculator.

EXAMPLE 35   Compute $26^{-0.024}$ on a calculator.

Solution: Press **26, y^x, 0.024, +/−, =.** The display reads 0.924785.   ∎

Notice that taking roots of *negative numbers* gives an error message. Try taking $(-2)^{1/2}$ on your calculator. It will not work because the square root of a negative number, when multiplied by itself, would have to give the number. Since a "minus times a minus gives a plus," you cannot find two identical negative numbers that, when multiplied, give a negative number. For this reason, the square roots of negative numbers are called **imaginary numbers.**

You can easily take powers and roots of numbers in scientific notation. Let's look at $(8 \times 10^4)^2$. Clearing parentheses gives

$$8^2 \times (10^4)^2$$

What does $(10^4)^2$ mean? Working it out we have

$$(10^4)^2 = 10^4 \times 10^4 = 10^8$$

so

$$(8 \times 10^4)^2 = 8^2 \times (10^4)^2 = 64 \times 10^8$$

In general, for any exponents m and n we have

$$(10^m)^n = 10^{m \times n} \text{ or } 10^{mn}$$

and

$$(10^m)^{1/n} = 10^{m/n}$$

EXAMPLE 36   Compute $(10^6)^3$.

Solution:

$$(10^6)^3 = 10^{6 \times 3} = 10^{18}$$ ∎

EXAMPLE 37   Compute $(10^{10})^{1/2}$.

Solution:

$$(10^{10})^{1/2} = 10^{10/2} = 10^5 \quad \blacksquare$$

EXAMPLE 38   Compute $(9 \times 10^8)^{1/2}$.

Solution:

$$(9 \times 10^8)^{1/2} = 9^{1/2} \times 10^{8 \times 1/2} = 9^{1/2} \times 10^{8/2} = 3 \times 10^4 \quad \blacksquare$$

EXAMPLE 39   Compute $(8 \times 10^{15})^{1/3}$.

Solution:

$$(8 \times 10^{15})^{1/3} = 8^{1/3} \times 10^{15 \times 1/3} = 8^{1/3} \times 10^{15/3} = 2 \times 10^5 \quad \blacksquare$$

Sometimes you must "adjust" the exponent of the number so that the answer does not have a fractional exponent in it. This would only be necessary if you had a calculator that didn't have scientific notation capabilities.

EXAMPLE 40   Compute $(4.6 \times 10^7)^{1/2}$.

Solution: Since 7 is not exactly divisible by 2, we can write

$$(4.6 \times 10^7)^{1/2} = (46 \times 10^6)^{1/2} = 46^{1/2} \times 10^{6 \times 1/2} = 6.8 \times 10^3$$

where we have multiplied 4.6 by 10 and divided 10^7 by 10. The value of the number hasn't changed. By this procedure, we avoid fractional exponents in the answer. We found $46^{1/2}$ by using a calculator. $\blacksquare$

EXAMPLE 41   Compute $(9.3 \times 10^{-5})^{1/2}$.

Solution:

$$(9.3 \times 10^{-5})^{1/2} = (93 \times 10^{-6})^{1/2} = 93^{1/2} \times 10^{-6 \times 1/2}$$
$$= 93^{1/2} \times 10^{-6/2} = 9.6 \times 10^{-3} \quad \blacksquare$$

EXAMPLE 42   Compute $(10^9)^{1/4}$.

Solution:

$$(10^9)^{1/4} = (10 \times 10^8)^{1/4} = 10^{1/4} \times 10^{8 \times 1/4}$$
$$= 10^{1/4} \times 10^{8/4} = 10^{0.25} \times 10^{8/4} = 1.8 \times 10^2 \quad \blacksquare$$

As mentioned before, if you use a scientific calculator to find powers and roots of numbers, you don't have to adjust the exponent. The calculator will do all the work for you. However, for simple cases, the procedure outlined in Examples 36 through 42 works very well. (Of course, we used a calculator to find $46^{1/2}$, $93^{1/2}$, and $10^{0.25}$.)

EXAMPLE 43   Compute $(6.62 \times 10^{11})^{1/3}$.

Solution: Press **6.62**, **EE**, **11**, **y^x**, **0.3333** . . . , **=** or **6.62**, **EE**, **11**, **INV**, **y^x**, **3**, **=** or **6.62**, **EE**, **11**, **y^x**, **(**, **1**, **÷**, **3**, **)**, **=**. The display will read 8.7154 03, which in handwritten scientific notation would be 8.7154×10^3. The **INV**, **y^x** operation is the same as the **y$^{1/x}$** or **$\sqrt[x]{y}$** button found on some calculators. If you want to take $4^{1/3}$, for example, you can do it in up to four ways, depending on which calculator you have. Push: **4**, **y^x**, **0.333** . . . , **=** or **4**, **INV**, **y^x**, **3**, **=** or **4**, **y$^{1/x}$**, **3**, **=** or **4**, **y^x**, **(**, **1**, **÷**, **3**, **)**, **=**. The answer is 1.5874. $\blacksquare$

EXAMPLE 44   Compute $(4.58 \times 10^{-22})^{3.62}$.

Solution: Press **4.58**, **EE**, **+/−**, **22**, **y^x**, **3.62**, **=**. The display reads 5.6537 −78 or 5.65×10^{-78}, where we have rounded to three digits. $\blacksquare$

One final word about using calculators. It is important that you know how to do simple calculations without a calculator. After all, your calculator may malfunction at a critical time. In addition, you should have a feeling for what a reasonable answer is when using a calculator. Otherwise, you have no chance at all to spot a wrong answer caused by pushing the wrong buttons. There is an old rule that computer programmers use that applies to calculators also. It is the **GIGO** rule:

Garbage **I**n, **G**arbage **O**ut.

In other words, don't turn off your brain when you turn on your calculator.

PROBLEMS

KEYED PROBLEMS

1. Write 50,000,000 as a 5 multiplied by a 10 with some exponent.

2. Write the product $2 \times 5 \times 10$ in all the six possible ways that the numbers can be mixed up. Calculate the answer for each way.

3. In the multiplication $6 \times 7 \times 10$ put parentheses in reasonable places and multiply.

4. Write the multiplication $\frac{1}{4} \times 16 \times 5$ with parentheses in reasonable places and multiply.

5. Multiply the following after rearranging as was done in Example 5:

$$20 \times \tfrac{1}{4}$$

6. Convert 38,423 to scientific notation.

7. Convert 25.6 to scientific notation.

8. Convert 7,360,000 to scientific notation.

9. Convert 8.2×10^5 to regular notation.

10. Convert 5.72×10^8 to regular notation.

11. Convert 58.85×10^7 to regular notation.

12. Multiply $10^6 \times 10^{13}$.

13. Multiply 8×10^5 by 3×10^{15}.

14. Multiply the following and put the answer into scientific notation: $(5 \times 10^{12})(7 \times 10^6)$.

15. Perform the following division:

$$\frac{10^7}{10^5}$$

16. Compute

$$\frac{4 \times 10^8}{3 \times 10^5}$$

17. Compute

$$\frac{(5 \times 10^5)(8 \times 10^8)}{(9 \times 10^2)(3 \times 10^4)}$$

18. Divide

$$\frac{10^7}{10^8}$$

19. Divide

$$\frac{10^9}{10^9}$$

20. Compute

$$\frac{8 \times 10^{-6}}{6 \times 10^5}$$

21. Compute

$$\frac{4 \times 10^8}{3 \times 10^{-7}}$$

22. Compute

$$\frac{7 \times 10^{-4}}{3 \times 10^{-9}}$$

23. Compute

$$\frac{9.3 \times 10^{-15}}{3}$$

24. Compute

$$\frac{8.34 \times 10^{-12}}{10^{-9}}$$

25. Compute

$$\frac{9}{4 \times 10^{15}}$$

26. Convert 0.72 into scientific notation.

27. Convert 0.00048 into scientific notation.

28. Convert 4.25×10^{-3} into regular notation.

29. Calculate $(3 \times 27)^{1/2}$.

30. Calculate $(8 \times 2)^2$.

31. Using a calculator, find $5^{1/2}$.

32. What is $27^{2/3}$?

33. Compute $7^{2/3}$ on a calculator.

34. Compute $5^{6.71}$ on a calculator.

35. Compute $17^{-0.055}$ on a calculator.

36. Compute $(10^4)^5$.

37. Compute $(10^{16})^{1/4}$.

38. Compute $(25 \times 10^{12})^{1/2}$.

39. Compute $(64 \times 10^{21})^{1/3}$.

40. Compute $(7.2 \times 10^9)^{1/2}$.

41. Compute $(3.6 \times 10^{-7})^{1/2}$.

42. Compute $(10^{15})^{1/4}$.

43. Compute $(5.73 \times 10^{15})^{1/5}$.

44. Compute $(7.47 \times 10^{-16})^{5.41}$.

SUPPLEMENTAL PROBLEMS

45. Convert the following numbers to scientific notation.
    a. 3264
    b. 582
    c. 0.043
    d. 0.000572
    e. 4,670,000
    f. 0.000009
    g. 6,000,000,000
    h. 7001

46. Convert the following numbers to regular notation.
    a. 3.7×10^2
    b. 4.89×10^7
    c. 5.1×10^{-2}
    d. 8.92×10^{-9}
    e. 5.117×10^5
    f. 32.4×10^3
    g. 0.01×10^{-2}
    h. 0.32×10^4

47. Compute the following. Leave the answers in scientific notation.
    a. $(4.2 \times 10^2)(3.6 \times 10^8)$
    b. $(8 \times 10^{15})(6 \times 10^{23})$
    c. $(5.3 \times 10^{-2})(6 \times 10^5)$
    d. $(3.1 \times 10^{-5})(2 \times 10^{-10})$
    e. $(4.9 \times 10^6)(8 \times 10^{-12})$
    f. $(3 \times 10^{-10})(4)$

48. Compute the following. Leave the answers in scientific notation.
    a. $\dfrac{8 \times 10^5}{4 \times 10^2}$
    b. $\dfrac{6 \times 10^{-2}}{4 \times 10^7}$
    c. $\dfrac{4.7 \times 10^{-12}}{8.2 \times 10^{-15}}$
    d. $\dfrac{7.43 \times 10^{10}}{2 \times 10^{-4}}$
    e. $\dfrac{3.2}{4 \times 10^3}$
    f. $\dfrac{2.7 \times 10^{-2}}{4}$

49. Compute

$$\frac{(4.07 \times 10^{-8})(3.26 \times 10^{-5})}{8.99 \times 10^{-7}}$$

50. Compute

$$\frac{5.88 \times 10^5}{(3.16 \times 10^7)(7.02 \times 10^{-6})}$$

51. Compute

$$\frac{(3.27 \times 10^4)(8.53 \times 10^7)}{(5.55 \times 10^8)(7.76 \times 10^{-5})}$$

For Problems 52 through 54: To work these out without a calculator, convert all terms to numbers with the same exponent. Then you can add them. Example: $3 \times 10^3 + 2 \times 10^5 = 3 \times 10^3 + 200 \times 10^3 = 203 \times 10^3 = 2.03 \times 10^5$. Another way to work this example is $3 \times 10^3 + 2 \times 10^5 = 0.03 \times 10^5 + 2 \times 10^5 = 2.03 \times 10^5$.

If one number is much smaller than the other number in addition or subtraction, you can usually ignore the smaller one. For example, $3 \times 10^5 + 2 \times 10^{-4} = 3 \times 10^5$. We have ignored the 2×10^{-4} because it is so much smaller than 3×10^5. Let's write the numbers in regular notation to see this more clearly. Since $3 \times 10^5 = 300,000$ and $2 \times 10^{-4} = 0.0002$, when we add them we get $300,000 + 0.0002 = 300,000.0002$. Certainly we can ignore the 0002 after the decimal point. An analogy may be helpful in explaining this. If you are weighing an elephant and a flea lands on his ear, do you think the weight you get would be any different? I don't think so either!

52. Compute $7.61 \times 10^4 + 9.23 \times 10^5 + 4.61 \times 10^{-3} + 0.712$.

53. Compute $2.21 \times 10^{-5} - 8.90 \times 10^{-6}$.

54. Compute $4.88 \times 10^{-7} - 3.22 \times 10^{-2} + 5.66 \times 10^8$.

55. Use a calculator to find the following powers and roots:
    a. $(4.26)^{4/5}$
    b. $(8.99)^{1.26}$
    c. $(6.25)^{-0.011}$
    d. $(5.42 \times 10^{-11})^{-4.45}$
    e. $(4.77 \times 10^9)^{0.76}$
    f. $(3.2 \times 10^{-4})^{-0.015}$

6

SIGNIFICANT FIGURES

In science classes, students frequently measure objects and then divide numbers like 2.51 inches and 6.82 inches. If they use a calculator, they get

$$\frac{2.51}{6.82} = 0.3680352$$

Notice that 0.3680352 seems to have too many digits or figures in it to be reported as an answer. Why? This chapter discusses the way to determine how many digits in a number make sense. Reporting the proper number of digits in a scientific report is one aspect of honesty in science.

6-1
THE PERCENT UNCERTAINTY OF A MEASUREMENT IS
THE KEY TO UNDERSTANDING SIGNIFICANT FIGURES

Now back to our division. The numbers 2.51 and 6.82 were derived from an experiment—say, inches measured with a ruler graduated in tenths of an inch (see Figure 6-1). The students had to estimate the hundredths place, so 2.51 and 6.82 are not known too well, maybe to within ±1 in their last digit. (Read ±1 as "plus or minus one.") The last digit, the one in the hundredths place, is an estimated digit (see Figure 6-2). So 2.51 and 6.82 are known to within ±0.01, which is the same as saying that they are known to within ±1 in their last digit. Thus the 2.51 could be anywhere between 2.52 and 2.50, and 6.82 could be between 6.83 and 6.81.

Our two measurements are thus good to within about

$$\frac{0.01}{2.51} \times 100 = 0.4\% \quad \text{and} \quad \frac{0.01}{6.82} \times 100 = 0.1\%$$

(Opposite) A sundial. Compare the precision with which you can read a sundial and a digital watch or clock.

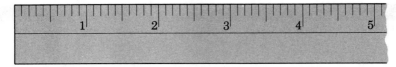

FIGURE 6-1 A ruler graduated in tenths of an inch.

(See Chapters 2 and 10 for discussions on percent.)

In these calculations, the 0.01 is the ±1 in the hundredths place, and a number with a percent sign after it is called the **percent uncertainty**. The division of the two numbers, 2.51 and 6.82, should be reported to no more than about 0.4%, which is the largest percent uncertainty of our two ruler measurements. Taking 0.4% of 0.3680352, we get 0.3680352 × 0.004 = 0.001. This means that 0.368052 should be reported to the nearest 0.001, or 0.368. The 8 in the thousandths place is known to within ±1, and our answer is in line with our measurements. You could express the answer as 0.368 ± 0.001, showing that the 8 is uncertain by ±1. Even if you don't write a ± after a number, people will assume that your number is good to within ±1 in the last place that you have written.

A good rule of thumb when dividing or multiplying two or more numbers is as follows. The percent uncertainty of the answer should be about the same as the largest percent uncertainty in any of the numbers that make up the problem.

EXAMPLE 1   You want to divide the measurement 6.3 feet by 2.7 feet. Each number is known to within ±1 in the tenths place. What should your answer be, to fall within the proper percent uncertainty?

Solution: On a calculator 6.3/2.7 = 2.3333333. The percent uncertainty of each number is 0.1/6.3 × 100 = 1.6% and 0.1/2.7 × 100 = 3.7%. The larger of the two is 3.7%. Taking 3.7% of 2.3333333, we get 2.3333333 × 0.037 = 0.086 or about 0.1. We should therefore report the answer is 2.3. If you wanted to be extra careful, you could report your answer as 2.3 ± 0.1, which specifically says that there is an uncertainty of ±1 in the tenths place.   ■

EXAMPLE 2   You multiply 4.68 × 7.36 on a calculator and get 34.4448. How should the answer be reported if 4.68 and 7.36 are good to within ±2 in the hundredths place?

Solution: Take percent uncertainties of each number:

$$\frac{0.02}{4.68} \times 100 = 0.43\% \qquad \frac{0.02}{7.36} \times 100 = 0.27\%$$

Now take 0.43% (the largest percent uncertainty) of 34.4448:

$$0.0043 \times 34.4448 = 0.15$$

The answer is good to within values between ± 1 and ± 2 in the tenths place (or you could say the answer is good to within values between 0.10 and 0.20, since 0.15 is between these two numbers). To be safe from overreporting your answer, you should use the larger uncertainty and report your answer as 34.4 $\pm$ 0.2. If you were to write the answer as 34.4, the reader would assume you meant 34.4 $\pm$ 0.1, which is a bit misleading about the uncertainty of the answer. ■

As you can see, this procedure is a bit cumbersome. Fortunately, there is a shortcut that works most of the time if all the numbers in the problem are good to within ± 1 in the last place: **When you are doing a multiplication or division, the number of digits reported in your answer should be the same as the smallest number of digits in any of the numbers of the problem.** Example 3 shows how easy the shortcut is to use.

EXAMPLE 3   In multiplying 4.73 × 14.47, what is a reasonable answer containing the proper number of digits?

Solution: 4.73 × 14.47 = 68.4431 on a calculator. The 4.73 has three digits, and the 14.47 has four. The answer should have three digits and should be reported as 68.4. ■

6-2
RECOGNIZING SIGNIFICANT FIGURES IS EASIEST WHEN NUMBERS ARE EXPRESSED IN SCIENTIFIC NOTATION

The **significant figures** of a number are the digits that are known with certainty *plus* one digit that is uncertain. In Example 1 the numbers 6.3 and 2.7 each have two significant figures. The 6 and 2 are known for sure and the 3 and 7 are in doubt. That's why we said that each number is known to within ± 0.1. The answer, 2.3, also has two significant figures. The 2 is known for sure, and the 3 is doubtful.

In Example 2 the numbers 4.68 and 7.36 each have three significant figures; their product, 34.4, is also expressed with three significant figures.

There is a problem in recognizing what digits are significant when the number contains zeros.

1. *Zeros in the middle of the number.* All zeros in the middle of a number are significant. An example would be 307.02. This number has five significant figures.

2. *Zeros after a decimal point.* An example is 32.00. The zeros are significant, and the number 32.00 has four significant figures.

3. *Zeros before the number.* An example is 0.00246. This number has three significant figures. The zero before the decimal point is for clarity only and tells you that a decimal point follows—we could have written the number as .00246. The two 0's between the decimal point and the 2 locate the decimal point. Zeros before a number that are used to locate the decimal point are not significant. A better way to see this is to convert 0.00246 into scientific notation:

$$0.00246 = 2.46 \times 10^{-3}$$

Now you can see that the number has only three significant figures. The best way to figure out significant figures is to convert numbers to scientific notation.

4. *Zeros after the number but before the decimal point.* An example would be 32,000. Now we have a problem in figuring out the significant figures. If we convert the number to scientific notation, we get four possibilities:

number:	3.2000×10^4	3.200×10^4	3.20×10^4	3.2×10^4
significant figures:	5	4	3	2

Which is correct? It depends on our original number. Were three zeros really significant? Was the measurement that obtained it good to within ±1 in 32,000? In other words, is only the last zero in doubt, and are all the others known for sure? If the answer is yes, then the percent uncertainty is

$$\frac{1}{32,000} \times 100 = 0.003\%$$

and in scientific notation we should write 3.2000×10^4. However, maybe the measurement was good to within only ±1000 in 32,000 or

$$\frac{1000}{32,000} \times 100 = 3.1\%$$

Then we should write the number as 3.2×10^4, since the 2 is doubtful. The measurement could have been good to within some other value— say, to within 100 or 10. Then we would write 3.20×10^4 or 3.200×10^4. So how we write the number depends on how good our measurement is. When you write a number in scientific notation, people assume that all the digits in the **mantissa** part (the part of the number before the

"×" sign) are significant. This is not always true if you write a number in regular notation. In regular notation you must somehow indicate which digits are significant when there is an ambiguity.

EXAMPLE 4   How many significant figures are in each of the following numbers? 400.01; 376.10; 0.00003317; 7,8$\bar{0}$0 (the zero with the bar over it is doubtful).

Solution:

400.01 has five significant figures.

376.10 has five significant figures.

$0.00003317 = 3.371 \times 10^{-5}$ has four significant figures.

$7.8\bar{0}0 = 7.80 \times 10^3$ has three significant figures (remember, a doubtful digit is significant). ∎

You could occasionally make a mistake in determining significant figures by using the shortcut in which you report in your answer the smallest number of significant figures that were in the problem. The example below illustrates this.

EXAMPLE 5   In the division 11.2/9.9, how many significant figures should there be in the answer?

Solution: 11.2/9.9 = 1.131313 on a calculator. Using our shortcut, we would report the answer as 1.1, since 9.9 has two significant figures. Let's calculate the percent uncertainty of each number in the problem:

$$\frac{0.1}{11.2} \times 100 = 0.89\% \qquad \frac{0.1}{9.9} \times 100 = 1.01\%$$

The answer, 1.1, has a percent uncertainty of

$$\frac{0.1}{1.1} \times 100 = 9.1\%$$

The data in the problem are better than those in the answer, so we are justified in reporting the answer as having about a 1% error. To do this we can add another digit and report the answer as

$$\frac{11.2}{9.9} = 1.13$$

The percent uncertainty is

$$\frac{0.01}{1.13} \times 100 = 0.88\%$$

This is very close to the largest percent uncertainty of 1.01%, and we can safely report three significant figures in our answer. ∎

To understand what happened in Example 5, compare 99 and 100. The percent uncertainties of these numbers are

$$\frac{1}{99} \times 100 = 1.01\% \quad \text{and} \quad \frac{1}{100} \times 100 = 1.00\%$$

The percent uncertainties are about the same, but 99 has two significant figures and 100 has three. What this discussion means, then, is that numbers like 99 and 98 can be considered as having three significant figures in calculations. And numbers between 980 and 999 can be considered as having four significant figures in calculations. Can you show that this is true?

6-3
ACCURACY AND PRECISION TELL US HOW
GOOD OUR MEASUREMENTS ARE

At this point we will discuss the factors that affect the percent uncertainty of a measurement.

When you make a measurement, there are two main things that determine how close your reading comes to the actual or true value. They are (1) to what uncertainty you can read the instrument; (2) how close the instrument reading is to the true value.

The uncertainty with which you can read the instrument is called the **precision** of the measurement. If you make a series of measurements with the same instrument, the agreement between the numbers is an indication of the precision of the measurement. This means that if you consistantly read an instrument carefully, your measurements will have a high precision.

Another way to look at precision is to consider the number of significant figures that you can read off an instrument. Assuming that you know how to read it, a ruler graduated in hundredths of an inch will give more significant figures than a ruler graduated in tenths of an inch. The rod in Figure 6-2 was measured to be 2.51 inches using a ruler graduated in tenths of an inch. If we had used a ruler graduated in hundredths of an inch, our rod might have been measured as being 2.514 inches. Clearly, 2.514 has more

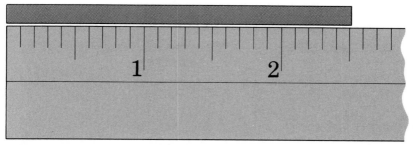

FIGURE 6-2 A magnified view of our ruler measuring a rod that is 2.51 inches long. Notice that the hundredth place is estimated to be "1" but could be a bit more or less, depending on how good you are at reading rulers.

significant figures than does 2.51. Since 2.514 has a smaller percent uncertainty than does 2.51, the measurement 2.514 inches is more precise. Thus, the more significant figures an instrument can give, the more precise the instrument. This assumes, of course, that you know how to read the instrument and that it is working properly. We might note here that instruments capable of high precision tend to be more expensive than lower-precision instruments.

How close the reading is to the true value is called the **accuracy** of the measurement. Figure 6-3 gives two pictorial representations of accuracy and precision for a series of measurements.

Now let's look at a problem you might have if your instrument is defective. Consider again the ruler discussed at the beginning of this chapter so that you can measure the rod that is 2.51 inches long. The percent uncertainty of your reading is $0.01/2.51 \times 100 = 0.4\%$. This is the **precision** of your measurement. Now suppose your ruler is warped (bent) and the rulings on it are closer together than they should be. Then you will read more of them in measuring the rod. Possibly the actual length of the rod is 2.45 inches, and because of the warping you read 2.51 inches. Your reading is off by 2.51 inches − 2.45 inches = 0.06 inches. The percent uncertainty is $0.06/2.51 \times 100 = 2.4\%$. This is the accuracy of your measurement. Thus you can be a very good ruler reader (high precision, 0.4%) and still get bad results (low accuracy, 2.4%) if you use a bad ruler.

The "bad ruler" problem is recognized by scientists who make every effort to calibrate their equipment. To calibrate our ruler, we would get a rod of known length (say, from the National Institute of Standards and Technology) and measure it with our ruler. We could then compensate for the warp and improve the accuracy of our measurements. The precision would remain the same. A better solution, of course, is to get a good ruler.

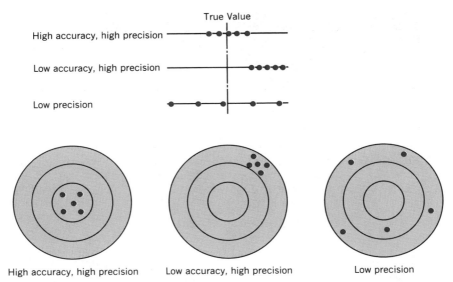

FIGURE 6-3 Two pictorial representations of accuracy and precision. The large dots represent the measurements. In the bottom diagrams, the "true value" is in the center of the smallest circle. Notice that in the case of low precision, we don't talk about the accuracy. Although the average of the low-precision values may *by chance* give the correct answer, it is not likely. It is thus meaningless to talk about the accuracy of measurements of low precision.

EXAMPLE 6   We will assume that the instructor in a lab class can read an instrument as well as it was designed to be read. In weighing a sample of magnesium oxide on an analytical balance that reads in grams (abbreviated "g"), the instructor gets 2.0342 g. A student, who is not yet an expert in reading the balance, weighs the same sample and gets 2.035 g. Who made the more precise weighing?

Solution: 2.0342 g has less percent uncertainty (and more significant figures) than does 2.035 g. The instructor's weight is more precise.   ◾

EXAMPLE 7   Refer to Example 6. An expert from the National Institute of Standards and Technology weighs the same magnesium oxide sample on a special balance known to be calibrated. The weight is 2.0349 g. Whose weight is more accurate, the instructor's or the student's? (Assume 2.0349 g is the true weight.)

Solution: The student's weight is more accurate because it is closer to the true value. This is probably due to luck. Since the instructor's reading was

An analytical balance in use. Balances of this type can be read with a precision of 0.1 milligram.

so far from the true value, either the instructor had a bad day or the balance is not reliable. ■

In general, when you make a measurement, your answer reflects a combination of three things: (1) the precision inherent (built-in) in the instrument; (2) the precision to which you read the instrument; and (3) the inherent accuracy of the instrument. Good technical skill in reading an instrument will improve the precision of your measurements. Careful attention to calibration (comparison to a known standard) will improve your accuracy. Using high-quality instruments will improve both your accuracy and precision.

6-4
DETERMINING SIGNIFICANT FIGURES IN ADDITION
AND SUBTRACTION IS DIFFERENT FROM THAT
IN MULTIPLICATION AND DIVISION

Determining significant figures in addition and subtraction is different from doing so in multiplication and division. The following will illustrate why. Suppose you want to measure 123.46 mL of water.[1] You could do it in either

[1]Note that mL stands for milliliters (there are about 5 mL in a teaspoon). A liter (L) is a measure of volume that is equal to about a quart. One milliliter equals one-thousandth of a liter. In other words, there are 1000 mL in 1 L.

of two ways using (1) a 50-mL buret accurate to within ±0.01 mL and (2) a 100-mL graduated cylinder, accurate to within ±1 mL (see Figure 6-4 for a description). The two ways are the following.

1. Use the 50-mL buret to measure two 50.00-mL portions of water (the last two zeros are significant). Then use the buret to measure 23.46 mL of water. Add all the portions together to get 123.46 mL of water.

2. Use the 100-mL graduated cylinder to measure 100 mL of water. Use the 50-mL buret to measure 23.46 mL of water. Add all the portions to get 123.46 mL of water.

Procedure 1 really does give 123.46 mL of water if you are very good at using a buret. Procedure 2 doesn't. Why? Because the 100-mL graduated

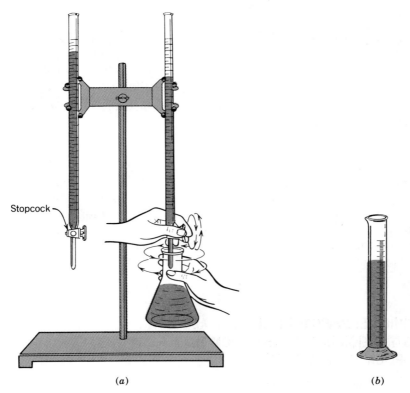

Stopcock

(a)   (b)

FIGURE 6-4 (a) A 50-mL buret. (b) A 100-mL graduated cylinder. A solution is dripped out of the buret slowly. It is easy to control the amount of fluid coming out of the buret by using the stopcock (a valve). A good operator can let out a fraction of a drop at a time. The graduated cylinder, on the other hand, is used to add large amounts of solution. Since the solution must be poured out, fine control is not easy.

cylinder can be read only to within ±1 mL. So if you add 23.46 mL of water to the 100 mL of water, you still have an uncertainty in your volume of ±1 mL. The best you can say is that you have 123 mL of water.

The least precise volume measure determines the uncertainty of the final volume. In procedure 2 we would add the volumes as follows:

$$
\begin{array}{r}
100. \\
+ \ 23.46 \\
\hline
123.46
\end{array}
$$

The 4 and 6 in the answer are not significant because one of the numbers, here the 100, doesn't have any digits to the right of the decimal point. Thus we must write the sum as 123 mL.

Some authors use a "?" to represent the unknown digits:

$$
\begin{array}{r}
100.?? \\
+ \ 23.46 \\
\hline
123.??
\end{array}
$$

The rule is that a "?" number plus a known number gives a "?" number, and "?" numbers are not significant.

EXAMPLE 8   Add 236.1 and 3.247.

Solution:

$$
\begin{array}{r}
236.1 \\
+ \ 3.247 \\
\hline
239.347
\end{array}
\qquad \text{Using "?" we see that} \qquad
\begin{array}{r}
236.1?? \\
+ \ 3.247 \\
\hline
239.3??
\end{array}
$$

so we must report the sum as 239.3.   ■

EXAMPLE 9   Compute 47.32 − 3.1.

Solution:

$$
\begin{array}{r}
47.32 \\
- \ 3.1? \\
\hline
44.2?
\end{array}
$$

We must report the sum as 44.2.   ■

To see why this method of retaining significant figures is reasonable, let's look at some percent uncertainties. Take the expression 100 mL + 23.46 mL. The percent uncertainties are

$$\frac{1}{100} \times 100 = 1\% \quad \text{and} \quad \frac{0.01}{23.46} \times 100 = 0.04\%$$

For the two possible answers, we have

$$\frac{1}{123} \times 100 = 0.8\% \quad \text{and} \quad \frac{0.01}{123.46} \times 100 = 0.008\%$$

Clearly, the sum of 100 mL and 23.46 mL must be reported as 123 mL and not as 123.46 mL, because 0.8% is a bigger percent uncertainty than 0.008%. Moreover, 0.8% is close to 1%, which is the largest percent uncertainty of our two numbers, 100 mL and 23.46 mL. *Remember, it is the largest percent uncertainty of our original numbers which determines the percent uncertainty of our final answer.*

6-5
ROUNDING A NUMBER PRESERVES THE PROPER NUMBER OF SIGNIFICANT FIGURES

In our discussion about measuring out 123.46 mL of water, we rounded 123.46 mL to 123 mL. How do we round numbers?

The convention for rounding numbers depends on the digits to the right of the last digit we will keep. When we rounded 123.46 mL to three significant figures, we looked at the 46. This is less than 50, so we dropped it and left the 3 alone. If the volume had been 123.58 mL, we would have rounded it to 124 mL because 58 is larger than 50.

So we can use the following rule: **If the digits to the right of the ones we want to keep are greater than 5 or 50 or 500 or 5000 or a 5 with as many zeros as we need, the last digit we keep is increased by 1. If the digits to the right are less than 5 or 500 or 5000, and so on, the last digit is left unchanged.**

EXAMPLE 10   Round 57.337 to three significant figures.

Solution: We will keep the 57.3 as is and discard the 37, which is less than 50.   ■

EXAMPLE 11   Round 3.986 to three significant figures.

Solution: Since the last digit, a 6, is larger than 5, we increase the 8 to a 9 and get 3.99. ■

EXAMPLE 12   Round 15.9932 to three significant figures.

Solution: Since 932 is larger than 500, we increase 15.9 by 0.1 to get 16.0. ■

If the digits to the right of the ones we want to keep are exactly 5 or 50 or 500, and so on, most scientists follow this convention: **If the last digit to be kept is even, leave it unchanged. If the last digit to be kept is odd, increase it by 1. Zero is considered an even number.** Can you figure out why? (The answer is given after Example 14.)

EXAMPLE 13   Round 25.550 to three significant figures.

Solution: In 25.$\overline{5}$50 the digit with the bar is odd, so we increase it by 1 and the number becomes 25.6. ■

EXAMPLE 14   Round 125.00 to two significant figures.

Solution: In 1$\overline{2}$5.00 the digit with the bar is even, so we leave it unchanged. The final answer is 120 or, even better, 1.2×10^2. ■

The reason for the procedure shown in Examples 13 and 14 above is this: In rounding a series of numbers, the last significant digit will probably be even about half the time and odd the other half. So about half the time we increase the digit, and half the time we leave it unchanged. That's why we consider zero to be an even number. There are then five even digits (0, 2, 4, 6, 8) and five odd digits (1, 3, 5, 7, 9). In this way, our final answer hopefully won't be biased by the rounding procedure.

6-6
EXACT NUMBERS HAVE AN INFINITE NUMBER OF SIGNIFICANT FIGURES

Sometimes talking about significant figures doesn't mean very much. Suppose you have three apples. You would write 3 apples. It wouldn't make much sense to write 3.0 or 3.00 apples because you have *exactly* three apples. The zeros after the decimal point do not mean anything if you're talking about whole apples. Of course, if you wanted to tell someone that you had 3.50 apples, you would mean that you had $3\frac{1}{2}$ apples to three significant figures.

Thus, when we express numbers of objects that come as unique pieces, significant figures do not apply. The numbers we use are called **exact numbers** and have an infinite number of significant figures.

For a chemical illustration of exact numbers, consider the chemical formula H_2O. The subscripts indicate exactly 2 atoms of H and 1 atom of O.

6-7
THE MORAL OF THIS CHAPTER IS TO ALWAYS PRESENT YOUR RESULTS IN THE PROPER NUMBER OF SIGNIFICANT FIGURES

In science as in life, you want to be honest but you don't want to be foolish. Presenting the proper number of significant figures in an answer is honest. Presenting more than the proper number is certainly misleading and possibly dishonest. Presenting less than the proper number of significant figures is foolish—you should always make yourself look as good as honesty allows.

PROBLEMS

KEYED PROBLEMS

1. You are dividing 8.4 feet by 3.6 feet. Each number is known to within ± 1 in the tenths place. What should your answer be?

2. You are multiplying 6.84×8.76 on a calculator. If each number is good to within ± 2 in the hundredths place, how should the answer be reported?

3. In multiplying 2.13×27.86, what is a reasonable answer?

4. How many significant figures are in the following: 2.001; 97.300; 0.001161?

5. In the division 1.23/0.98, what is the answer containing the proper number of significant figures?

6. Which of the following two weights is more precise, 1.0342 g or 1.134 g?

7. If the "true" weight is 1.1328 g, which of the weights in Problem 6 is more accurate?

8. Add 429.6 g and 7.685 g.

9. Compute $82.47 - 6.5$.

10. Round 28.941 to three significant figures.

11. Round 4.888 to three significant figures.

12. Round 33.996 to three significant figures.

13. Round 96.750 to three significant figures.

14. Round 445.00 to two significant figures.

SUPPLEMENTAL PROBLEMS

15. Round the following numbers to three significant figures.
    a. 4.268
    b. 25.144
    c. 0.03055
    d. 3.987×10^{12}
    e. 9.231×10^{-4}
    f. 32650

16. In the following multiplications, round your answer to three significant figures.
    a. 4.23×6.41
    b. 25.2×87.6
    c. 125.1×9.66
    d. 0.342×0.768
    e. $(4.31 \times 10^6)(9.32 \times 10^4)$
    f. $(2.86 \times 10^{-4})(6.68 \times 10^{-8})$

17. In the following divisions, round your answer to three significant figures.
    a. 8.32/4.66
    b. 32.1/25.21
    c. 562/3.25
    d. 0.167/0.0876
    e. $(6.37 \times 10^{-5})/(5.42 \times 10^6)$
    f. $(2.11 \times 10^{-3})/(6.68 \times 10^{-2})$

18. In the following multiplications, round your answer to the appropriate number of significant figures.
    a. 3.2×4.6
    b. 5.11×15.32
    c. 9.86×3.20
    d. 0.030×2.61
    e. $(3.216 \times 10^3)(4.23 \times 10^5)$
    f. $(1.2 \times 10^{-2})(3.13 \times 10^{-4})$

19. In the following divisions, round your answer to the appropriate number of significant figures.
    a. 4.3/9.6
    b. 1.67/26.38
    c. 2.20/1.86
    d. 0.312/0.0026
    e. $(4.6 \times 10^3)/(2.33 \times 10^4)$
    f. $(2.671 \times 10^{-2})/(7.32 \times 10^{-6})$

20. Perform the following additions and subtractions and round your answer to the appropriate number of significant figures.
    a. $3.26 + 4.1$
    b. $5.77 + 6.00$
    c. $251.6 + 1.167$
    d. $25.44 - 13.1$
    e. $187.5 - 57.92$
    f. $200.00 - 3.00$

21. In the molecular interpretation of the chemical equation $2H_2 + O_2 \rightarrow 2H_2O$, do the coefficients have only one significant figure or are they exact numbers? Explain.

7
UNITS AND
UNIT CONVERSIONS

Now that we have covered some of the essentials of atoms, molecules, and scientific notation, we can begin a systematic study of the arithmetic of chemistry. This chapter will cover units and unit conversions.

7-1
UNITS AND THEIR ABBREVIATIONS ARE USED
TO DESCRIBE PHYSICAL QUANTITIES

It is important to keep track of the units (sometimes called dimensions) used in a problem. **Units** indicate how a physical quantity is measured. For instance, *3 inches* is the number *3* with the unit *inches* attached. The unit part sets the scale or basic multiple; thus, 3 *inches* is not 3 *feet* or 3 *pounds* or 3 *seconds* or 3 *oranges,* even though the 3 is the same in all expressions.

It is common to abbreviate the units, with singular and plural forms having the same abbreviation and being used interchangeably. Abbreviations for some common units are as follows:

foot or feet = ft
inch or inches = in.
pound or pounds = lb
gram or grams = g
kilogram or kilograms = kg
calorie or calories = cal

(Opposite) Skiers in action. The length of skis is measured in centimeters. For instance, a skier who is 6 feet tall might use a pair of skis that are 205 centimeters long, which is about 6 feet 9 inches long.

> second or seconds = s
> meter or meters = m
> millimeter or millimeters = mm
> centimeter or centimeters = cm
> milligram or milligrams = mg
> liter or liters = L
> milliliter or milliliters = mL

NOTE: A period is *not* put after an abbreviation unless it happens to come at the end of a sentence. The exception to this rule is the abbreviation for inch or inches (in.), which might be confused for the word "in" if a period wasn't used.

7-2
"ONE" IS THE IDENTITY ELEMENT IN MULTIPLICATION

If any number is multiplied by 1, the number is not changed. We say that 1 is the **identity element** in multiplication. For example,

$$7 \times 1 = 7$$

$$\tfrac{4}{3} \times 1 = \tfrac{4}{3}$$

$$2.36 \times 1 = 2.36$$

Generally, if n stands for any number, it follows that

$$n \times 1 = n$$

The number "1" can take many forms. For instance,

$$\frac{3}{3} = 1 \qquad \frac{9.46}{9.46} = 1 \qquad \frac{xy}{xy} = 1 \qquad \frac{\frac{7}{3}}{\frac{7}{3}} = 1$$

where we have used the fact that *any* number (except zero) divided by itself equals 1.

Just as was true of numbers alone, numbers with units attached don't change when they are multiplied by 1 in any form. For example, 3 ft × 1 = 3 ft. If the 1 should happen to be 9.46/9.46, say, then

$$3 \text{ ft} \times \frac{9.46}{9.46} = 3 \text{ ft}$$

7-3
UNITS DIVIDE IN A WAY SIMILAR TO ANY ALGEBRAIC QUANTITY

It seems reasonable that 3 ft/3 ft = 1, since any quantity divided by itself equals 1. Thus it must be true that both 3/3 = 1 and ft/ft = 1. To convince yourself of this, think of the following example. If my lab bench is 20 ft long and your lab bench is 10 ft long, my bench is *twice* as long as your bench:

$$\frac{20 \text{ ft}}{10 \text{ ft}} = 2$$

Notice that my bench is *not* 2 ft longer than your bench—it is twice as long. It should now be clear that **units can be divided in a way similar to numbers or any algebraic quantity.** Thus in this example we have the following:

$$\frac{20}{10} = 2 \quad \text{and} \quad \frac{\text{ft}}{\text{ft}} = 1$$

EXAMPLE 1   My experiment needs 10 mL of methanol as a solvent. Yours needs 2 mL. How many times more methanol do I need than you need? (Methanol is also called wood alcohol. It is jellied to make sterno and is very poisonous.)

Solution:

$$\frac{10 \text{ mL}}{2 \text{ mL}} = 5$$

Thus I need five times as much methanol as you do.   ■

7-4
IDENTITIES AND THEIR RATIOS ALLOW US TO
CONVERT FROM ONE UNIT TO ANOTHER

The fact that units have many of the properties of algebraic quantities and can be operated on separately from the numbers that they come with is very useful. Consider the identity[1]

$$12 \text{ in.} = 1 \text{ ft}$$

---

[1]The numbers 12 and 1 are exact numbers because the relationship between feet and inches is a definition.

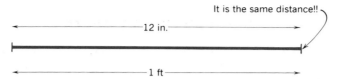

FIGURE 7-1 A line has the same length, no matter what you call it.

This is called an **identity** because both the 12 in. and the 1 ft represent a distance that is *physically* the same. The line drawn in Figure 7-1, which represents 1 ft or 12 in., is the same length no matter what you call it.

If both sides of the identity, 12 in. = 1 ft, are divided by 12 in., we get

$$\frac{12\ \text{in.}}{12\ \text{in.}} = \frac{1\ \text{ft}}{12\ \text{in.}}$$

Since 12 in./12 in. = 1, it must be true that

$$\frac{1\ \text{ft}}{12\ \text{in.}} = 1$$

If we instead divide both sides of the identity 12 in. = 1 ft by 1 ft, we get

$$\frac{12\ \text{in.}}{1\ \text{ft}} = \frac{1\ \text{ft}}{1\ \text{ft}}$$

Since 1 ft/1 ft = 1, it must again be true that

$$\frac{12\ \text{in.}}{1\ \text{ft}} = 1$$

Two forms of "1" have been obtained:

$$\frac{12\ \text{in.}}{1\ \text{ft}} \quad \text{and} \quad \frac{1\ \text{ft}}{12\ \text{in.}}$$

These quantities are called **ratio identities** because they are ratios made from an identity. The use of ratio identities allows us to easily convert from one unit to another.

7-5
CONVERTING FROM ONE UNIT TO ANOTHER USES RATIO IDENTITIES

Suppose that we want to convert 36 in. into feet. Then we simply multiply 36 in. by the ratio identity that divides the inches and leaves feet. In other words, **put the unit you want to end up with on top and the unit you want to divide on the bottom:**

$$36 \text{ in.} \times \frac{1 \text{ ft}}{12 \text{ in.}} = \frac{36 \text{ in.}}{1} \times \frac{1 \text{ ft}}{12 \text{ in.}} = \frac{36 \,\cancel{\text{in.}}}{12\,\cancel{\text{in.}}} \times \frac{1 \text{ ft}}{1} = 3.0 \times 1 \text{ ft} = 3.0 \text{ ft}$$

Doing this in one step we have

$$36 \,\cancel{\text{in.}} \times \frac{1 \text{ ft}}{12\,\cancel{\text{in.}}} = 3.0 \text{ ft}$$

(Notice that we are being careful about significant figures.) If the other ratio identity had been used by mistake, we would have gotten

$$36 \text{ in.} \times \frac{12 \text{ in.}}{1 \text{ ft}} = \frac{432 \text{ in.}^2}{1 \text{ ft}} = 432 \frac{\text{in.}^2}{\text{ft}}$$

where in.^{2} = in. × in. Obviously, this is undesirable because we wanted the unit feet.

Now let's do this problem in the reverse direction. Convert 3.0 ft to inches. For this, the ratio identity 12 in./1 ft is used:

$$3.0 \,\cancel{\text{ft}} \times \frac{12 \text{ in.}}{1\,\cancel{\text{ft}}} = 36 \text{ in.}$$

To help you to convert easily from one unit to another, we will outline a definite procedure that you can use. To help you spot the *given* and *asked for* in a problem, we will put a (G) after the given and an (AF) after the asked for. This procedure will also be followed for most of the problems in the next two chapters, Chapters 8 and 9. After that, you should have enough practice in solving problems to determine the given and asked for by yourself.

For instance, in the problem

How many pounds (AF) are there in 200 g (G)?

the 200 g are given and the pounds are asked for.

Let's solve this problem using the procedure given in the previous paragraph.

PROBLEM   How many pounds (AF) are there in 200 g (G)?

Steps in the solution:

1. Determine the given and the asked for. In this problem we have indicated them by a (G) and an (AF).
2. Write down an identity relating the *units* of the given and asked for. In this problem the identity is 1 lb = 453.6 g.
3. Write down the given, put a "×" sign after it, draw a line, and follow it with an "=" sign:

$$200 \text{ g} \times \underline{\hspace{2cm}} =$$

4. From the identity in step 2, put the *asked-for unit on top of the line* and the *given unit under the line:*

$$200 \text{ g} \times \frac{1 \text{ lb}}{453.6 \text{ g}} =$$

5. Compute the answer. Be *sure* that the given unit divides, thus leaving the asked-for unit:

$$200 \cancel{\text{ g}} \times \frac{1 \text{ lb}}{453.6 \cancel{\text{ g}}} = \frac{200}{453.6} \text{ lb} = 0.441 \text{ lb} \quad \blacksquare$$

Remember, the only thing that converting from one unit to another does is to change the *name* we give to a certain quantity. It doesn't change that quantity physically. Thus all ratio identities must equal 1 so that they only change the *name* of the quantity, not the size of it.

Many of the identities that are used in this book are based on the metric system. Since scientists use the metric system almost exclusively, now is a good time to get familiar with it.

7-6
THE METRIC SYSTEM USES PREFIXES IN FRONT OF A UNIT TO INDICATE A POWER OF 10

In 1790, the French Academy of Sciences presented a system of measurement called the **metric system** to the French National Assembly. In 1960, an international committee agreed on a particular choice of metric units. This set of metric units is called the **International System** and is abbreviated **SI**, after the French name *Le Système International d'Unités*. In the SI system,

TABLE 7-1

FIVE SI BASE UNITS

PHYSICAL QUANTITY	UNIT	SYMBOL OR ABBREVIATION
Length	meter	m
Mass	kilogram	kg
Time	second	s
Temperature	kelvin	K
Amount of substance	mole	mol

a prefix, or group of letters, is placed in front of the SI unit to indicate a power of 10.

Each physical quantity has an **SI base unit**. The five SI base units that are used in this book are listed in Table 7-1.

The prefixes in the SI system are listed in Table 7-2. The prefixes that are in boldface type are the ones that are used in this book or that are so commonly used by chemists that you should be familiar with them.

To show you how to use Table 7-2, let's choose the prefix *milli*. *Milli* means 10^{-3} or one-thousandth. Therefore we can write the following identities:

Length: 10^{-3} meter = 1 millimeter   or 10^{-3} m = 1 mm

Mass: 10^{-3} gram = 1 milligram   or 10^{-3} g = 1 mg

Time: 10^{-3} second = 1 millisecond   or 10^{-3} s = 1 ms

TABLE 7-2

SI PREFIXES

MULTIPLE	PREFIX	SYMBOL	MULTIPLE	PREFIX	SYMBOL
10^{18}	exa	E	10^{-1}	**deci**	**d**
10^{15}	peta	P	10^{-2}	**centi**	**c**
10^{12}	tera	T	10^{-3}	**milli**	**m**
10^{9}	giga	G	10^{-6}	**micro**	**μ**[b]
10^{6}	**mega**[a]	**M**	10^{-9}	**nano**	**n**
10^{3}	**kilo**	**k**	10^{-12}	**pico**	**p**
10^{2}	hecto	h	10^{-15}	femto	f
10	deca	da	10^{-18}	atto	a

[a]The prefixes in boldface type are used in this book or are commonly used by chemists.

[b]The letter μ is Greek "mu" and is read "micro" in this case.

We can multiply both sides of each of these identities by 10^3. Then we obtain the following identities, which are equivalent to those just given but look slightly different:

$$\text{Length:} \quad 1 \text{ m} = 10^3 \text{ mm}$$

$$\text{Mass:} \quad 1 \text{ g} = 10^3 \text{ mg}$$

$$\text{Time:} \quad 1 \text{ s} = 10^3 \text{ ms}$$

To show that both sets of identities are equivalent to one another, let's use both of them to convert 58 g to milligrams. Using the identity 10^{-3} g = 1 mg, we obtain

$$58 \text{ g} \times \frac{1 \text{ mg}}{10^{-3} \text{ g}} = 58 \times 10^3 \text{ mg} = 5.8 \times 10^4 \text{ mg}$$

Using the identity 1 g = 10^3 mg, we get

$$58 \text{ g} \times \frac{10^3 \text{ mg}}{1 \text{ g}} = 58 \times 10^3 \text{ mg} = 5.8 \times 10^4 \text{ mg}$$

There is an exception to using the prefixes. When we measure time that is longer than a few hundred seconds, even scientists still use minutes and hours.

Table 7-3 presents some useful identities involving both the metric system and the everyday system used in the United States, which uses feet, inches, quarts, and so on.

7-7
WHEN CONVERTING FROM ONE UNIT TO ANOTHER, BE SURE THAT THE UNITS DIVIDE OUT IN THE CORRECT WAY

This section presents many examples of unit conversions, most of which involve the metric system.

EXAMPLE 2   Convert 100 g (G) into pounds (AF).

Solution: The identity is 453.6 g = 1 lb.

$$100 \text{ g} \times \frac{1 \text{ lb}}{453.6 \text{ g}} = \frac{100}{453.6} \text{ lb} = 0.220 \text{ lb} \quad \blacksquare$$

TABLE 7-3

SOME USEFUL IDENTITIES

QUANTITY	ABBREVIATION
Length	
1 m = 100 cm	m = meter
1 m = 1000 mm	mm = millimeter
1 cm = 10 mm	cm = centimeter
1 m = 10^9 nm	nm = nanometer
1 km = 1000 m	km = kilometer
1 m = 39.37 in.	in. = inch
1 in. = 2.54 cm	mi = mile
1 mi = 5280 ft	ft = foot
12 in. = 1 ft	
Volume	
1 L = 1000 mL	L = liter
1 L = 1.057 qt	mL = milliliter
1 mL = 1 cm^3	cm^3 = cubic centimeter or cc
1 gal = 4 qt	qt = quart
	gal = gallon
Mass	
1 kg = 1000 g	kg = kilogram
1 lb = 453.6 g	g = gram
1000 mg = 1 g	lb = pound
10^6 μg = 10^6 mcg = 1 g	μg = mcg = microgram^a
Time	
1 min = 60 s	min = minute
1 h = 60 min	s = second
	h = hour
Energy	
1 cal = 4.184 J	cal = calorie
1 kcal = 4.184 kJ	J = joule
1 kJ = 1000 J	kJ = kilojoule
1 kcal = 1 Cal = 1000 cal	kcal or Cal = kilocalorie

^{a}The letter μ is Greek "mu" and is read "micro" in this case. Sometimes μg is written as mcg, especially in nonscientific uses such as on the labels of nutritional supplements.

EXAMPLE 3   Convert 2.00 lb (G) into grams (AF).

Solution: The identity is 453.6 g = 1 lb.

$$2.00 \cancel{\text{lb}} \times \frac{453.6 \text{ g}}{1 \cancel{\text{lb}}} = 907 \text{ g} \quad \blacksquare$$

EXAMPLE 4   How many seconds (AF) are there in 10 min (G)?

Solution: The identity is 60 s = 1 min.

$$10 \text{ min} \times \frac{60 \text{ s}}{1 \text{ min}} = 600 \text{ s} = 6.0 \times 10^2 \text{ s} \quad \blacksquare$$

EXAMPLE 5   How many centimeters (AF) are there in 5.00 in. (G)?

Solution: The identity is 2.54 cm = 1 in.

$$5.00 \text{ in.} \times \frac{2.54 \text{ cm}}{1 \text{ in.}} = 12.7 \text{ cm} \quad \blacksquare$$

EXAMPLE 6   How many inches (AF) are there in 100 cm (G)?

Solution: The identity is 2.54 cm = 1 in.

$$100 \text{ cm} \times \frac{1 \text{ in.}}{2.54 \text{ cm}} = 39.4 \text{ in.} \quad \blacksquare$$

EXAMPLE 7   How many millimeters (AF) are there in 10 m (G)?

Solution: The identity is 1 m = 1000 mm.

$$10 \text{ m} \times \frac{1000 \text{ mm}}{1 \text{ m}} = 10,000 \text{ mm} = 1.0 \times 10^4 \text{ mm} \quad \blacksquare$$

EXAMPLE 8   How many kilograms (AF) are there in 500 g (G)?

Solution: The identity is 1000 g = 1 kg.

$$500 \text{ g} \times \frac{1 \text{ kg}}{1000 \text{ g}} = 0.500 \text{ kg} \quad \blacksquare$$

EXAMPLE 9   How many liters (AF) are there in 250 mL (G)?

Solution: The identity is 1000 mL = 1 L.

$$250 \text{ mL} \times \frac{1}{1000 \text{ mL}} = 0.250 \text{ L} \quad \blacksquare$$

Sometimes you don't have the identity needed to make a conversion. However, by using two or more identities from Table 7-3 (or some other source), you can make the conversion. Suppose you go to a gas station that sells gas by the liter. It takes 40.0 L to fill the tank of your car. How many gallons is this? Looking at Table 7-3, you don't see an identity relating liters and gallons. But you do see two identities that will do the job. They are 1 L = 1.057 qt and 4 qt = 1 gal. So write down the given (40.0 L) and draw two horizontal lines, two "×" signs, and an "=" sign. The conversion is done in two steps that are combined into one: liters → quarts → gallons.

$$40.0 \text{ L} \times \underline{\hspace{2cm}} \times \underline{\hspace{2cm}} =$$

You want gallons after the equals sign, so insert the appropriate ratio identities to divide the liters and the quarts, giving gallons:

$$40.0 \, \cancel{L} \times \frac{1.057 \, \cancel{qt}}{1 \, \cancel{L}} \times \frac{1 \text{ gal}}{4 \, \cancel{qt}} = 10.6 \text{ gal}$$

EXAMPLE 10   How many seconds (AF) are there in 1 h (G)?

Solution: The identities from Table 7.1 are 60 s = 1 min and 60 min = 1 h.

$$1 \, \cancel{hr} \times \frac{60 \, \cancel{min}}{1 \, \cancel{hr}} \times \frac{60 \text{ s}}{1 \, \cancel{min}} = 3600 \text{ s}$$

If you've studied Chapter 6, which is on significant figures, you might ponder about how many significant figures 3600 s has. ■

EXAMPLE 11   A wine bottle contains 750 mL (G) of wine. How many gallons (AF) is this?

Solution: The three identities we need from Table 7.1 are

$$1 \text{ L} = 1000 \text{ mL} \qquad 1 \text{ L} = 1.057 \text{ qt} \qquad 1 \text{ gal} = 4 \text{ qt}$$

$$750 \, \cancel{mL} \times \frac{1 \, \cancel{L}}{1000 \, \cancel{mL}} \times \frac{1.057 \, \cancel{qt}}{1 \, \cancel{L}} \times \frac{1 \text{ gal}}{4 \, \cancel{qt}} = 0.198 \text{ gal} \quad ■$$

It is interesting that 0.198 gal is almost 0.200 gal or one-fifth of a gallon. Before liquor and wine manufacturers converted to the metric system, a "fifth" was exactly one-fifth of a gallon. But they rounded the "fifth" to 750 mL, thereby saving a few milliliters. If you do Problems 11 and 24 at the end of this chapter, you will see how much wine or liquor they saved, thus increasing their profits.

A final word about units. If a number has a unit attached to it and you don't write the unit, nobody except you knows what you mean. For instance, if you write only an 8 when you mean 8 kg, how is the reader supposed to know what you are talking about?

PROBLEMS

KEYED PROBLEMS

1. I drive 20 miles to work every day. You drive 5 miles to work every day. How many times farther is my job than your job?

2. Convert 200 g into pounds.

3. Convert 5.0 lb into grams.

4. How many seconds are there in 40 min?

5. How many centimeters are there in 8.0 in.?

6. How many inches are there in 50 cm?

7. How many millimeters are there in 200 m?

8. How many kilograms are there in 200 g?

9. How many liters are there in 725 mL?

10. How many seconds are there in a day? (*Hint:* 1 day = 24 h.)

11. How many milliliters are there in 0.200 gal?

SUPPLEMENTAL PROBLEMS

12. The recommended dietary allowance (RDA) of vitamin C is 60 mg. How many grams is this?

13. Many people believe that high doses of vitamin C should be taken, and they take around 3 g a day. How many milligrams is this?

14. The RDA for vitamin B_{12} is 3 μg. How many grams is this?

15. The distance from New York City to Seattle (Washington) is about 3.0×10^3 miles. How many kilometers is this?

16. A chemistry student weighs 150 lb. How many kilograms does the student weigh?

17. A solid rectangular object measures 38 in. × 26 in. × 34 in. on each side. How many centimeters does the object measure?

18. The New York City marathon race is 26 miles, 385 yards. How many inches is this?

19. You have been waiting on line for 1000 s to buy your textbooks. How many hours have you been waiting?

20. The label on a can of beer reads: volume = 300 mL. How many quarts is this?

21. One milligram of natural vitamin E equals 1.49 international units (IU). How many milligrams of natural vitamin E are there in a 400-IU capsule?

22. About 0.10 lb of matter is turned into energy in the explosion of a medium-sized atomic bomb. How many grams is this?

23. How many liters are there in a half-gallon bottle of soda?

24. A bottle of bourbon whiskey contains 750 mL. This is the size that used to be called a "fifth." A "fifth" was one-fifth of a gallon. How much less bourbon are you getting now than you were getting before they switched over to metric labeling of bottles?

25. You drive into a gas station and fill up with 48 L of gas. How many gallons is this?

26. The average male college student probably consumes about 2500 kcal/day. How many kilojoules is this? How many joules?

27. The average American diet contains about 18% of its energy in the form of refined sugar. If the average daily diet contains 9200 kJ of energy, how many kilocalories from refined sugar does the average American consume each day?

28. The yellow light from a sodium vapor lamp has a wavelength of 589 nm. How many centimeters is this?

29. Adults under 55 years of age with a fasting serum cholesterol above 240 mg/dL are probably at increased risk for coronary heart disease. How many millimoles per liter is this if 1 mg/dL = 0.02586 mmol/L? (From Table 7.2 we see that dL is the abbreviation for deciliter.)

8

AVOGADRO'S NUMBER, THE MOLE, AND MOLECULAR WEIGHT

Consider a golf ball that weighs 100 g and a bowling ball that weighs 8000 g. The bowling ball weighs 80 times more than the golf ball.

Now consider 10 golf balls and 10 bowling balls, as is done in Figure 8-1. The 10 golf balls weigh 1000 g, and the 10 bowling balls weigh 80,000 g. But the bowling balls still weigh 80 times more than the golf balls.

What have we learned from this discussion? We have learned the following: *If one bowling ball weighs 80 times more than one golf ball, then 10 bowling balls weigh 80 times more than 10 golf balls.*

If we had a million golf balls and a million bowling balls, the bowling balls would still weigh 80 times more than the golf balls.

We could take *any number* of bowling balls and an equal number of golf balls. As long as we take the *same number* of each, the bowling balls still weigh 80 times more than the golf balls. The same principle holds when we consider atoms.

8-1
ATOMIC WEIGHTS ARE RELATED TO NUMBERS OF ATOMS

In Chapter 2 we discussed the relative atomic weights of atoms. For instance, we learned that the relative atomic weight of hydrogen is about 1 u and that the relative atomic weight of oxygen is about 16 u. Oxygen atoms

(Opposite) A New York City fireboat. The molecular formula of water was a matter of great debate among scientists at the beginning of the nineteenth century. Most thought the formula was HO. In 1811, Avogadro realized that the formula of water was actually H_2O.

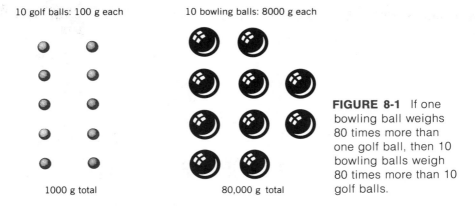

10 golf balls: 100 g each   10 bowling balls: 8000 g each

1000 g total   80,000 g total

FIGURE 8-1 If one bowling ball weighs 80 times more than one golf ball, then 10 bowling balls weigh 80 times more than 10 golf balls.

weigh 16 times more than hydrogen atoms. (Ignore the different isotopes here, because they don't affect the discussion.)

If we take 10 hydrogen atoms and 10 oxygen atoms, the oxygen atoms still weigh 16 times more than the hydrogen atoms; and if we take a million of each, the oxygen atoms still weigh 16 times more than the hydrogen atoms.

Suppose now that we want to take a billion hydrogen atoms and a billion oxygen atoms. What would be the weight of each batch of atoms? Without even knowing how we are going to measure weight, we know that we need 16 times more oxygen by weight than hydrogen.

So we have discovered the following: *To take equal numbers of atoms of hydrogen and oxygen, we would weigh out 16 times more oxygen than hydrogen.*

EXAMPLE 1   The relative atomic weight of bromine is 80 u, and the relative atomic weight of helium is 4 u. If you wanted to take an equal number of Br atoms and He atoms, what would be the ratio of their masses that you would need?

Solution: Each Br atom weighs 20 times more than each He atom, because 80 u/4 u = 20. You would have to weigh out 20 times more Br than He. ∎

8-2
THE GRAM IS USED AS THE UNIT OF ATOMIC WEIGHT

Now we have to settle on a *unit of mass* that we will use to weigh atoms. The atomic mass unit is too small to be practical. As you might suspect, scientists use the metric system, and the unit of mass they use to weigh atoms is the **gram**. Technically, the gram is a unit of mass. But nobody ever will ask you to "mass" something—you will always "weigh" something.

So from now on we will speak of 1.01 g of hydrogen atoms and 16.0 g of oxygen atoms and 79.9 g of bromine atoms when referring to their atomic weights.

When we attach grams to the relative atomic weight number, it is customary to drop the word "relative." So we will now speak of the "atomic weight of an element."

Since we already found that equal numbers of H atoms and O atoms have a mass ratio of 1.01 to 16.0, then a mass ratio of H atoms to O atoms of 1.01 to 16.0 should have *equal numbers* of H atoms and O atoms. Thus *1.01 g of hydrogen atoms has the same number of atoms as 16.0 g of oxygen atoms*. And 1.01 g of hydrogen atoms has the same number of atoms as 79.9 g of bromine. And 16.0 g of oxygen has the same number of atoms as 79.9 g of bromine.

Since the atomic weights of the elements are in the same ratio as the weights of the individual atoms, **one atomic weight of any element contains the same number of atoms as one atomic weight of any other element.**

EXAMPLE 2   What are the atomic weights of the following elements, all of which contain the same number of atoms? (Use a Table of Atomic Weights.) The elements are C, Kr, U, and P.

Solution: The masses of each element that contain the same number of atoms are just the atomic weights in grams:

$$C = 12.0 \text{ g} \qquad Kr = 83.8 \text{ g} \qquad U = 238 \text{ g} \qquad P = 31.0 \text{ g} \quad \blacksquare$$

8-3
ONE ATOMIC WEIGHT OF ATOMS CONTAINS ONE AVOGADRO'S NUMBER OF ATOMS

The number of atoms in one atomic weight of an element is found, by experiment, to be 6.0221367 × 10²³. This number is called **Avogadro's number**, named in honor of Amadeo Avogadro (1776–1856),[1] an Italian physicist who, in 1811, made some *very* significant contributions to our understanding of the concept of atoms and molecules. We will write Avogadro's number to three significant figures, or 6.02 × 10²³. So, for instance, one atomic weight of oxygen, or 16.0 g of oxygen, contains 6.02 × 10²³ atoms of oxygen.

EXAMPLE 3   How many atoms are there in 132.9 g of cesium?

Solution: There are 6.02 × 10²³ atoms of cesium, since 132.9 g of cesium is one atomic weight of cesium.   ■

[1]His full name was Lorenzo Romano Amadeo Carlo Avogadro di Quaregna e Ceretto!

We can also find the number of grams in 6.02×10^{23} atoms of an element.

EXAMPLE 4   How many grams does 6.02×10^{23} atoms of silicon weigh?

Solution: 6.02×10^{23} atoms of silicon corresponds to one atomic weight of silicon, or 28.1 g of silicon.   ■

8-4
A MOLE IS ONE ATOMIC WEIGHT OR ONE AVOGADRO'S NUMBER OF ATOMS

In previous sections of this chapter, we discussed two ways of looking at a certain quantity of an element.

1. One atomic weight.
2. Avogadro's number, 6.02×10^{23}.

One way refers to a certain *mass* of an element; the other way refers to a certain *number* of atoms of an element.

Chemists have a name for this amount of an element. It is called a **mole** of an element. **One mole of an element is both one atomic weight of an element and one Avogadro's number of atoms of an element.**

EXAMPLE 5   How much does one mole of chlorine atoms weigh?

Solution: One mole of chlorine atoms weighs one atomic weight of chlorine, which is 35.453 g.   ■

EXAMPLE 6   How many atoms are in a mole of chlorine (Cl) atoms?

Solution: A mole of chlorine atoms, like a mole of anything, contains 6.02×10^{23} atoms of chlorine.   ■

Another way to look at a mole[2] is this. From a pile of atoms of an element, count out 6.02×10^{23} atoms. Put them on a balance that reads in

---

[2]The word *mole* seems to come from the first four letters of the German word for molecular weight, *molekulargewicht*. The word *molecule* comes from the Latin word *molecula*, which is the diminutive of the Latin word *moles*, meaning mass.

grams. The balance will read the weight of one mole of the element or one atomic weight. The heavier the atoms, of course, the greater the atomic weight.

EXAMPLE 7   If you could count one atom per second, how many years would it take you to count 6.02×10^{23} atoms?

(**NOTE:** this example tries to give you an idea of how large Avogadro's number is. It is more difficult than the other examples in this chapter.)

Solution: First we must calculate the number of seconds in a year. Since 60 s = 1 min, 60 min = 1 h, 24 h = 1 day, and 365 days = 1 year, we have

$$1 \text{ year} \times \frac{365 \text{ days}}{1 \text{ year}} \times \frac{24 \text{ hr}}{1 \text{ day}} \times \frac{60 \text{ min}}{1 \text{ hr}} \times \frac{60 \text{ s}}{1 \text{ min}}$$

$$= 365 \times 24 \times 60 \times 60 = 31,536,000 \text{ s}$$

where we have used four ratio identities in a row to convert from years to seconds. Thus there are about 3.2×10^7 s in 1 year.

Since we can count 1 atom per second, we can say that 1 atom = 1 s. Thus we have the identities

$$1 \text{ atom} = 1 \text{ s}$$

$$3.2 \times 10^7 \text{ s} = 1 \text{ year}$$

Now all we have to do is convert 6.02×10^{23} atoms to the equivalent number of years,

$$6 \times 10^{23} \text{ atoms} \times \frac{1 \text{ s}}{1 \text{ atom}} \times \frac{1 \text{ year}}{3.2 \times 10^7 \text{ s}} = \frac{6 \times 10^{23}}{3.2 \times 10^7} \text{ years}$$

$$= 1.9 \times 10^{16} \text{ years}$$

or about 2×10^{16} years. Since 1 billion years = 10^9 years, we have

$$2 \times 10^{16} \text{ years} \times \frac{1 \text{ billion years}}{10^9 \text{ years}}$$

$$= 2 \times 10^7 \text{ billion years} = 20,000,000 \text{ billion years}$$

Because the universe is only about 10 or 20 billion years old, it is hardly likely that any human beings would ever have actually been able to count as high as Avogadro's number. ∎

8-5
CALCULATING MOLES AND NUMBERS OF ATOMS USES ATOMIC WEIGHTS AND AVOGADRO'S NUMBER

A very common and useful thing to be able to do is to calculate the number of moles and number of atoms in a sample of an element. To do this kind of calculation, we will use the methods developed in Chapter 7.

Consider the element calcium (symbol Ca). The atomic weight of Ca is 40.1 g. This corresponds to one mole of Ca as well as 6.02×10^{23} atoms of Ca. So we can write the following equality (the abbreviation of "mole" is "mol"):

$$40.1 \text{ g of Ca} = 1 \text{ mol of Ca} = 6.02 \times 10^{23} \text{ atoms of Ca}$$

It is customary to leave out the "of" in this type of expression as well as to write "Ca atoms" instead of "atoms of Ca":

$$40.1 \text{ g Ca} = 1 \text{ mol Ca} = 6.02 \times 10^{23} \text{ Ca atoms}$$

If we want to, we can write this expression as three identities:

$$40.1 \text{ g Ca} = 1 \text{ mol Ca}$$

$$40.1 \text{ g Ca} = 6.02 \times 10^{23} \text{ Ca atoms}$$

$$1 \text{ mol Ca} = 6.02 \times 10^{23} \text{ Ca atoms}$$

We can use these identities to help us solve problems like Examples 8 through 13. What we do is select the identity that has both the given (G) and asked-for (AF) units in it. Then we form a ratio identity that divides the given unit and leaves the asked-for unit. Refer to Chapter 7 for more details if necessary.

Notice that in problems involving moles and grams, units divide in the same way that they did in Chapter 7. Thus

$$\frac{\text{mol Ca}}{\text{mol Ca}} = 1 \quad \text{and} \quad \frac{\text{g Ca}}{\text{g Ca}} = 1 \quad \text{and so on.}$$

EXAMPLE 8   How many grams of Ca (AF) are there in 2.0 mol of Ca (G)?

Solution: The identity is 40.1 g = 1 mol Ca. We want to convert mol Ca to g Ca. First write

$$2.0 \text{ mol Ca} \times \text{———} =$$

and then write

$$2.0 \text{ mol Ca} \times \frac{40.1 \text{ g Ca}}{1 \text{ mol Ca}} = 80 \text{ g Ca} \quad \blacksquare$$

EXAMPLE 9   How many moles of Ca (AF) are there in 20 g of Ca (G)?

Solution: The identity is 40.1 g Ca = 1 mol Ca. We want to convert g Ca to mol Ca:

$$20 \text{ g Ca} \times \frac{1 \text{ mol Ca}}{40.1 \text{ g Ca}} = 0.50 \text{ mol Ca} \quad \blacksquare$$

EXAMPLE 10   How many atoms of Ca (AF) are there in 20 g of Ca (G)?

Solution: The identity is 40.1 g Ca = 6.02×10^{23} Ca atoms. We want to convert g Ca to Ca atoms:

$$20 \text{ g Ca} \times \frac{6.02 \times 10^{23} \text{ Ca atoms}}{40.1 \text{ g Ca}} = \frac{20}{40.1} \times 6.02 \times 10^{23} \text{ Ca atoms}$$

$$= 3.0 \times 10^{23} \text{ Ca atoms} \quad \blacksquare$$

EXAMPLE 11   How many atoms of Ca (AF) are there in 2.0 mol of Ca (G)?

Solution: The identity is 1 mol Ca = 6.02×10^{23} Ca atoms. We want to convert mol Ca to Ca atoms:

$$2.0 \text{ mol Ca} \times \frac{6.02 \times 10^{23} \text{ Ca atoms}}{1 \text{ mol Ca}} = 12 \times 10^{23} \text{ Ca atoms}$$

$$= 1.2 \times 10^{24} \text{ Ca atoms} \quad \blacksquare$$

EXAMPLE 12   How many moles of Ca (AF) are there in 3×10^{23} atoms of Ca (G)?

Solution: The identity is 1 mol Ca = 6.02×10^{23} Ca atoms. We want to convert Ca atoms to mol Ca:

$$3 \times 10^{23} \text{ Ca atoms} \times \frac{1 \text{ mol Ca}}{6.02 \times 10^{23} \text{ Ca atoms}} = \frac{3 \times 10^{23}}{6.02 \times 10^{23}} \text{ mol Ca}$$

$$= 0.5 \text{ mol Ca} \quad \blacksquare$$

EXAMPLE 13   How many grams of Ca (AF) are there in 1.2×10^{24} atoms of Ca (G)?

Solution: The identity is $40.1 \text{ g Ca} = 6.02 \times 10^{23}$ Ca atoms. We want to convert Ca atoms to g Ca:

$$1.2 \times 10^{24} \text{ Ca atoms} \times \frac{40.1 \text{ g Ca}}{6.02 \times 10^{23} \text{ Ca atoms}}$$

$$= \frac{1.2 \times 10^{24}}{6.02 \times 10^{23}} \times 40.1 \text{ g Ca} = 80 \text{ g Ca} \quad \blacksquare$$

8-6
ONE MOLE OF MOLECULES CONTAINS AN AVOGADRO'S NUMBER OF MOLECULES WHOSE MASS IS ONE MOLECULAR WEIGHT

As we indicated in our footnote about the derivation of the word "mole," you might expect that chemists would talk about moles of molecules. They do. **A mole of molecules is defined as 6.02×10^{23} molecules,** or

$$1 \text{ mol of molecules} = 6.02 \times 10^{23} \text{ molecules}$$

This is reasonable, since one mole of *any* substance contains 6×10^{23} particles of that substance.

The mass (in grams) of one mole of a molecule is called the **molecular weight** of that molecule, just as the mass (in grams) of one mole of atoms is called the **atomic weight** of that atom.

The question now is: How do we calculate molecular weights? There are no tables of molecular weights readily available, because it is easy to calculate the molecular weight of a molecule from the atomic weights of the atoms that make up that molecule. As the following examples show, you just add up the individual atomic weights.

EXAMPLE 14   What is the molecular weight of the diatomic molecule O_2?

Solution: The atomic weight of oxygen is 16.0 g; and since O_2 has two O atoms, the molecular weight of O_2 is $2 \times 16.0 \text{ g} = 32.0 \text{ g}$. $\blacksquare$

EXAMPLE 15   What is the molecular weight of water, H_2O?

Solution: The atomic weights are H = 1.01 g and O = 16.0 g. The molecular weight of water is $1.01 \text{ g} + 1.01 \text{ g} + 16.0 \text{ g} = 18.0 \text{ g}$. $\blacksquare$

EXAMPLE 16   Calculate the molecular weight of glucose, $C_6H_{12}O_6$.

Solution: The atomic weights are C = 12.0 g, H = 1.01 g, O = 16.0 g. Then:

$$
\begin{array}{lrl}
\text{For C we have} & 6 \times 12.0\,\text{g} = & 72.0\,\text{g} \\
\text{For H we have} & 12 \times 1.01\,\text{g} = & 12.1\,\text{g} \\
\text{For O we have} & 6 \times 16.0\,\text{g} = & \underline{96.0\,\text{g}} \\
\text{The molecular weight is} & & 180.1\,\text{g} \quad \blacksquare
\end{array}
$$

EXAMPLE 17   Calculate the molecular weight of $Ca_3(PO_4)_2$.

Solution: The atomic weights are Ca = 40.1 g, P = 31.0 g, O = 16.0 g. Then:

$$
\begin{array}{lrl}
\text{For Ca we have} & 3 \times 40.1\,\text{g} = & 120.3\,\text{g} \\
\text{For P we have} & 2 \times 31.0\,\text{g} = & 62.0\,\text{g} \\
\text{For O we have} & 8 \times 16.0\,\text{g} = & \underline{128.0\,\text{g}} \\
\text{The molecular weight is:} & & 310.3\,\text{g} \quad \blacksquare
\end{array}
$$

8-7
CALCULATING MOLES AND NUMBERS OF MOLECULES USES MOLECULAR WEIGHTS AND AVOGADRO'S NUMBER

As we did for atoms, we can write three identities for a molecule that will help us to convert between grams, moles, and number of molecules. For the water molecule, H_2O, we have

$$18.0\,\text{g } H_2O = 1 \text{ mol } H_2O = 6.02 \times 10^{23} \, H_2O \text{ molecules}$$

We can also express this as three separate identities:

$$18.0\,\text{g } H_2O = 1 \text{ mol } H_2O$$
$$18.0\,\text{g } H_2O = 6.02 \times 10^{23} \, H_2O \text{ molecules}$$
$$1 \text{ mol } H_2O = 6.02 \times 10^{23} \, H_2O \text{ molecules}$$

EXAMPLE 18   Using the above identities, calculate how many moles and molecules of H_2O (AF) there are in 9.0 g of H_2O (G).

Solution: To get moles, the identity is 18.0 g H_2O = 1 mol H_2O. We want to convert g H_2O to mol H_2O:

$$9.0 \;\cancel{g\,H_2O} \times \frac{1 \text{ mol } H_2O}{18.0 \;\cancel{g\,H_2O}} = 0.50 \text{ mol } H_2O$$

To get molecules, the identity is 18.0 g H_2O = 6.02×10^{23} H_2O molecules. We want to convert g H_2O to H_2O molecules:

$$9.0 \;\cancel{g\,H_2O} \times \frac{6.02 \times 10^{23} \text{ } H_2O \text{ molecules}}{18.0 \;\cancel{g\,H_2O}} = 3.0 \times 10^2 \text{ } H_2O \text{ molecules} \quad \blacksquare$$

EXAMPLE 19 How many moles of glucose (AF) are there in 135 g of glucose (G)?

Solution: The molecular weight of glucose is 180 g. The identity is 180 g $C_6H_{12}O_6$ = 1 mol $C_6H_{12}O_6$. We want to convert g $C_6H_{12}O_6$ to mol $C_6H_{12}O_6$:

$$135 \;\cancel{g\,C_6H_{12}O_6} \times \frac{1 \text{ mol } C_6H_{12}O_6}{180 \;\cancel{g\,C_6H_{12}O_6}} = 0.750 \text{ mol } C_6H_{12}O_6 \quad \blacksquare$$

8-8
CALCULATING MOLES OF ATOMS IN MOLECULES USES THE MOLECULAR FORMULA

It is useful to know how to calculate the number of moles of atoms in a mole of molecules. This can be done by setting up the proper identities. For example, each molecule of methane, CH_4, has one atom of C and four atoms of H. And 1 mol of CH_4 contains 1 mol of C and 4 mol of H. We can write the following identities:

$$1 \text{ } CH_4 \text{ molecule} = 1 \text{ C atom} = 4 \text{ H atoms}$$

$$1 \text{ mol } CH_4 = 1 \text{ mol C} = 4 \text{ mol H}$$

Of course, 1 molecule of CH_4 doesn't actually equal 1 atom of C or 4 atoms of H. But we can use the equals sign here if we think of it as reading "contains." The same goes for the second identity involving moles.

EXAMPLE 20 Using the identities just given, how many moles of C atoms and moles of H atoms (AF) are there in 5.0 mol of CH_4 (G)?

Solution: The identities are

$$1 \text{ mol } CH_4 = 1 \text{ mol } C \quad \text{and} \quad 1 \text{ mol } CH_4 = 4 \text{ mol } H$$

To get moles of C, we want to convert mol CH_4 to mol C:

$$5.0 \text{ mol } CH_4 \times \frac{1 \text{ mol } C}{1 \text{ mol } CH_4} = 5.0 \text{ mol } C$$

To get moles of H, we want to convert mol CH_4 to mol H:

$$5.0 \text{ mol } CH_4 \times \frac{4 \text{ mol } H}{1 \text{ mol } CH_4} = 20 \text{ mol } H \quad \blacksquare$$

EXAMPLE 21   How many atoms of C (AF) and atoms of H (AF) are in 5 molecules of CH_4 (G)?

Solution: The identities are

$$1 \text{ } CH_4 \text{ molecule} = 1 \text{ C atom} \quad \text{and} \quad 1 \text{ } CH_4 \text{ molecule} = 4 \text{ H atoms}$$

To get atoms of C, we want to convert CH_4 molecules to C atoms:

$$5 \text{ } CH_4 \text{ molecules} \times \frac{1 \text{ C atom}}{1 \text{ } CH_4 \text{ molecule}} = 5 \text{ C atoms}$$

To get atoms of H, we want to convert CH_4 molecules to H atoms:

$$5 \text{ } CH_4 \text{ molecules} \times \frac{4 \text{ C atoms}}{1 \text{ } CH_4 \text{ molecule}} = 20 \text{ H atoms} \quad \blacksquare$$

EXAMPLE 22   How many atoms of H (AF) are there in 2.0 mol of pentane, C_5H_{12} (G)?
   There are two ways to solve this problem. See the discussion after this example for some details of what we have done.

Solution 1: This approach takes the following route:

$$\text{mol pentane} \rightarrow \text{mol H} \rightarrow \text{H atoms}$$

The identities are

$$1 \text{ mol pentane} = 12 \text{ mol H} \quad \text{and} \quad 1 \text{ mol H} = 6.02 \times 10^{23} \text{ atom H}$$

$$2.0 \text{ mol pentane} \times \frac{12 \text{ mol H}}{1 \text{ mol pentane}} \times \frac{6.02 \times 10^{23} \text{ H atoms}}{1 \text{ mol H}}$$

$$= 1.4 \times 10^{25} \text{ H atoms}$$

Solution 2: This approach takes the following route:

$$\text{mol pentane} \rightarrow \text{pentane molecules} \rightarrow \text{H atoms}$$

The identities are

$$1 \text{ mol pentane} = 6 \times 10^{23} \text{ pentane molecules} \quad \text{and} \quad 1 \text{ pentane molecule}$$

$$= 12 \text{ H atoms}$$

$$2.0 \text{ mol pentane} \times \frac{6.02 \times 10^{23} \text{ pentane molecules}}{1 \text{ mol pentane}}$$

$$\times \frac{12 \text{ H atoms}}{1 \text{ pentane molecule}} = 1.4 \times 10^{25} \text{ H atoms} \quad \blacksquare$$

In Example 22, we used two identities to convert 2.0 mol of pentane into 1.4×10^{25} H atoms. We could, of course, have worked the solution in two steps using one identity at a time. For example, referring to solution 1 in Example 22, we would have

$$2 \text{ mol pentane} \times \frac{12 \text{ mol H}}{1 \text{ mol pentane}} = 24 \text{ mol H}$$

and

$$24 \text{ mol H} \times \frac{6.02 \times 10^{23} \text{ H atoms}}{1 \text{ mol H}} = 1.4 \times 10^{25} \text{ H atoms}$$

It is clearly shorter to solve the problem in one step. Again, using solution 1 in Example 22 as the example, we do the following.

1. Write down the two identities:

$$1 \text{ mol pentane} = 12 \text{ mol H}$$
$$1 \text{ mol H} = 6.02 \times 10^{23} \text{ H atoms}$$

2. Write down the given, a "×" sign, a horizontal line, a "×" sign, a horizontal line, and an "=" sign:

$$2.0 \text{ mol pentane} \times \underline{\hspace{3cm}} \times \underline{\hspace{3cm}} =$$

3. Put in one identity so that the given unit is divided:

$$2.0 \cancel{\text{ mol pentane}} \times \frac{12 \text{ mol H}}{1 \cancel{\text{ mol pentane}}} \times \underline{\hspace{3cm}} =$$

4. Put in the other identity to divide the "mol H" and give the answer:

$$2.0 \cancel{\text{ mol pentane}} = \frac{12 \cancel{\text{ mol H}}}{1 \cancel{\text{ mol pentane}}} \times \frac{6.02 \times 10^{23} \text{ H atoms}}{1 \cancel{\text{ mol H}}}$$

$$= 1.4 \times 10^{25} \text{ H atoms}$$

At this point, you should work solution 2 of Example 22 in separate steps as was done for solution 1.

In your study of the previous examples, you may have noticed that sometimes the name and sometimes the chemical formula were used to identify the unit. We could have written "mol C_5H_{12}" or "mol pentane." Both are correct—you just have to be sure to identify the unit. Some authors will use an abbreviation to identify the unit—such as "mol G" for "moles of glucose." In solving problems, you can make the choice as long as everyone knows what you are talking about.

EXAMPLE 23   How many atoms of oxygen (AF) are there in 49 g of sulfuric acid, H_2SO_4 (G)?

Solution: The molecular weight of H_2SO_4 is 98.0 g. The identities are

$$1 \text{ mol } H_2SO_4 = 98.0 \text{ g } H_2SO_4$$

$$1 \text{ mol } H_2SO_4 = 4 \text{ mol O}$$

$$1 \text{ mol O} = 6.02 \times 10^{23} \text{ O atoms}$$

Our conversion should take the route g $H_2SO_4 \rightarrow$ mol $H_2SO_4 \rightarrow$ mol O $\rightarrow$ O atoms:

$$49 \cancel{\text{ g } H_2SO_4} \times \frac{1 \cancel{\text{ mol } H_2SO_4}}{98.0 \cancel{\text{ g } H_2SO_4}} \times \frac{4 \cancel{\text{ mol O}}}{1 \cancel{\text{ mol } H_2SO_4}} \times \frac{6.02 \times 10^{23} \text{ O atoms}}{1 \cancel{\text{ mol O}}}$$

$$= 1.2 \times 10^{24} \text{ O atoms} \quad \blacksquare$$

At this point, it might not be clear to you which identities you need to use to solve a problem. As is always the case, you will find that the more problems you solve, the more skillful you will become. Even experts have to think carefully about what identities to use if they haven't solved a certain kind of problem before, so don't get discouraged. Later chapters in this book will give you much more practice in problem solving.

8-9
THE MOLE IS DEFINED IN TERMS OF A
CERTAIN QUANTITY OF A SUBSTANCE

As we have already implied, the definition of the mole can be expanded to include things other than atoms and molecules. The extended definition of the mole, the one chemists actually use, is

A mole is 6.02×10^{23} particles (of anything).[3]

EXAMPLE 24 How many electrons (AF) are in a mole of electrons (G)?

Solution: There are 6.02×10^{23} electrons in 1 mol of electrons. ■

EXAMPLE 25 How many photons (light "rays") (AF) are in 3.0 mol of photons (G)?

Solution: The identity is 1 mol photons = 6.02×10^{23} photons. We want to convert moles of photons to number of photons:

$$3.0 \ \cancel{\text{mol photons}} \times \frac{6.02 \times 10^{23} \text{ photons}}{1 \ \cancel{\text{mol photons}}} = 1.8 \times 10^{24} \text{ photons} \quad ■$$

We will close this chapter with a discussion of the mole interpretation of chemical equations.

---

[3]The official definition of the mole is this:

1. The mole is the amount of substance of a system which contains as many elementary entities as there are atoms in 0.012 kilogram (or 12 g) of carbon-12; its symbol is "mol."
2. When the mole is used, the elementary entities must be specified and may be atoms, molecules, ions, electrons, or other particles, or specified groups of such particles.

8-10
THE MOLE INTERPRETATION OF CHEMICAL EQUATIONS USES THE COEFFICIENTS OF THE SUBSTANCES IN THE BALANCED EQUATION

We can now interpret chemical equations in a different way from that done in Chapter 3. In Chapter 3, we said that a chemical equation such as

$$2H \rightarrow H_2$$

can be interpreted as "Two atoms of H react to give one molecule of H_2."

Now we can look at the equation in terms of moles. From this point of view,

$$2H \rightarrow H_2$$

is interpreted as "Two moles of H react to give one mole of H_2."

All we have done is increase the quantity. Instead of talking about atoms and molecules, we are now talking about moles. We have gone from talking about a few atoms and molecules to talking about many atoms and molecules (on the order of Avogadro's number, 6.02×10^{23}).

This approach is very useful because it allows us to do calculations using reasonable amounts of material. Mole-sized masses can be weighed out on ordinary balances; a few atoms or molecules weigh so little that it would be impossible to weigh them.

EXAMPLE 26   Interpret the equation $2H_2 + O_2 \rightarrow 2H_2O$ in terms of moles.

Solution: This is interpreted as "Two moles of H_2 react with one mole of O_2 to form two moles of H_2O."   ∎

EXAMPLE 27   Interpret the equation $6CO_2 + 6H_2O \rightarrow C_6H_{12}O_6 + 6O_2$ in terms of moles.

Solution: This is interpreted as "Six moles of CO_2 react with six moles of H_2O to form one mole of $C_6H_{12}O_6$ and six moles of O_2."   ∎

PROBLEMS

KEYED PROBLEMS

1. The relative atomic weight of calcium is 40 u, and the relative atomic weight of carbon is 12 u. If you wanted to take an equal number of Ca atoms and C atoms, what would be the ratio of their masses that you would need?

2. What are the atomic weights of the following elements, all of which contain the same number of atoms? The elements are Na, Xe, W, and Pb.

3. How many atoms are there in 55.847 g of iron?

4. How many grams does 6.02×10^{23} atoms of palladium weigh?

5. How much does 1 mol of nickel weigh?

6. How many atoms are in 1 mol of nickel?

7. How many atoms would have been counted to today's date if Avogadro had started counting one atom per second on 12:01 A.M. of January 1, 1811? Make any reasonable assumptions you like about leap years, and so on. (This problem is more difficult than the others.)

8. How many grams of silicon are there in 2.0 mol of silicon?

9. How many moles of silicon are there in 14 g of silicon?

10. How many atoms of silicon are there in 14 g of Si?

11. How many atoms of Si are there in 2.0 mol of Si?

12. How many moles of Si are there in 3.0×10^{23} atoms of Si?

13. How many grams of Si are there in 1.2×10^{24} atoms of Si?

14. What is the molecular weight of the diatomic molecule F_2?

15. What is the molecular weight of ammonia, NH_3?

16. Calculate the molecular weight of ethanol, C_2H_6O.

17. Calculate the molecular weight of $Al_2(SO_4)_3$.

18. Using the following identities, calculate how many moles and molecules of NH_3 there are in 34 g of NH_3. First we write one identity containing all the information needed, and then we split it up into three identities. As one identity we have

$$17.0 \text{ g } NH_3 = 1 \text{ mol } NH_3 = 6.02 \times 10^{23} \text{ } NH_3 \text{ molecules}$$

As three separate identities we have

$$17.0 \text{ g } NH_3 = 1 \text{ mol } NH_3$$
$$17.0 \text{ g } NH_3 = 6.02 \times 10^{23} \text{ } NH_3 \text{ molecules}$$
$$1 \text{ mol } NH_3 = 6.02 \times 10^{23} \text{ } NH_3 \text{ molecules}$$

NOTE: Only two of these three identities are needed to solve this problem.

19. How many moles of ethanol are there in 23 g of ethanol, C_2H_6O?

20. Using the identities at the beginning of Section 8-8, how many moles of C and H atoms are there in 7 mol of CH_4?

21. How many atoms of C and atoms of H are there in 5 molecules of ethane, C_2H_6?

22. How many atoms of H are there in 8.0 mol of hexane, C_6H_{14}?

23. How many atoms of oxygen are there in 49 g of phosphoric acid, H_3PO_4?

24. How many electrons are there in 2.0 mol of electrons?

25. How many photons are there in 5.0 mol of photons?

26. Interpret the equation $2NO + O_2 \rightarrow 2NO_2$ in terms of moles.

27. Interpret the equation $Mg(OH)_2 + 2HCl \rightarrow MgCl_2 + 2H_2O$ in terms of moles.

SUPPLEMENTAL PROBLEMS

28. How many atoms are there in 19 g of fluorine?

29. Calculate the molecular weight of calcium sulfate, $CaSO_4$.

30. How many moles are there in 25 g of manganese dioxide, MnO_2?

31. How many grams are there in 4 mol of sodium thiosulfate, $Na_2S_2O_3$?

32. How many atoms of sulfur and oxygen are there in 3.0 mol of sulfur trioxide, SO_3?

33. How many atoms of chlorine and fluorine are there in 50 g of chlorine trifluoride, ClF_3?

34. How many moles are there in 86 g of hydrazoic acid, HN_3?

35. How many grams are there in 9.50×10^{24} molecules of CO_2?

36. How many grams are there in 3.20×10^{21} molecules of diethyl ether, $C_4H_{10}O$?

37. The formula of vitamin C, ascorbic acid, is $C_6H_8O_6$. How many moles of ascorbic acid are there in 4.0 g of ascorbic acid?

38. The formula of vitamin B_{12} is $C_{63}H_{90}CoN_{14}O_{14}P$. How many atoms of carbon are there in 1.0×10^{-10} mol of vitamin B_{12}?

39. Thyroxine, $C_{15}H_{11}I_4NO_4$, is produced by the thyroid gland. Normal blood levels in the human being are about 5 μg in 100 mL of blood serum. How many moles of thyroxine is this?

NOTE: T_4 is the common abbreviation for thyroxine because it contains four iodine atoms per molecule.

40. A birth control pill contains 1.0 mg of progesterone, $C_{21}H_{30}O_2$. How many moles of progesterone is this?

41. Some people on the liquid protein diet have died, possibly from a potassium deficiency. Normal blood serum contains about 4×10^{-3} mol of potassium in 1 L. How many grams of potassium is this?

42. Interpret the following equation in terms of moles (this reaction is part of the Ostwald process, which is used in making nitric acid):

$$4NH_3 + 5O_2 \rightarrow 4NO + 6H_2O$$

43. Interpret the following equation in terms of moles:

$$PBr_3 + 3HOH \rightarrow P(OH)_3 + 3HBr$$

44. What is the mass, in grams, of a single gold atom?

9

STOICHIOMETRY

The last item discussed in Chapter 8 was the mole interpretation of chemical equations. In this chapter, we shall use this interpretation to solve numerical problems involving balanced chemical equations. The solution of these kinds of problems is called **stoichiometry** [from the Greek *stoicheion* (element) and *metry* (measurement)].

9-1
IDENTITIES FROM THE BALANCED CHEMICAL EQUATION USE
THE COEFFICIENTS OF THE SUBSTANCE IN THE EQUATION

Consider again the equation $2H_2 + O_2 \rightarrow 2H_2O$. The mole interpretation of this balanced equation is: 2 mol of H_2 react with 1 mol of O_2 to give 2 mol of H_2O. We could also write this as

$$2 \text{ mol } H_2 + 1 \text{ mol } O_2 \rightarrow 2 \text{ mol } H_2O$$

Now if we replace the "+" and the "$\rightarrow$" signs with "=" signs, we get

$$2 \text{ mol } H_2 = 1 \text{ mol } O_2 = 2 \text{ mol } H_2O$$

This expression looks very much like the identities that we wrote in Chapters 7 and 8. However, as was true of some of the identities in Chapter 8, this mole identity must be interpreted carefully. *It is meaningless without reference to the balanced chemical equation.* After all, 2 mol of H_2 do not really equal 1 mol of O_2, but 2 mol of H_2 do react with 1 mol of O_2 according to the balanced chemical equation. The reason that we like to write an

(Opposite) The space shuttle Columbia taking off on April 12, 1981. This was the first shuttle flight. Chemists must understand stoichiometry in order to calculate the fuel–oxygen ratio for maximum power.

identity between the substances of a balanced chemical equation is that it allows us to easily solve problems.

One story that might make the use of the "=" sign here a bit clearer is the story of the two cavemen, Charlie and Ogg. Charlie has three clams, and he wants some of the delicious dates that Ogg has. So Ogg agrees to trade eight dates for the three clams. Thus, in the cave barter system, three clams are equivalent to eight dates, or

$$3 \text{ clams} = 8 \text{ dates}$$

But you must be aware that if clams get scarce, things could change—maybe then you could get eight dates for only one clam. The trade would then be

$$1 \text{ clam} = 8 \text{ dates}$$

This is something like the substances in the balanced chemical equation. Only if you know the balanced chemical equation that relates the substances can you write the mole identity.

EXAMPLE 1   Write the mole identity between the carbon and the oxygen in the reaction $C + O_2 \rightarrow CO_2$. (There is no need to consider the CO_2 for this problem.)

Solution: The mole identity is 1 mol C = 1 mol O_2. ■

EXAMPLE 2   Write the identity between carbon and oxygen in the reaction $2C + O_2 \rightarrow 2CO$. (Again, ignore the CO for this problem.)

Solution: The identity is 2 mol C = 1 mol O_2. ■

As you can see from Examples 1 and 2, the mole identity changes if the balanced chemical equation changes, even though the substances are the same. You might wonder why carbon reacts in two different ways with oxygen. Only CO_2 is formed when there is a lot of oxygen present, such as in open flames. Some CO is formed when the oxygen supply is limited, such as in an internal combustion engine.

EXAMPLE 3   Write the three mole identities from the reaction $2H_2 + O_2 \rightarrow 2H_2O$.

Solution: The identities are 2 mol H_2 = 1 mol O_2, 2 mol H_2 = 2 mol H_2O, and 1 mol O_2 = 2 mol H_2O. ■

9-2
MOLE RELATIONSHIPS FROM THE BALANCED CHEMICAL EQUATION
USE THE COEFFICIENTS OF THE SUBSTANCES IN THE EQUATION

We can use identities like the ones in Example 3 to relate moles of different substances in a chemical reaction.

EXAMPLE 4   How many moles of O_2 (AF) are needed to completely burn 5.0 mol of H_2 (G) to form water?

Solution: The balanced equation for the reaction is $2H_2 + O_2 \rightarrow 2H_2O$. The identity relating moles H_2 and moles O_2 is 2 mol H_2 = 1 mol O_2. Then we convert 5.0 mol H_2 into moles O_2:

$$5.0 \ \text{mol } H_2 \times \frac{1 \ \text{mol } O_2}{2 \ \text{mol } H_2} = 2.5 \ \text{mol } O_2 \quad \blacksquare$$

EXAMPLE 5   How many moles of ammonia, NH_3, (AF) can be produced from 12 mol of H_2 (G) in the reaction $3H_2 + N_2 \rightarrow 2NH_3$?

Solution: The identity relating moles NH_3 and moles H_2 is 3 mol H_2 = 2 mol NH_3. Now we convert 12 mol H_2 to moles of NH_3:

$$12 \ \text{mol } H_2 \times \frac{2 \ \text{mol } NH_3}{3 \ \text{mol } H_2} = 8 \ \text{mol } NH_3 \quad \blacksquare$$

9-3
MASS RELATIONSHIPS FROM THE BALANCED CHEMICAL EQUATION
USE BOTH THE COEFFICIENTS AND THE ATOMIC
OR MOLECULAR WEIGHTS

Mass relationships between substances in a balanced equation are very important. This is so because chemicals are commonly weighed; therefore, if you want to know how much reactant to use or how much product you can get, you must know how to do mass calculations.

It is the mole identity that allows us to convert from moles of one substance to moles of another. We can then convert from grams to moles and from moles to grams by using the atomic or molecular weights.

Suppose that we are interested in two substances in a chemical reaction. We are given grams of one substance and asked to find grams of the other. The following conversions would be needed,

$$
\begin{array}{ccccccc}
\text{grams} & \xrightarrow{\ \text{A}\ } & \text{moles} & \xrightarrow{\ \text{B}\ } & \text{moles} & \xrightarrow{\ \text{C}\ } & \text{grams} \\
\text{substance 1} & & \text{substance 1} & & \text{substance 2} & & \text{substance 2}
\end{array}
$$

or simply

$$\text{grams} \overset{\text{A}}{\to} \text{moles} \overset{\text{B}}{\to} \text{moles} \overset{\text{C}}{\to} \text{grams}$$

Conversions **A** and **C** use the atomic or molecular weights in the identity. Conversion **B** uses the appropriate mole relationship from the balanced equation.

Let's look at the following problem in detail. How many grams of O_2 (AF) are needed to burn 8.0 g of H_2 (G) in the reaction $2H_2 + O_2 \to 2H_2O$? We are given 8.0 g of H_2. We want to find grams of O_2. We can do the following.

A. *Convert g H_2 to moles H_2:* The identity is 1 mol H_2 = 2.02 g H_2. Therefore, we have

$$8.0 \, \cancel{g \, H_2} \times \frac{1 \text{ mol } H_2}{2.02 \, \cancel{g \, H_2}} = 4.0 \text{ mol } H_2$$

B. *Convert the 4.0 mol H_2 to mol O_2.* The identity is 2 mol H_2 = 1 mol O_2. Therefore, we have

$$4.0 \, \cancel{\text{mol } H_2} \times \frac{1 \text{ mol } O_2}{2 \, \cancel{\text{mol } H_2}} = 2.0 \text{ mol } O_2$$

C. *Convert the 2.0 mol O_2 into grams O_2.* The identity is 1 mol O_2 = 32.0 g O_2. Therefore, we have

$$2.0 \, \cancel{\text{mol } O_2} \times \frac{32.0 \text{ g } O_2}{1 \, \cancel{\text{mol } O_2}} = 64 \text{ g } O_2$$

Thus it takes 64 g of oxygen to burn 8.0 g of hydrogen.

This problem could have been done in one step using all three identities at once. The three identities are

A. 1 mol H_2 = 2.02 g H_2.
B. 2 mol H_2 = 1 mol O_2.
C. 1 mol O_2 = 32.0 g O_2.

The solution, in one step, is

$$8.0 \, \cancel{g \, H_2} \times \frac{1 \, \cancel{\text{mol } H_2}}{2.02 \, \cancel{g \, H_2}} \times \frac{1 \, \cancel{\text{mol } O_2}}{2 \, \cancel{\text{mol } H_2}} \times \frac{32.0 \text{ g } O_2}{1 \, \cancel{\text{mol } O_2}} = 64 \text{ g } O_2$$

$$\qquad\qquad\qquad \uparrow \qquad\qquad \uparrow \qquad\qquad \uparrow$$

Conversion: **A**   **B**   **C**

Let's look closely at the procedure used to solve this problem in one step.

1. Write down the given and the asked for (G) = 8.0 g H_2, (AF) = ? g O_2.
2. Write down the three identities:

   A. *To convert grams given to moles given:* 1 mol H_2 = 2.02 g H_2.

   B. *To convert moles given to moles asked for:* 2 mol H_2 = 1 mol O_2.

   C. *To convert moles asked for to grams asked for:* 1 mol O_2 = 32.0 g O_2.

3. Write down the given, a "×", a horizontal line, a "×", a horizontal line, a "×", a horizontal line, and an "=":

   $$8.0 \text{ g } H_2 \times \underline{\hspace{2cm}} \times \underline{\hspace{2cm}} \times \underline{\hspace{2cm}} = ? \text{ g } O_2$$

4. Fill the lines with the appropriate ratio identities and compute the answer:

$$8.0 \text{ g } H_2 \times \frac{1 \text{ mol } H_2}{2.02 \text{ g } H_2} \times \frac{1 \text{ mol } O_2}{2 \text{ mol } H_2} \times \frac{32.0 \text{ g } O_2}{1 \text{ mol } O_2} = 64 \text{ g } O_2$$

$$\text{Conversion:} \quad\quad\quad \underset{\uparrow}{A} \quad\quad\quad \underset{\uparrow}{B} \quad\quad\quad \underset{\uparrow}{C}$$

The following examples should illustrate the method of solving these kinds of problems.

EXAMPLE 6   How many grams of O_2 (AF) are needed to burn completely 81 g of Al (G) in the following reaction:

$$4Al + 3O_2 \rightarrow 2Al_2O_3$$

Solution: The identities needed are **(A)** 1 mol Al = 27.0 g Al, **(B)** 4 mol Al = 3 mol O_2, and **(C)** 1 mol O_2 = 32.0 g O_2. Our conversion should take the route g Al → mol Al → mol O_2 → g O_2:

$$81 \text{ g Al} \times \frac{1 \text{ mol Al}}{27.0 \text{ g Al}} \times \frac{3 \text{ mol } O_2}{4 \text{ mol Al}} \times \frac{32.0 \text{ g } O_2}{1 \text{ mol } O_2} = 72 \text{ g } O_2 \quad \blacksquare$$

$$\text{Conversion:} \quad\quad\quad \underset{\uparrow}{A} \quad\quad\quad \underset{\uparrow}{B} \quad\quad\quad \underset{\uparrow}{C}$$

EXAMPLE 7   How many grams of water (AF) are produced from the combustion of 0.50 mol of methane, CH_4 (G)? The reaction is

$$CH_4 + 2O_2 \rightarrow CO_2 + 2H_2O$$

Solution: This problem is a bit shorter than Example 6 because we are given the moles of methane. Therefore, conversion **A**, where grams are converted to moles, can be eliminated.

The identities are **(B)** 1 mol CH_4 = 2 mol H_2O and **(C)** 1 mol H_2O = 18.0 g H_2O. Our conversion should take the route mol $CH_4 \rightarrow$ mol $H_2O \rightarrow$ g H_2O:

$$0.50 \text{ mol CH}_4 \times \frac{2 \text{ mol H}_2O}{1 \text{ mol CH}_4} \times \frac{18.0 \text{ g H}_2O}{1 \text{ mol H}_2O} = 18 \text{ g H}_2O \quad \blacksquare$$

Conversion: **B**   **C**

EXAMPLE 8   How many moles of CO_2 (AF) are produced by the oxidation of 90 g of glucose, $C_6H_{12}O_6$ (G)? The reaction is

$$C_6H_{12}O_6 + 6O_2 \rightarrow 6CO_2 + 6H_2O$$

Solution: Again, this problem only uses two identities. They are **(A)** 1 mol glucose = 180 g glucose and **(B)** 1 mol glucose = 6 mol CO_2. Our conversion should take the route g glucose $\rightarrow$ mole glucose $\rightarrow$ mole CO_2:

$$90 \text{ g glucose} \times \frac{1 \text{ mol glucose}}{180 \text{ g glucose}} \times \frac{6 \text{ mol CO}_2}{1 \text{ mol glucose}} = 3.0 \text{ mol CO}_2 \quad \blacksquare$$

Conversion: **A**   **B**

EXAMPLE 9   How many grams of calcium carbonate, $CaCO_3$ (AF), are needed to completely neutralize (react with) 109.5 g of hydrochloric acid, HCl (G)? The reaction is $CaCO_3 + 2HCl \rightarrow CaCl_2 + H_2O + CO_2$. (Note: Hydrochloric acid is a water solution of hydrogen chloride, HCl. It is a strong acid. $CaCO_3$ acts as a base. The reaction between an acid and a base is called **neutralization.** $CaCO_3$ is the main ingredient in TUMS antacid tablets, and hydrochloric acid is "stomach acid.")

Solution: This solution requires all three conversions. The identities are **(A)** 1 mol HCl = 36.46 g HCl, **(B)** 2 mol HCl = 1 mol $CaCO_3$, and **(C)** 1 mol $CaCO_3$ = 100.1 g $CaCO_3$. Our conversion should take the route g HCl $\rightarrow$ mol HCl $\rightarrow$ mol $CaCO_3 \rightarrow$ g $CaCO_3$:

$$109.5 \text{ g HCl} \times \frac{1 \text{ mol HCl}}{36.46 \text{ g HCl}} \times \frac{1 \text{ mol CaCO}_3}{2 \text{ mol HCl}} \times \frac{100.1 \text{ g CaCO}_3}{1 \text{ mol CaCO}_3}$$

$$= 150.3 \text{ g CaCO}_3 \quad \blacksquare$$

Conversion: **A**   **B**   **C**

9-4
THE LIMITING REAGENT IS THE REACTANT THAT IS USED UP FIRST IN A CHEMICAL REACTION

So far, the stoichiometry problems that we have solved assumed that all the reactants were used up. Frequently, one of the reactants will be in excess.

An example from everyday life should illustrate what we mean. Suppose you are in charge of making hero sandwiches (subs, torpedoes, grinders, hoagies, po' boys) in the school cafeteria. For each hero, you need one piece of hero bread, one slice of bologna, and one slice of cheese. Assume that mustard, mayonnaise, lettuce, and tomato are in great excess. You have 100 pieces of hero bread, 80 slices of bologna, and 60 slices of cheese. How many heros can you make? Only 60—the cheese is the limiting ingredient. It runs out first. The bread and the bologna are in excess and will be left over.

Consider again the reaction $2H_2 + O_2 \rightarrow 2H_2O$. This says that 2 mol of H_2 react with 1 mol of O_2. But suppose you are given a flask with 2 mol of H_2 and 2 mol of O_2 in it. After the reaction, what would be left in the flask? Well, the 2 mol of H_2 would react with 1 mol of O_2 to form 2 mol of H_2O. *And 1 mol of O_2 would remain unreacted.* So after the reaction there would be 1 mol of O_2 and 2 mol of H_2O in the flask. The H_2 would have reacted completely. The H_2 is the **limiting reagent.** The limiting reagent is completely used up in a chemical reaction. It is the reactant that "limits" the reaction because when it is used up, the reaction stops. Thus, the limiting reagent determines how much of the products are made. Let's make a table to illustrate this:

balanced equation:	$2H_2$	$+ O_2$	$\rightarrow 2H_2O$
moles before reaction:	2	2	0
moles that react:	-2	-1	$+2$
moles after reaction:	0	1	2

In the row "moles that react," a minus sign indicates that a substance is being used up, and a plus sign indicates that a substance is being formed. H_2 and O_2 are being used up; H_2O is being formed. The last row, "moles after reaction," is just the sum of the first two rows.

EXAMPLE 10   For the reaction $C + O_2 \rightarrow CO_2$, you are given the reactants 2.0 mol of C and 1.5 mol of O_2. What substances are left after reaction, and how many moles of each are there?

Solution: From the balanced equation we know that 1 mol of C reacts with 1 mol of O_2. Therefore, 1.5 mol of O_2 must react with 1.5 mol of C. Let's make a table:

balanced equation:	C	+	O_2	$\rightarrow$	CO_2
moles before reaction:	2.0		1.5		0.0
moles that react:	-1.5		-1.5		$+1.5$
moles after reaction:	0.5		0.0		1.5

Therefore, after reaction, 0.5 mol of C and 1.5 mol of CO_2 remain. The O_2 is the limiting reagent, and there is no O_2 left.   ■

There is a simple way to determine the limiting reagent. Consider the following imaginary reaction: $3A + 7B \rightarrow 2C + D$, where A, B, C, and D represent different substances. The balanced equation indicates that

$$3 \text{ mol A} = 7 \text{ mol B} = 2 \text{ mol C} = 1 \text{ mol D}$$

The "3 mol A", "7 mol B", "2 mol C", and "1 mol D" *can each be considered as a unit,* and each will be called a **reaction equivalent** (abbreviated REQ). Thus we have that

$$3 \text{ mol A} = 1 \text{ REQ A} \qquad 7 \text{ mol B} = 1 \text{ REQ B}$$

$$2 \text{ mol C} = 1 \text{ REQ C} \qquad 1 \text{ mol D} = 1 \text{ REQ D}$$

only for the reaction $3A + 7B \rightarrow 2C + D$.
   You can see that 1 REQ of any reactant will completely react with 1 REQ of any other reactant (to give, of course, 1 REQ of each product). Thus 1 REQ A reacts completely with 1 REQ B. Therefore, if the moles of each reacting substance are converted to REQs, *the substance with the smallest REQ is the limiting reagent.*
   To convert from moles to REQs, we just use the appropriate identity.

EXAMPLE 11   In the reaction $3A + 7B \rightarrow 2C + D$, if 5 mol of A and 8 mol of B are mixed, which reactant will be completely used up after reaction (i.e., which reactant is the limiting reagent)?

Solution: The identities needed are 3 mol A = 1 REQ A and 7 mol B = 1 REQ B. We have

$$5 \text{ mol A} \times \frac{1 \text{ REQ A}}{3 \text{ mol A}} = 1.67 \text{ REQ A}$$

and

$$8 \text{ mol B} \times \frac{1 \text{ REQ B}}{7 \text{ mol B}} = 1.14 \text{ REQ B}$$

Since there are fewer REQs of B, reactant B will be completely used up. Reactant B is the limiting reagent. ∎

If you examine Example 11 closely, you will see that all we really did to find the REQs was to divide the given moles of A and B by their respective coefficients from the balanced chemical equation:

$$\frac{5 \text{ mol A}}{3} = 1.67 \qquad \frac{8 \text{ mol B}}{7} = 1.14$$

(We have left off most units because in this one case they serve no useful purpose.)

So we have a "shortcut" to determine the limiting reagent. **To determine the limiting reagent, divide the given moles of each reactant by its respective coefficient from the balanced chemical equation; the reactant with the smallest quotient is the limiting reagent.** We don't have to mention the REQ anymore—it was just introduced to derive the preceding rule.

EXAMPLE 12   You have 10 mol of Al powder (G) and 10 mol of ferric oxide, Fe_2O_3 (G). When these react according to the equation $2Al + Fe_2O_3 \rightarrow Al_2O_3 + 2Fe$, which reactant will be completely used up (i.e., which reactant will be the limiting reagent (AF))?

NOTE: This reaction is called the **thermite reaction** and produces molten iron and a great shower of sparks.

Solution: Just divide the given moles by the respective coefficients:

$$\text{For Al:} \quad 10/2 = 5 \qquad \text{For } Fe_2O_3: \quad 10/1 = 10$$

We have left off all the units this time. Since Al gives the smallest quotient, 5, the Al will be all used up. Al is the limiting reagent. ∎

Usually you will be given grams of reactants and asked to determine the limiting reagent. Then you must first convert all grams of reactants to moles of reactants.

EXAMPLE 13   In the presence of plenty of oxygen, methane burns to give carbon dioxide and water. However, if methane burns with a shortage of oxygen, some carbon monoxide, CO, will be produced according to the reaction $2CH_4 + 3O_2 \rightarrow 2CO + 4H_2O$. If 80 g CH_4 and 96 g O_2 (G) are ignited, which reactant is the limiting reagent (AF)?

Solution: Convert all gram quantities to moles. The identities needed are 1 mol CH_4 = 16.0 g CH_4 and 1 mol O_2 = 32.0 g O_2. Thus we have

$$80 \text{ g } CH_4 \times \frac{1 \text{ mol } CH_4}{16.0 \text{ g } CH_4} = 5.0 \text{ mol } CH_4$$

$$96 \text{ g } O_2 \times \frac{1 \text{ mol } O_2}{32.0 \text{ g } O_2} = 3.0 \text{ mol } O_2$$

To find the limiting reagent, divide the moles by the respective coefficients from the balanced equation.

For CH_4:   5.0/2 = 2.5   For O_2:   3.0/3 = 1

Oxygen is the limiting reagent. Methane is in excess and cannot burn efficiently, thus giving carbon monoxide instead of carbon dioxide.   ∎

Sometimes the problem will ask you to calculate not only the limiting reagent but also how much of each reactant was left over and how much of each product was formed. To solve this kind of problem, you must use all the techniques presented in this chapter.

EXAMPLE 14   Referring to Example 13, calculate the number of grams of CH_4 (AF) left after the reaction; also calculate the number of grams of CO and H_2O (AF) formed.

NOTE: 80 g CH_4 and 96 g O_2 are the (G) in this problem.

Solution: We found in Example 13 that O_2 is the limiting reagent. Therefore, O_2 determines how much CH_4 is burned and how much CO and H_2O are formed. Thus the problem is similar to the following one. How many grams of CH_4 (AF) are needed to react with 96 g of O_2 (G), and how many grams of CO and H_2O (AF) are formed?

We will break the solution into five parts.

1. Calculate moles of CH_4. The identities needed are 1 mol O_2 = 32.0 g O_2 and 3 mol O_2 = 2 mol CH_4:

$$96 \text{ g } O_2 \times \frac{1 \text{ mol } O_2}{32.0 \text{ g } O_2} \times \frac{2 \text{ mol } CH_4}{3 \text{ mol } O_2} = 2.0 \text{ mol } CH_4$$

2. Calculate moles of CO formed. The identities needed are 1 mol O_2 = 32.0 g O_2 and 3 mol O_2 = 2 mol CO:

$$96 \ \cancel{g \ O_2} \times \frac{1 \ \cancel{mol \ O_2}}{32.0 \ \cancel{g \ O_2}} \times \frac{2 \ mol \ CO}{3 \ \cancel{mol \ O_2}} = 2.0 \ mol \ CO$$

3. Calculate moles of H_2O formed. The identities needed are 1 mol O_2 = 32.0 g O_2 and 3 mol O_2 = 4 mol H_2O:

$$96 \ \cancel{g \ O_2} \times \frac{1 \ \cancel{mol \ O_2}}{32.0 \ \cancel{g \ O_2}} \times \frac{4 \ mol \ H_2O}{3 \ \cancel{mol \ O_2}} = 4.0 \ mol \ H_2O$$

4. Calculate moles of all substances present after the reaction. A table is useful:

balanced equation:	$2CH_4$ +	$3O_2$ →	$2CO$ +	$4H_2O$
moles before reaction:	5.0	3.0	0.0	0.0
moles that react:	−2.0	−3.0	+2.0	+4.0
moles after reaction:	3.0	0.0	2.0	4.0

NOTE: This table gives the number of moles of CH_4 and O_2 before reaction (from 80 g CH_4 and 98 g O_2), as calculated in Example 13.

5. Calculate the number of grams of all substances present after the reaction. The table in step 4 (in the row labeled "moles after reaction") gives the number of moles of each substance after the reaction. The identities needed are 1 mol CH_4 = 16.0 g CH_4, 1 mol CO = 28.0 g CO, and 1 mol H_2O = 18.0 g H_2O:

$$3.0 \ \cancel{mol \ CH_4} \times \frac{16.0 \ g \ CH_4}{1 \ \cancel{mol \ CH_4}} = 48 \ g \ CH_4$$

$$2.0 \ \cancel{mol \ CO} \times \frac{28.0 \ g \ CO}{1 \ \cancel{mol \ CO}} = 56 \ g \ CO$$

$$4.0 \ \cancel{mol \ H_2O} \times \frac{18.0 \ g \ H_2O}{1 \ \cancel{mol \ H_2O}} = 72 \ g \ H_2O$$

There is no O_2 present after the reaction because it was the limiting reagent. ■

If you are ever in doubt whether a problem involves a limiting reagent, remember this. If the mass of more than one reactant is given in the problem, you can be pretty sure it's a limiting reagent problem.

PROBLEMS

KEYED PROBLEMS

1. Write the mole identity between sulfur and oxygen in the reaction $S + O_2 \rightarrow SO_2$.

2. Write the mole identity between sulfur dioxide and oxygen in the reaction $2SO_2 + O_2 \rightarrow 2SO_3$.

3. Write the three mole identities from the reaction $4Al + 3O_2 \rightarrow 2Al_2O_3$.

4. How many moles of O_2 are needed to burn 5.0 mol of Al in the reaction of Example 3 above?

5. How many moles of HCl can be produced from 10 mol of Cl_2 by the reaction $H_2 + Cl_2 \rightarrow 2HCl$?

6. How many grams of O_2 are needed to react completely with 128 g of SO_2 in the reaction $2SO_2 + O_2 \rightarrow 2SO_3$?

7. How many grams of water are produced from the combustion of 0.50 mol of octane, C_8H_{18}? The reaction is $C_8H_{18} + \frac{25}{2}O_2 \rightarrow 8CO_2 + 9H_2O$.

8. How many moles of CO_2 are produced by the oxidation of 23 g of ethanol, C_2H_5OH? The reaction is $C_2H_5OH + 3O_2 \rightarrow 2CO_2 + 3H_2O$.

9. How many grams of magnesium hydroxide (milk of magnesia), $Mg(OH)_2$, are needed to neutralize (or react with) 40.0 g of HCl? The reaction is $Mg(OH)_2 + 2HCl \rightarrow MgCl_2 + 2HOH$.

10. For the reaction $S + O_2 \rightarrow SO_2$, you are given the reactants 4 mol of S and 3 mol of O_2. What substances are left after the reaction, and how many moles of each are there?

11. In the reaction $3A + 7B \rightarrow 2C + D$, if 10 mol of A and 8 mol of B are mixed, which reactant will be completely used up after the reaction (i.e., which reactant is the limiting reagent)?

12. You have 40 mol of Al powder and 15 mol of ferric oxide, Fe_2O_3. When these react according to the equation $2Al + Fe_2O_3 \rightarrow Al_2O_3 + Fe$, which reactant will be completely used up—that is, which reactant will be the limiting reagent?

13. When methane burns with excess oxygen, carbon dioxide and water are produced: $CH_4 + 2O_2 \rightarrow CO_2 + 2H_2O$. If 8.0 g of CH_4 and 64 g of O_2 are ignited, which reactant is the limiting reagent?

14. Referring to Problem 13, calculate the number of grams of O_2 left after the reaction; also calculate the number of grams of CO_2 and H_2O formed.

SUPPLEMENTAL PROBLEMS

15. Certain recipes use a mixture of baking soda (sodium bicarbonate, $NaHCO_3$) and vinegar (a 5% solution of acetic acid, $HC_2H_3O_2$) as a substitute for baking powder. How many grams of $NaHCO_3$ should be added to react completely with 20 g of $HC_2H_3O_2$? The reaction is $NaHCO_3 + HC_2H_3O_2 \rightarrow NaC_2H_3O_2 + CO_2 + H_2O$.

16. Aspirin, $C_9H_8O_4$, is synthesized from salicylic acid, $C_7H_6O_3$, and acetic anhydride, $C_4H_6O_3$, in the reaction $C_7H_6O_3 + C_4H_6O_3 \rightarrow C_9H_8O_4 + HC_2H_3O_2$. How many grams of aspirin can be made from 100 g of salicylic acid?

17. The Haber process for the synthesis of ammonia uses the reaction $N_2 + 3H_2 \rightarrow 2NH_3$. How many grams of hydrogen, H_2, are required to react completely with 7.0 g of N_2?

18. The antacid tablet TUMS contains $CaCO_3$ as the active ingredient. How many grams of hydrochloric acid, HCl, can be neutralized by 1.0 g $CaCO_3$ (about two TUMS tablets)? The reaction is $2HCl + CaCO_3 \rightarrow CaCl_2 + CO_2 + H_2O$.

19. The modern Hall process for producing aluminum was invented in the 1880s; prior to that, aluminum was more expensive than gold. Aluminum was made by the reduction of an aluminum salt with metallic sodium; $3Na + AlCl_3 \rightarrow Al + 3NaCl$. If 10 g Na and 10 g $AlCl_3$ are mixed together, how many grams of Al metal will be produced?

20. The "fixer" in photography (also known as "hypo") is sodium thiosulfate, $Na_2S_2O_3$. It removes undeveloped silver bromide (AgBr) from the film by the reaction $AgBr + 2Na_2S_2O_3 \rightarrow Na_3[Ag(S_2O_3)_2] + NaBr$. How many grams of $Na_2S_2O_3$ is required to remove 0.50 g of AgBr?

    (**NOTE:** Ignore the square brackets in the formula to the right of the arrow.)

21. Calcium oxide (CaO, quicklime) reacts with water to form calcium hydroxide ($Ca(OH)_2$, slaked lime) by the reaction $CaO + H_2O \rightarrow Ca(OH)_2$. If 5.0 g of CaO are reacted with 10 g of H_2O, what is the limiting reagent? How many grams of each substance will be left over after the reaction is completed?

22. Propyne, C_3H_4, can react with H_2 in the presence of a nickel catalyst to produce propane, C_3H_8: $C_3H_4 + 2H_2 \rightarrow C_3H_8$. (This is an example of a process called **hydrogenation**.) If 25 g C_3H_4 and 20 g H_2 are mixed together, how many grams of C_3H_8 will be produced?

23. Acetylene, C_2H_2, can be made by the reaction $CaC_2 + 2H_2O \rightarrow C_2H_2 + Ca(OH)_2$. CaC_2 is called calcium carbide and is used in the "carbide cannon" (which makes a loud boom used to start sailing races). How many grams of acetylene can be made from 50 g of CaC_2?

24. Ethyl alcohol, C_2H_5OH, can be made by fermenting sugar (such as glucose, $C_6H_{12}O_6$) according to the reaction $C_6H_{12}O_6 \rightarrow 2C_2H_5OH + 2CO_2$. This process can be used to make alcohol for drinking as well as for use in gasohol. How many grams of glucose are needed to make 100 g of ethyl alcohol?

25. Aluminum reacts with bromine to form aluminum bromide according to the reaction $2Al + 3Br_2 \rightarrow 2AlBr_3$. How many grams of bromine are needed to react completely with 250 g Al?

10

PERCENT COMPOSITION, EMPIRICAL FORMULAS, AND MOLECULAR FORMULAS

This chapter discusses certain mass and percentage topics involving molecules. We begin with a review of percent.

10-1
PERCENT IS BASED ON 100

If you get 95% on an exam, this means that out of 100 possible points, you got 95 points. Percent (abbreviated "%") means parts per hundred and is used mostly for convenience. It is easier to say that you got 95% (meaning 95 out of a possible 100) on an exam than to say that you got 0.95 (out of a possible 1).[1]

EXAMPLE 1   A student gets 82 questions out of 100 correct on an exam. What is the fraction correct and the percent correct?

Solution: The fraction correct is 82/100 = 0.82. The percent correct is 82/100 × 100 = 82%.   ■

---

[1]A baseball player's batting average is another example of a decimal number. It is based on 1000 instead of 100. So when Joe DiMaggio hit .348, it really meant that he averaged 348 hits per 1000 times at bat. The decimal is calculated from 348/1000 = 0.348. His percent of hits would be 348/1000 × 100 = 34.8%.

(Opposite) A magnet floating above a superconducting disk that has been cooled with liquid nitrogen. The formula of one type of material that is superconducting at the temperatures of liquid nitrogen (-195.8 °C) is $YBa_2Cu_3O_7$.

The difference between the fraction correct and the percent correct is simply a factor of 100. Multiplying the fraction correct by 100 gives the percent correct.

Sometimes an instructor will give an exam that doesn't have 100 points. A percent correct can still be calculated, as Example 2 shows.

EXAMPLE 2   A student takes an exam consisting of 80 short-answer questions. The student gets 67 questions correct. What is the percent correct on this exam?

Solution: The percent correct is $67/80 \times 100 = 84\%$.   ■

A formula for calculating the percent correct (or the grade) on an exam is

$$\frac{\text{number of questions correct}}{\text{total number of questions}} \times 100 = \% \text{ correct}$$

This formula assumes that all the questions have the same value. If questions have different values, you must calculate your grade from the following formula:

$$\frac{\text{number of points you received}}{\text{total number of points on exam}} \times 100 = \% \text{ of points correct}$$

Another thing you should realize is that

$$\% \text{ points correct} + \% \text{ points incorrect} = 100\%$$

EXAMPLE 3   On an exam, a student lost 7 points out of a possible 30. What percentage did the student score?

Solution: The number of points the student received was $30 - 7 = 23$. The grade on the exam was $23/30 \times 100 = 77\%$.   ■

In Example 3 the percent points incorrect on the exam was $7/30 \times 100 = 23\%$. The sum of the % points correct + % points incorrect = 77% + 23% = 100%.

10-2
PERCENT COMPOSITION IS THE MASS PERCENT OF THE ELEMENTS IN A COMPOUND

The percent by mass of all the elements in a compound is called the **percent composition.** Each element's percent by mass is found in a way that is similar to calculating exam scores. Take the mass of an element in the compound, divide by the total mass of the compound, and multiply by 100:

$$\frac{\text{element's mass}}{\text{total mass of compound}} \times 100 = \% \text{ by mass of the element}$$

In practice, the atomic and molecular weights are used in this calculation.

EXAMPLE 4   What is the percent by mass of carbon in CO_2?

Solution: Atomic and molecular weights are easily available numbers that we can use. The atomic weight of carbon is 12.0 g, and the molecular weight of CO_2 is 44.0 g. Therefore,

$$\frac{12.0 \text{ g C}}{44.0 \text{ g CO}_2} \times 100 = 27.3\% \text{ carbon} \quad \blacksquare$$

EXAMPLE 5   What is the percent by mass of oxygen in CO_2?

Solution: The atomic weight of oxygen is 16.0 g. There are two oxygen atoms in a molecule of CO_2, so the total weight of oxygen in 1 mol (44.0 g) of CO_2 is 32.0 g. Thus

$$\frac{32.0 \text{ g O}}{44.0 \text{ g CO}_2} \times 100 = 72.7\% \text{ oxygen} \quad \blacksquare$$

Notice that when we combine the percent by mass from Example 4 with that from Example 5, our total is 100%:

$$27.3\% \text{ carbon} + 72.7\% \text{ oxygen} = 100\% \text{ total}$$

A listing of all the percents by mass of the elements in a compound is called the percent composition of that compound.

EXAMPLE 6   What is the percent composition of CO_2?

Solution: From the two previous examples, the percent composition of CO_2 is 27.3% carbon and 72.7% oxygen.   ■

EXAMPLE 7   What is the percent composition of water, H_2O?

Solution: The atomic weights are H = 1.01 g and O = 16.0 g. The molecular weight of H_2O is 18.0 g. The percent composition is

$$\frac{2 \times 1.01 \text{ g H}}{18.0 \text{ g } H_2O} \times 100 = 11.1\% \text{ hydrogen}$$

$$\frac{1 \times 16.0 \text{ g O}}{18.0 \text{ g } H_2O} \times 100 = 88.9\% \text{ oxygen}$$

The "2" in 2 × 1.01 g H is needed because of the mole identity 2 mol H = 1 mol H_2O. The "1" in 1 × 16.0 g O shows that 1 mol O = 1 mol H_2O.   ■

EXAMPLE 8   What is the percent composition of glucose, $C_6H_{12}O_6$?

Solution: The atomic weights are C = 12.0 g, H = 1.01 g, and O = 16.0 g. The molecular weight of glucose is 180 g. The percent composition is

$$\frac{6 \times 12.0 \text{ g C}}{180 \text{ g glucose}} \times 100 = 40.0\% \text{ carbon}$$

$$\frac{12 \times 1.01 \text{ g H}}{180 \text{ g glucose}} \times 100 = 6.73\% \text{ hydrogen}$$

$$\frac{6 \times 16.0 \text{ g O}}{180 \text{ g glucose}} \times 100 = 53.3\% \text{ oxygen}$$ ■

If you are given a quantity of a substance and you know the percent composition, you can calculate the mass of each element in the sample.

EXAMPLE 9   What is the mass of each element in 30.0 g of glucose?

Solution: From Example 8 we know the percent composition of glucose. Thus the mass of each element in 30.0 g of glucose is (remember to convert percents into the equivalent decimal form):

mass of carbon:	$30.0 \text{ g} \times 0.400 = 12.0$ g carbon	
mass of hydrogen:	$30.0 \text{ g} \times 0.0673 = 2.02$ g hydrogen	
mass of oxygen:	$30.0 \text{ g} \times 0.533 = 16.0$ g oxygen	■

10-3
AN EMPIRICAL FORMULA AND A MOLECULAR FORMULA
CAN BE CALCULATED FROM THE PERCENT COMPOSITION
AND MOLECULAR WEIGHT OF THE COMPOUND

A chemical formula that gives the relative number of each type of atom in a molecule of a substance is called the **empirical formula** of the substance. It is the simplest formula of a substance. Notice the words "relative number" in this definition. The empirical formula is not necessarily the same as the actual molecular formula. Take the case of glucose, whose actual molecular formula is $C_6H_{12}O_6$. The empirical formula of glucose is CH_2O. This is the simplest formula that has the correct ratio of carbon, hydrogen, and oxygen atoms. The empirical weight of CH_2O is 30.0 g. **Empirical weight** is the term we shall use for the mass represented by the empirical formula. If we take six empirical formulas, we have $6 \times CH_2O$ or $C_6H_{12}O_6$ with a molecular weight of $6 \times 30.0 \text{ g} = 180$ g.

EXAMPLE 10   The empirical formula of ethylene is CH_2. The molecular weight of ethylene is 28.0 g. What is the molecular formula of ethylene?

Solution: The empirical weight of CH_2 is 14.0 g. If we divide 28.0 g by 14.0 g, we find the number of empirical formula units in the molecular formula. Since 28.0 g/14.0 g = 2, the molecular formula of ethylene is $2 \times CH_2 = C_2H_4$. ■

EXAMPLE 11   The empirical formula of hydrogen peroxide is HO. The molecular weight is 34.0 g. What is the molecular formula of hydrogen peroxide?

Solution: The empirical weight of HO is 17.0 g. Since 34.0 g/17.0 g = 2, the molecular formula of hydrogen peroxide is $2 \times HO = H_2O_2$. ■

Sometimes the empirical formula and the molecular formula are the same. An example would be the empirical formula of sulfur dioxide, SO_2, whose empirical weight is 64.1 g. The molecular weight of sulfur dioxide is also 64.1 g, and thus the molecular formula of sulfur dioxide is SO_2.

It is important for chemists to know the molecular formulas of substances. Since the percent composition and the molecular weights can be determined experimentally, we can use these data to determine molecular formulas.

The simplest way to approach the problem is to assume that we have 100 g of the substance. This is just a convenience—any amount would do—but taking percents of 100 is very easy, and everybody does it this way.

EXAMPLE 12   The percent composition of a hydrocarbon is 75.0% carbon and 25.0% hydrogen. The molecular weight is 16.0 g. What is the molecular formula?

Solution: To calculate the empirical formula, we use percent composition. Taking 100 g of hydrocarbon, we see that it has 75.0 g carbon and 25.0 g hydrogen. Since the ratio of atoms in the molecule is the same as the ratio of moles of atoms in the substance, we now calculate the number of moles of each element in these masses:

$$75.0 \text{ g C} \times \frac{1 \text{ mol C}}{12.0 \text{ g C}} = 6.25 \text{ mol C}$$

$$25.0 \text{ g H} \times \frac{1 \text{ mol H}}{1.01 \text{ g H}} = 24.8 \text{ mol H}$$

Writing the empirical formula as $C_{6.25}H_{24.8}$ is not very simple—we must simplify it. We can do this by dividing both subscripts by the smallest one. This will give a "1" as the smallest subscript:

$$6.25/6.25 = 1 \qquad 24.8/6.25 = 3.97$$

We round 3.97 to 4 because we need a whole number of atoms in our empirical formula. The difference between 3.97 and 4 is due to experimental error and/or rounding error. The empirical formula is thus C_1H_4 or CH_4. The empirical weight is 16.0 g. Since the molecular weight is also 16.0 g, and 16.0 g/16.0 g = 1, the molecular formula is CH_4.   ■

EXAMPLE 13   The percent composition of a hydrocarbon is found to be 82.7% carbon and 17.4% hydrogen. The molecular weight is 58.1 g. What is the empirical formula and the molecular formula?

Solution: Taking 100 g of hydrocarbon, we see that it has 82.7 g carbon and 17.4 g hydrogen. The number of moles of each element is

$$82.7 \text{ g C} \times \frac{1 \text{ mol C}}{12.0 \text{ g C}} = 6.89 \text{ mol C}$$

$$17.4 \text{ g H} \times \frac{1 \text{ mol H}}{1.01 \text{ g H}} = 17.2 \text{ mol H}$$

Again, writing the empirical formula as $C_{6.89}H_{17.2}$ is not the simplest way, and we must divide both subscripts by the smallest one:

$$6.89/6.89 = 1 \qquad 17.2/6.89 = 2.5$$

An empirical formula of $CH_{2.5}$ is still not the simplest, since we want whole number subscripts. If we double both subscripts, we get C_2H_5. This is a fine empirical formula whose empirical weight is 29.1 g. Since the molecular weight of our hydrocarbon is 58.1 g, we divide 58.1 g by 29.1 g and get 58.1 g/29.1 g = 2; the molecular formula of our hydrocarbon is $2 \times C_2H_5 = C_4H_{10}$. ■

Example 13 illustrates an interesting point. If the empirical formula turns out to have noninteger (noninteger means not a whole number) subscripts, try multiplying the subscripts by 2 or 3 or some other small whole number. This will usually turn the subscripts into whole numbers.

EXAMPLE 14   In Example 8 we calculated the percent composition of glucose from the formula $C_6H_{12}O_6$. Let's work backward from the percent composition and the molecular weight to arrive at the molecular formula of $C_6H_{12}O_6$. Glucose is 40.0% carbon, 6.73% hydrogen, and 53.3% oxygen and has a molecular weight of 180 g.

Solution: Assuming we are given a sample of 100 g glucose, we have 40.0 g carbon, 6.73 g hydrogen, and 53.3 g oxygen. The moles of each element are

$$40.0 \text{ g C} \times \frac{1 \text{ mol C}}{12.0 \text{ g C}} = 3.33 \text{ mol C}$$

$$6.73 \text{ g H} \times \frac{1 \text{ mol H}}{1.01 \text{ g H}} = 6.66 \text{ mol H}$$

$$53.3 \text{ g O} \times \frac{1 \text{ mol O}}{16.0 \text{ g O}} = 3.33 \text{ mol O}$$

Dividing each number by the smallest, 3.33, we get

$$3.33/3.33 = 1 \qquad 6.66/3.33 = 2 \qquad 3.33/3.33 = 1$$

The empirical formula is $C_1H_2O_1$ or CH_2O, which has an empirical weight of 30.0 g. Since 180 g/30.0 g = 6, the molecular formula is 6 × CH_2O = $C_6H_{12}O_6$. ■

EXAMPLE 15   A gaseous compound has the following percent composition: 5.05% hydrogen and 95.0% fluorine. Its molecular weight is 120 g. What is the molecular formula of the compound?

Solution: Assuming we are given 100 g of gas, we have 5.05 g hydrogen and 95.0 g fluorine. The number of moles of each element is

$$5.05 \text{ g H} \times \frac{1 \text{ mol H}}{1.01 \text{ g H}} = 5.00 \text{ mol H}$$

$$95.0 \text{ g F} \times \frac{1 \text{ mol F}}{19.0 \text{ g F}} = 5.00 \text{ mol F}$$

Dividing by the smallest number we get

$$5.00/5.00 = 1 \qquad 5.00/5.00 = 1$$

The empirical formula is HF, and the empirical weight is 20.0 g. Since 120 g/20.0 g = 6, the molecular formula is 6 × HF = H_6F_6.

NOTE: The formula for hydrogen fluoride is commonly written as HF; however, in the gas phase, HF molecules stick together and on the average form "clumps" of six. Thus, in the gas phase, we can write the molecular formula of hydrogen fluoride as H_6F_6. Another way to write it would be $(HF)_6$. ■

10-4
THE FORMULA WEIGHT OF A GIANT ARRAY IS THE MASS OF THE SIMPLEST RATIO OF ATOMS IN THE ARRAY

If a substance is a "giant array" of atoms (see Chapter 4) such as NaCl, the empirical formula would be NaCl. It wouldn't make much sense to talk about the molecular formula of NaCl, since there are no separate molecules in the NaCl crystal. The empirical weight of NaCl is 58.4 g and is the mass of the simplest ratio of atoms in the NaCl crystal. Many authors call the mass of the simplest ratio of atoms in the giant array the **formula weight.** Thus the formula weight of NaCl is 58.4 g.

PROBLEMS

KEYED PROBLEMS

1. A student gets 88 questions out of 100 correct on an exam. What is the fraction correct and percent correct?

2. A student takes an exam consisting of 89 short-answer questions. The student gets 72 questions correct. What is the percent correct on this exam?

3. On an exam, you lost 12 points out of a possible 40 points. What was your grade?

4. What is the percent by mass of sulfur in sulfur dioxide, SO_2?

5. What is the percent by mass of oxygen in sulfur dioxide, SO_2?

6. What is the percent composition of SO_2?

7. What is the percent composition of hydrogen peroxide, H_2O_2?

8. What is the percent composition of diethyl ether, $C_4H_{10}O$?

9. What is the mass of each element in 50.0 g of diethyl ether, $C_4H_{10}O$?

10. The empirical formula of benzene is CH. The molecular weight of benzene is 78.1 g. What is the molecular formula of benzene?

11. The empirical formula of an oxide of nitrogen is NO_2. The molecular weight is 92.0 g. What is the molecular formula of the oxide of nitrogen?

12. A smelly liquid is 15.78% carbon and 84.22% sulfur. The molecular weight is 76.13 g. What is the molecular formula of the liquid?

13. The percent composition of a hydrocarbon is found to be 85.6% carbon and 14.4% hydrogen. The molecular weight is 56.1 g. What is the empirical formula and the molecular formula?

14. The percent composition of butyric acid is 54.5% carbon, 36.3% oxygen, and 9.15% hydrogen. The molecular weight of butyric acid is 88.1 g. What is the molecular formula of butyric acid?

15. A sulfur–fluorine compound has the following percent composition: 25.2% sulfur and 74.8% fluorine. Its molecular weight is 254 g. What is the molecular formula of the compound?

SUPPLEMENTAL PROBLEMS

16. Calculate the percent composition of the following interhalogen compounds: ClF_3, BrF_5, IF_7. (An interhalogen compound consists only of a combination of halogens, which are the elements F, Cl, Br, I, and At.)

17. Calculate the percent composition of plaster of Paris, $(CaSO_4)_2 \cdot H_2O$. What is the percent of H_2O in plaster of Paris? The "·" in the formula $(CaSO_4)_2 \cdot H_2O$ is read "dot" and signifies that the H_2O is "water of hydration." This water of hydration is loosely held to the $CaSO_4$ and can be removed by heating.

18. Calculate the percent composition of copper sulfate pentahydrate, which appears as deep blue crystals. The formula is $CuSO_4 \cdot 5H_2O$, where the $5H_2O$ is the water of hydration (see Problem 17). If the water is removed by heating, the crystals turn white.

19. Calculate the percent composition of calcium nitride, Ca_3N_2.

20. The percent composition of ascorbic acid (vitamin C) is 40.9% carbon, 54.5% oxygen, and 4.55% hydrogen. The molecular weight is 176 g. What is the molecular formula of ascorbic acid?

21. The percent composition of a strong acid is 2.04% hydrogen, 32.65% sulfur, and 65.31% oxygen. What is the molecular formula of the strong acid if the molecular weight is 98.1 g?

22. An unpleasant gaseous compound is 22.54% phosphorus and 77.46% chlorine. The molecular weight is 137.3 g. What is the molecular formula of the compound?

23. Upon analysis, the black coating found on tarnished silver contains 87.1% silver and 12.9% sulfur. The formula weight is 248 g. What is the chemical formula?

24. This problem is more difficult: When 5.000 g of a certain hydrocarbon is burned, 13.72 g of CO_2 and 11.23 g of H_2O are produced. What is the empirical formula of the hydrocarbon? (Hint: First calculate the percent of C in CO_2 and the percent of H in H_2O. Second, calculate the number of grams of C and H in the hydrocarbon. Finally, calculate the percent of C and H in the hydrocarbon and then determine the empirical formula.)

25. Calculate the formula weight of the following substances that consist of giant arrays of atoms: KBr, SiC.

26. This is a question about the differences and similarities of the terms "molecular weight" and "formula weight." Are all molecular weights also formula weights? Are all formula weights also molecular weights?

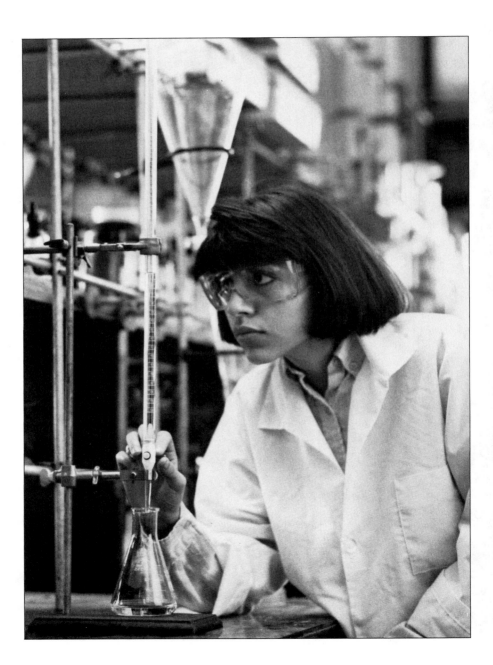

11

MOLARITY
AND SOLUTION STOICHIOMETRY

In this chapter we will discuss the kinds of solutions in which a solid or a liquid is dissolved in a liquid. An example of a solid dissolved in a liquid is a salt–water solution. A solid, salt, is dissolved in a liquid, water. An example of a liquid dissolved in a liquid is grain alcohol dissolved in water. Vodka is an example of such a solution.

In a solution in which a solid has been dissolved in a liquid, the solid is called the **solute.** The liquid is called the **solvent.** In a solution of a liquid in a liquid, the liquid present in smaller amounts is usually called the solute; however, the distinction is sometimes not clear.

EXAMPLE 1   If you prepare a solution by mixing water and sugar, which component is the solute and which is the solvent?

Solution: Water, a liquid, is the solvent; sugar, a solid, is the solute.   ■

11-1
MOLARITY IS A UNIT OF CONCENTRATION
INVOLVING MOLES AND LITERS

From your study of stoichiometry in Chapter 9, you might realize that we can do stoichiometric calculations with solutions. To perform such calculations, we must have a concentration unit that involves moles. One such

(Opposite) A student performing a titration with a buret.

concentration unit that is very useful is called **molarity.** A solution with a molarity of one is defined as follows:

> If one mole of a solute is dissolved in enough solvent to make one liter of solution, the resulting solution has a **molarity of one.** The abbreviation of molarity is M, so this solution would be referred to as 1 M and would read "one molar."

Of course, we don't always have to make up solutions that are 1 M. We can make solutions of any molarity, provided the solute can be dissolved in the solvent. If two moles of solute are dissolved in enough solvent to make one liter of solution, we would have a 2 molar solution (2 M). Figure 11-1 shows the procedure for making up one liter of a 0.500 M water solution of

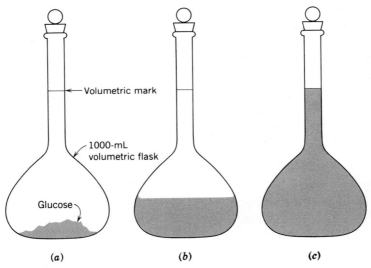

(a)   (b)   (c)

FIGURE 11-1 The procedure for making 1 L of 0.500 M glucose solution. (a) Weigh out 0.500 mol of glucose. This would be 90.1 g of glucose. Place the glucose in a 1000-mL volumetric flask. (b) Fill the flask about half full with water. Shake the flask to dissolve the glucose. If the flask is filled up to the volumetric mark before the glucose is dissolved, it will be difficult to dissolve the glucose. The neck is too narrow to allow good mixing, and the volume after mixing might turn out to be more than 1000 mL. However, the narrow neck has a purpose. If an error is made in filling the flask to the volumetric mark, the error in the volume will be small. (c) Fill the flask to the volumetric mark and mix completely. Notice that a volumetric flask has only one volumetric mark on it—in this case 1000 mL. It can only be used to make 1000 mL of solution. If you want to make a different volume of solution, you must use a volumetric flask of a different size.

glucose, $C_6H_{12}O_6$. Notice that in making up the solution, we add enough water to make up one liter of solution. We don't, as a rule, add one liter of water.

It seems reasonable that we could use *any quantity of solute and solvent* to make our solutions. Since molarity is the number of moles of solute in a liter of solution, molarity is calculated by dividing the number of moles of solute by the number of liters (or fraction of a liter) of solution:

$$\text{molarity} = \frac{\text{number of moles of solute}}{\text{number of liters of solution}}$$

$$= \text{number of moles of solute per liter of solution}$$

Since the number of moles refers to "moles of solute," and the liters always refers to "liters of solution," we can simplify things by writing (remember that the abbreviation of mole is mol)

$$\text{molarity} = \frac{\text{moles}}{\text{liter}} = \frac{\text{mol}}{\text{L}} = \text{M}$$

So to calculate molarity, just divide moles of solute by liters of solution.

EXAMPLE 2   A solution contains 2.0 mol of NaOH dissolved in 8.0 L of solution. What is the molarity of the solution?

Solution: Since molarity = mol/L, we have

$$\text{molarity} = \frac{2.0 \text{ mol NaOH}}{8.0 \text{ L}} = 0.25 \frac{\text{mol NaOH}}{\text{L}} = 0.25 \text{ M NaOH}$$

Notice that a 0.25 molar NaOH solution is abbreviated 0.25 M NaOH. ■

EXAMPLE 3   A solution contains 0.10 mol of H_2SO_4 dissolved in a total volume of 0.20 L. What is the molarity of the solution?

Solution:

$$\text{molarity} = \frac{\text{mol}}{\text{L}} = \frac{0.10 \text{ mol } H_2SO_4}{0.20 \text{ L}} = 0.50 \text{ M } H_2SO_4 \quad ■$$

EXAMPLE 4   A solution contains 0.300 mol of glucose in 250 mL of solution. What is the molarity of the solution?

Solution: First we must convert 250 mL to liters:

$$250 \text{ mL} \times \frac{1 \text{ L}}{1000 \text{ mL}} = 0.250 \text{ L}$$

$$\text{molarity} = \frac{\text{mol}}{\text{L}} = \frac{0.300 \text{ mol glucose}}{0.250 \text{ L}} = 1.20 \text{ M glucose} \quad \blacksquare$$

Most of the time, the amount of solute is given in grams, since that's the unit of mass that we read on our balances. Then you must convert from grams to moles in order to calculate the molarity of the solution.

EXAMPLE 5   A solution contains 15 g of $FeCl_3$ in 500 mL of solution. What is the molarity of the solution?

Solution: First calculate the number of moles in 15 g of $FeCl_3$. The formula weight of $FeCl_3$ is 162 g, so 1 mol weighs 162 g. (Remember from Section 10-4 that the term "formula weight" is used instead of "molecular weight" when we want to refer to the weight of the simplest formula of a giant array.)

$$15 \text{ g } FeCl_3 \times \frac{1 \text{ mol } FeCl_3}{162 \text{ g } FeCl_3} = 0.092 \text{ mol } FeCl_3$$

Then convert 500 mL to liters:

$$500 \text{ mL} \times \frac{1 \text{ L}}{1000 \text{ mL}} = 0.500 \text{ L}$$

Now calculate the molarity:

$$\text{molarity} = \frac{\text{mol}}{\text{L}} = \frac{0.092 \text{ mol } FeCl_3}{0.500 \text{ L}} = 0.18 \text{ M } FeCl_3$$

We can do this problem in one step:

$$15 \text{ g } FeCl_3 \times \frac{1 \text{ mol } FeCl_3}{162 \text{ g } FeCl_3} \times \frac{1}{500 \text{ mL}} \times \frac{1000 \text{ mL}}{1 \text{ L}} = 0.18 \text{ M } FeCl_3$$

Notice that the product of the first two terms is 0.092 mol Fe, whereas the product of the last two terms is 1/0.500 L. Thus the product of all four terms is 0.092 mol Fe/0.500 L, just as when we use separate steps. $\blacksquare$

It is useful to devise a general formula to help us solve problems involving molarity. Since

$$M = \frac{mol}{L}$$

we can write

$$M \times L = mol$$

which is the same as

$$\frac{mol}{L} \times L = mol$$

where you must remember that the "mol/L" term stands for molarity.

EXAMPLE 6   How many grams of NaCl must be taken to make 700 mL of an 0.20 M NaCl solution?

Solution: The number of liters is 700 mL $\times$ 1 L/1000 mL = 0.700 L. Using the equation

$$\frac{mol}{L} \times L = mol$$

we can calculate the number of moles of NaCl needed:

$$0.20 \frac{mol\ NaCl}{L} \times 0.700\ L = 0.14\ mol\ NaCl$$

Remember that the first term in this equation is the same as 0.20 M NaCl. To figure out the number of grams of NaCl, use the formula weight of NaCl:

$$0.14\ mol\ NaCl \times \frac{58.5\ g\ NaCl}{1\ mol\ NaCl} = 8.2\ g\ NaCl \quad \blacksquare$$

Examples 5 and 6 show the relationship between grams, moles, liters, and molarity, which we can write out as

molarity × liters = moles

moles × molecular (or formula) weight = grams

EXAMPLE 7   Using 7.3 g of $AgNO_3$, how many liters of solution are needed to make a 0.12 M solution?

Solution: The number of moles of $AgNO_3$ is

$$7.3 \text{ g AgNO}_3 \times \frac{1 \text{ mol AgNO}_3}{169.9 \text{ g AgNO}_3} = 0.043 \text{ mol AgNO}_3$$

Using the formula mol/L × L = mol and remembering that 0.12 M = 0.12 mol/L,

We will solve for this "L" which represents the number of liters of solution we need.
$$\downarrow$$

$$0.12 \frac{\text{mol AgNO}_3}{\text{L}} \times \text{L} = 0.043 \text{ mol AgNO}_3$$

$$\text{L} = \frac{0.043 \text{ mol AgNO}_3}{0.12 \dfrac{\text{mol AgNO}_3}{\text{L}}} = 0.36 \text{ L} \quad \blacksquare$$

Notice that in a complex fraction like

$$\frac{\text{mol}}{\dfrac{\text{mol}}{\text{L}}}$$

we can clear the layers by multiplying numerator and denominator by "L" (which is the same as multiplying the three-layered fraction by L/L = 1):

$$\frac{\cancel{\text{mol}} \times \text{L}}{\dfrac{\cancel{\text{mol}}}{\cancel{\text{L}}} \times \cancel{\text{L}}} = \text{L}$$

11-2
SOLUTION STOICHIOMETRY USES MOLARITY

A good way to visualize what is involved in solution stoichiometry is to understand a laboratory procedure called **titration**. It is performed using the apparatus shown in Figure 11-2. A buret is filled to a known level with one solution, say 0.1 M NaOH (0.1 M sodium hydroxide). (A buret was described in Figure 6-4.) In a flask placed below the buret, a known quantity (say 25 mL) of another solution (e.g., hydrochloric acid, HCl) is added. The

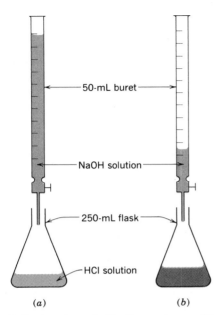

(a)   *(b)*

FIGURE 11-2 A titration using a 50-mL buret. (*a*) The setup before any NaOH solution has been added to the HCl solution in the flask. (*b*) The NaOH solution is slowly added to the flask until the titration is completed. We know when to stop adding the NaOH solution, because an indicator that we added to the HCl solution will change color, telling us that the solution has become neutral. Actually, the solution in the flask has really become very slightly basic—a fraction of a drop of extra NaOH solution past the neutral point must be added to change the indicator's color. This point in the titration is called the **end point**. The extra drop of NaOH we had to add to cause the indicator to change color introduced an error into our titration. This error is called the **titration error**. It is usually small and can often be ignored. A "student grade" 50-mL buret can deliver up to 50 mL of solution with an error of about ±0.1 mL. If 25 mL of NaOH is delivered during the titration, the percent uncertainty for each reading of the buret will be $0.1/25 \times 100 = 0.4\%$. When we carry out a titration using a buret, we make two readings: one reading before the solution is delivered and one after the titration is finished. Thus the total possible error is 0.2 mL, since the errors could combine if we were unlucky. The percent uncertainty for the titration is $0.2/25 \times 100 = 0.8\%$. This precision is sufficient for use in most general chemistry courses.

molarity of this solution is not known. We will find out what it is by doing the titration. The NaOH solution is added dropwise (a drop at a time) into the flask until an indicator that has been previously added to the HCl tells us that all the HCl has been neutralized (completely reacted with) by the NaOH. The **indicator** is a dye that changes color when the HCl is neutralized. The reaction is

$$HCl + NaOH \rightarrow NaCl + HOH$$

By reading the level of NaOH in the buret at this point, we know what volume of NaOH has been added to the HCl. From this we can calculate the molarity of the HCl solution. The following discussion will show you how to do this type of calculation.

Let's look at an actual titration of HCl with 0.100 M NaOH. We will take 25.00 mL of the HCl solution, mainly because an accurate pipet called a **volumetric pipet** (commonly available in chemistry labs) delivers 25.00 mL of solution (see Figure 11-3). The HCl solution is put into a 250-mL flask (again, it's convenient and commonly available), and an indicator called **phenolphthalein** is added. Phenolphthalein is colorless in acidic solutions and pink in basic solutions. A 50-mL buret is filled with 0.100 M NaOH solution, and the titration is performed. We find that 23.2 mL of 0.100 M NaOH is needed to turn the phenolphthalein pink. We have reached the **end**

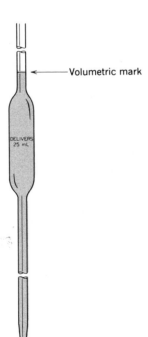

Volumetric mark

DELIVERS 25 mL

FIGURE 11-3 A 25-mL volumetric pipet. It can be read easily to about 0.06 mL. A volume of 25 mL is large enough so that if an error of ±0.06 mL is made in the reading, the percent uncertainty will be 0.06/25 × 100 = 0.24%. This precision is sufficient for most lab work. Notice that a 25-mL pipet has only one volumetric mark, namely, 25 mL. It can only be used to add 25 mL of solution.

point of the titration as described in Figure 11-2. In actual practice, a very small amount of NaOH (a drop or less) over the amount needed to neutralize the HCl is added to make the solution basic. Then the phenolphthalein turns a pale pink color. The volume of NaOH needed to neutralize (or completely react with) all the HCl is called the **equivalence point**. Notice that the equivalence point volume is usually *less* than the end-point volume. The difference is called the **titration error.** It is usually small and we will ignore it.

The number of moles of NaOH used in the titration is

$$0.100 \text{ M} \times 0.0232 \text{ L} = 0.00232 \text{ mol NaOH}$$

Since the chemical equation for this reaction is HCl + NaOH → NaCl + HOH, we see that 1 mol HCl = 1 mol NaOH. Thus the number of moles of HCl neutralized is

$$0.00232 \text{ mol NaOH} \times \frac{1 \text{ mol HCl}}{1 \text{ mol NaOH}} = 0.00232 \text{ mol HCl}$$

The molarity of the HCl solution is (remember we took 25.00 mL of HCl)

$$M = \frac{mol}{L} = \frac{0.00232 \text{ mol HCl}}{0.02500 \text{ L}} = 0.0928 \text{ M HCl}$$

EXAMPLE 8   What is the molarity of 25.0 mL of an H_2SO_4 (sulfuric acid) solution that requires 29.3 mL of 0.250 M NaOH (sodium hydroxide) to neutralize it?

Solution: The number of moles of NaOH is

$$0.250 \frac{\text{mol NaOH}}{L} \times 0.0293 \text{ L} = 0.00733 \text{ mol NaOH}$$

The balanced chemical equation for the reaction is

$$H_2SO_4 + 2NaOH \rightarrow Na_2SO_4 + 2HOH$$

Thus

$$1 \text{ mol } H_2SO_4 = 2 \text{ mol NaOH}$$

The number of moles of H_2SO_4 reacting with the NaOH is

$$0.00733 \text{ mol NaOH} \times \frac{1 \text{ mol } H_2SO_4}{2 \text{ mol NaOH}} = 0.00367 \text{ mol } H_2SO_4$$

The molarity of the H_2SO_4 solution is

$$M = \frac{mol}{L} = \frac{0.00367 \text{ mol } H_2SO_4}{0.0250 \text{ L}} = 0.147 \text{ M } H_2SO_4 \quad \blacksquare$$

EXAMPLE 9   How many milliliters of 0.15 M KOH (potassium hydroxide) solution is needed to completely neutralize (react with) 100 mL of 0.20 M H_3PO_4 (phosphoric acid)?

Solution: The number of moles of H_3PO_4 is

$$0.20 \; \frac{\text{mol } H_3PO_4}{L} \times 0.10 \text{ L} = 0.020 \text{ mol } H_3PO_4$$

The balanced equation is

$$3KOH + H_3PO_4 \rightarrow K_3PO_4 + 3HOH$$

Thus

$$3 \text{ mol KOH} = 1 \text{ mol } H_3PO_4$$

The number of moles of KOH needed is

$$0.020 \text{ mol } H_3PO_4 \times \frac{3 \text{ mol KOH}}{1 \text{ mol } H_3PO_4} = 0.060 \text{ mol KOH}$$

The number of milliliters of 0.15 M KOH solution needed is

$$M = \frac{mol}{L}$$

$$0.15 \text{ M KOH} = \frac{0.060 \text{ mol KOH}}{L}$$

$$L = \frac{0.060 \text{ mol KOH}}{0.15 \; \dfrac{\text{mol KOH}}{L}} = 0.40 \text{ L}$$

Converting liters to milliliters, we get

$$0.40 \text{ L} \times \frac{1000 \text{ mL}}{1 \text{ L}} = 400 \text{ mL} = 4.0 \times 10^2 \text{ mL}$$

which is expressed to two significant figures.

We can also solve this problem in one step:

$$0.20 \text{ M } H_3PO_4 \times 0.10 \text{ L} \times \frac{3 \text{ mol KOH}}{1 \text{ mol } H_3PO_4} \times \frac{1}{0.15 \text{ M KOH}} \times \frac{1000 \text{ mL}}{1 \text{ L}}$$

$$= 4.0 \times 10^2 \text{ mL} \quad \blacksquare$$

EXAMPLE 10   The titration in acid solution of 25.0 mL of an oxalic acid ($H_2C_2O_4$) solution required 22.5 mL of 0.097 M $KMnO_4$ (potassium permanganate) for all the oxalic acid to be used up. The balanced equation for the reaction is

$$2KMnO_4 + 5H_2C_2O_4 + 6HCl \rightarrow 10CO_2 + 2MnCl_2 + 8H_2O + 2KCl$$

What is the molarity of the oxalic acid solution? In this problem you can ignore the HCl because it is not involved in the calculation.

Solution: The number of moles of $KMnO_4$ used is

$$0.097 \text{ M } KMnO_4 \times 0.0225 \text{ L} = 0.00218 \text{ mol } KMnO_4$$

where we converted 22.5 mL to 0.0225 L. From the balanced equation we see that 2 mol $KMnO_4$ = 5 mol $H_2C_2O_4$. The number of moles of $H_2C_2O_4$ that reacted is

$$0.00218 \text{ mol } KMnO_4 \times \frac{5 \text{ mol } H_2C_2O_4}{2 \text{ mol } KMnO_4} = 0.00545 \text{ mol } H_2C_2O_4$$

The molarity of the oxalic acid solution is

$$M = \frac{\text{mol}}{L} = \frac{0.00545 \text{ mol } H_2C_2O_4}{0.0250 \text{ L}} = 0.218 \text{ M } H_2C_2O_4$$

The problem can also be solved in one step:

$$0.097 \text{ M } KMnO_4 \times 0.0225 \text{ L} \times \frac{5 \text{ mol } H_2C_2O_4}{2 \text{ mol } KMnO_4} \times \frac{1}{0.0250 \text{ L}}$$

$$= 0.218 \text{ M } H_2C_2O_4$$

Based on the material in Chapter 6, which was on significant figures, you might think about the percent uncertainty in 0.097 M $KMnO_4$ as compared to the rest of the numbers in the problem.   ■

Notice that in these examples, the solution with the known molarity was in the buret. One can just as well put a solution of unknown molarity in the buret. Then, of course, the solution of known molarity would have to be in the flask. The method of calculating the unknown molarity is similar in either case.

11-3
WHEN A SOLUTION IS DILUTED, THE NUMBER OF MOLES OF SOLUTE DOESN'T CHANGE

When a solution is diluted with pure solvent (such as water), the number of moles of solute originally present doesn't change. All that has been added is solvent. For instance, if we add water to a sugar solution, the number of moles of sugar in the solution, before and after the water is added, remains the same.

Let's use the subscript "i" (standing for "initial") for the undiluted solution, and the subscript "f" (standing for "final") for the diluted solution. Then, since the number of moles of solute doesn't change, we can state that

moles of solute in undiluted solution = moles of solute in diluted solution

Let

$$\text{moles of solute in undiluted solution} = \text{moles}_i = \text{mol}_i$$

and

$$\text{moles of solute in diluted solution} = \text{moles}_f = \text{mol}_f$$

Since initial and final moles are equal, we can say that

$$\text{mol}_i = \text{mol}_f$$

Moreover, since molarity × liters = moles, we have

$$M_i \times L_i = \text{mol}_i \quad \text{and} \quad M_f \times L_f = \text{mol}_f$$

Thus

$$M_i \times L_i = M_f \times L_f$$

which is the same as saying $\text{mol}_i = \text{mol}_f$.

Remember that L_i and L_f refer to the initial and final volumes of the solution. L_f does *not* refer to the amount of solvent added when making the dilution.

EXAMPLE 11   You have 200 mL of a 0.5 M NaCl solution. If you add water to bring the final volume to 500 mL, what is the new molarity of the solution?

Solution: We use the formula $M_iL_i = M_fL_f$. Substituting values from the problem, we have

$$(0.5 \text{ M})(200 \text{ mL}) = M_f(500 \text{ mL})$$

Notice that we didn't bother to convert milliliters to liters. It is not necessary because the milliliter units divide out anyway. Continuing, we solve for M_f:

$$M_f = \frac{(0.5 \text{ M})(200 \text{ mL})}{500 \text{ mL}} = 0.2 \text{ M} \quad \blacksquare$$

Chemists frequently prepare a dilute solution from a concentrated stock solution. The next example shows how to do this.

EXAMPLE 12   A chemist needs to prepare 900 mL of 0.15 M H_2SO_4 (sulfuric acid) solution. Concentrated sulfuric acid is 18 M. How many milliliters of concentrated sulfuric acid and about how many milliliters of water do you need to make the diluted solution?

Solution: **CAUTION:** *Concentrated sulfuric acid is very dangerous. If you get it on your skin or clothing, it will cause burns or holes. You should never add water to the concentrated acid. Always pour the acid slowly into a large amount of water and stir. The reason is that concentrated sulfuric acid reacts with water to produce a great deal of heat. If a small amount of water is added to the concentrated sulfuric acid, the heat released may boil the water and spatter the acid.* Now, getting back to our example, we use the formula

$$M_iL_i = M_fL_f$$

$$(18 \text{ M})(L_i) = (0.15 \text{ M})(900 \text{ mL})$$

$$L_i = \frac{(0.15 \text{ M})(900 \text{ mL})}{18 \text{ M}} = 7.5 \text{ mL}$$

Therefore, you need 7.5 mL of concentrated sulfuric acid and about 900 mL − 7.5 mL = 892.5 mL of water.

Notice that the amount of water added may not be exactly 892.5 mL. It is possible that the volume is changed when two substances are mixed. This solution will also be warm after the mixing and would have a slightly greater volume than the room temperature solution. So for the greatest accuracy, the dilute H_2SO_4 solution is prepared as follows: 7.5 mL of 18 M H_2SO_4 is added to about 400 mL water in a 1000-mL graduated cylinder. This solution is allowed to cool to room temperature. Then additional water is added, with stirring, to bring the volume up to 900 mL. ■

PROBLEMS

KEYED PROBLEMS

1. In a salt solution of water and salt, which component is the solute and which is the solvent?

2. A solution contains 1.5 mol of sucrose dissolved in 5.0 L of solution. What is the molarity of the solution?

3. A solution contains 0.250 mol of $KMnO_4$ dissolved in 0.350 L of solution. What is the molarity of the solution?

4. A solution contains 0.400 mol of fructose dissolved in 400 mL of solution. What is the molarity of the solution?

5. A solution contains 8.00 g of $CaSO_4$ dissolved in 800 mL of solution. What is the molarity of the solution?

6. How many grams of potassium bromide, KBr, must be used to make 75 mL of a 0.12 M KBr solution?

7. How many liters of solution are needed to make a 0.050 M solution from 5.2 g of $MgCl_2$?

8. What is the molarity of 25.0 mL of an oxalic acid ($H_2C_2O_4$) solution that required 23.5 mL of 0.0950 M KOH to neutralize (or react completely with) it? The reaction is

$$H_2C_2O_4 + 2KOH \rightarrow K_2C_2O_4 + 2HOH$$

9. How many milliliters of a 0.250 M NaOH solution are needed to completely neutralize 175 mL of 0.110 M H_3PO_4?

10. The titration, in acid solution, of 25.0 mL of oxalic acid solution required 27.3 mL of 0.125 M $KMnO_4$ for all the oxalic acid to be used up. What is the molarity of the oxalic acid?

11. You have 100 mL of a 0.15 M HCl solution. If you add water to bring the final volume to 500 mL, what is the new molarity of the solution?

12. You want to prepare 600 mL of 0.23 M HNO_3 (nitric acid). You will have to dilute concentrated (16 M) HNO_3. How many milliliters of concentrated nitric acid and about how many milliliters of water do you need to make the diluted solution?

CAUTION: *Concentrated nitric acid is very corrosive. Don't get it on you. If you do, wash it off quickly with lots and lots of water.*

SUPPLEMENTAL PROBLEMS

13. How many grams of $CuSO_4 \cdot 5H_2O$ (copper(II) sulfate pentahydrate) are needed to prepare 250 mL of a 0.075 M $CuSO_4$ solution? The centered dot in $CuSO_4 \cdot 5H_2O$ indicates that the 5 H_2O molecules are in the crystal and must be counted in the formula weight. The water is called **water of hydration.**

14. How many milliliters of 0.1400 M $Ce(SO_4)_2$ (Ce is the element cerium, used in analytical chemistry) are needed to react with 50.00 mL of a 0.07362 M $FeSO_4$ solution? The reaction is

$$2Ce(SO_4)_2 + 2FeSO_4 \rightarrow Ce_2(SO_4)_3 + Fe_2(SO_4)_3$$

15. How many milliliters of 0.5125 M HCl are needed to react completely with 35.00 mL of 0.1018 M $KMnO_4$? The reaction is

$$16HCl + 2KMnO_4 \rightarrow 5Cl_2 + 2MnCl_2 + 8H_2O + KCl$$

16. How many milliliters of 0.102 M NaOH are needed to neutralize 9.34 mL of 0.0987 M H_2SO_4? The reaction is

$$2NaOH + H_2SO_4 \rightarrow Na_2SO_4 + 2HOH$$

17. How many milliliters of solution do you get when you prepare a 0.0850 M solution from 15.2 g $ZnCl_2$? About how many milliliters of water are used?

18. How would you make 1.0 L of 0.50 M HCl from concentrated (12 M) HCl?

19. You are given 300 mL of a 1.50 M glucose solution. What final volume (in milliliters) do you need in order to dilute the glucose to 0.350 M? About how many milliliters of water are added?

20. Intravenous saline solution (NaCl solution) contains 0.85% NaCl in water. This solution is isotonic, meaning that it can be injected into veins without causing damage to blood cells or vein walls. What is the molarity of a 0.85% NaCl solution?

(**NOTE:** A 0.85% NaCl solution contains 0.85 g NaCl in 100 mL of solution.)

21. What is the molarity of 10.00 mL of vinegar [a dilute aqueous (water) solution of acetic acid, $HC_2H_3O_2$] that is neutralized by 17.02 mL of 0.5103 M NaOH? The reaction is

$$HC_2H_3O_2 + NaOH \rightarrow NaC_2H_3O_2 + HOH$$

22. Sodium hypochlorite, NaClO, in dilute solution is a bleach and disinfectant. A common brand name is Clorox. It can be prepared by the following reaction:

$$Cl_2 + 2NaOH \rightarrow NaClO + NaCl + H_2O$$

How many grams of Cl_2 are needed to react with 253 mL of 0.600 M NaOH?

23. Zinc reacts with hydrochloric acid according to the following equation:

$$Zn + 2HCl \rightarrow ZnCl_2 + H_2$$

How many milliliters of 0.500 M HCl are needed to react with 1.32 g of Zn?

24. One way to determine the concentration of chloride ion, Cl^-, in seawater is to perform a Mohr titration according to the following reaction:

$$AgNO_3 + Cl^- \rightarrow AgCl(s) + NO_3^-$$

(**NOTE:** AgCl(s) means solid silver chloride. As the silver nitrate solution is added to the seawater during the titration, a white material appears in the flask with the seawater. This white material is the solid silver chloride. It is called a precipitate.) What is the molarity of Cl^- in a sample of seawater if 25.00 mL of seawater required 26.32 mL of 0.4216 M $AgNO_3$?

25. If we need 42.8 mL of 0.296 M H_3PO_4 (phosphoric acid) to just neutralize 24.9 mL of KOH (potassium hydroxide), what is the molarity of the potassium hydroxide solution? The reaction is

$$H_3PO_4 + 3KOH \rightarrow K_3PO_4 + 3HOH$$

12

GASES AND THE IDEAL GAS LAW

Sulfur dioxide is used to make sulfuric acid, gaseous ammonia to make fertilizer. Freon gases are used in refrigerators and air conditioners, helium and argon in scientific experiments and in some industrial processes such as welding. The air around us is a gas. Because of the importance of air and other gases, it is useful to study the properties of gases.

12-1
MOLECULES OF A GAS COLLIDE AND CHANGE THEIR SPEED AND DIRECTION

A gas such as air consists of molecules that are constantly moving around. At room temperature and pressure, these molecules are usually far apart, so that if you were the size of a molecule, another molecule would seem very far away most of the time. But every so often, another molecule would collide with you and bounce off. And if the gas you were a part of was in a container, occasionally you would collide with a wall of the container. A picture of what gas molecules might look like as they fly around is given in Figure 12-1. Notice that the molecules are traveling in different directions and that they have different speeds. This is reasonable, since speeds and directions can change in a collision. Figure 12-2 shows how this can happen in the case of a straight-line collision.

(Opposite) A high-altitude balloon expands as it goes higher because the outside air pressure decreases.

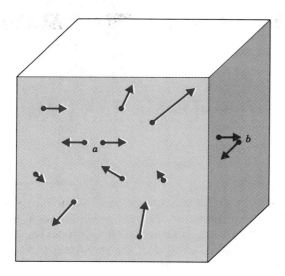

FIGURE 12.1 The dots with arrows represent gas molecules. The arrows show the direction of motion of the molecules. The length of an arrow is proportional to the speed of the molecule. Point *a* shows two molecules bouncing apart after colliding. Point *b* shows a molecule hitting a wall and bouncing off.

12-2
PRESSURE IS THE FORCE PER UNIT AREA

When gas molecules strike the walls of a container, they exert a force on the walls. This force is responsible for the pressure of the gas in the container. **Pressure** is defined as force per unit area. For example, at sea level, the air of the atmosphere exerts a pressure of 14.7 lb/in.^{2}. Read lb/in.^{2} as "pounds per square inch." This means that each square inch of surface has a force of 14.7 lb on it. See Figure 12-3 for a good way to understand atmospheric pressure.

Other units of pressure in addition to pounds per square inch are used. Figure 12-4 shows how 14.7 lb/in.^{2} of air pressure can support a column of mercury 760 mm high. The weight of the column of mercury is 14.7 lb for each square inch of cross section. The width of the column doesn't matter because a wider column will have more air pushing on it. Thus pressures are often reported in heights of a mercury column. A pressure of 760 mm of mercury (Hg) has been selected as "standard pressure" and is called one **atmosphere.** The unit "millimeters of mercury" is abbreviated **mmHg,** and the unit "atmosphere" is abbreviated **atm.**

Fast	Slow	Slow	Fast
(a)		(b)	

FIGURE 12-2 Speed change during collision. (a) Before collision: The fast molecule is catching up to the slow molecule. (b) After collision: The fast molecule has transferred some of its energy to the slow molecule. The fast molecule from part a is now the slow one, and the slow molecule from part a is now the fast one.

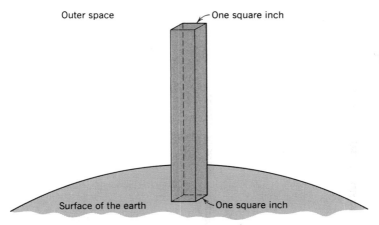

Outer space   One square inch

Surface of the earth   One square inch

FIGURE 12-3 Weight is the force resulting from the gravitational attraction between a mass and the earth. For example, if you weigh 150 lb, this means that there is a force of 150 lb attracting your mass to the earth. When we say that each square inch of the earth has a force of 14.7 lb on it because of the air of the atmosphere, we mean the following: All the air in a 1-in.^{2} column extending from the surface of the earth up to outer space weighs 14.7 lb.

From this discussion, we can write the following equality between different pressure units:

$$1 \text{ atm} = 760 \text{ mmHg} = 14.7 \text{ lb/in.}^2$$

The unit mmHg is often replaced by the unit **torr.** Thus 1 torr = 1 mmHg. The torr (there is no abbreviation for torr, and torr is both singular and plural) is named after the Italian scientist Evangelista Torricelli (1608–1647), who invented the barometer. We can write

$$1 \text{ atm} = 760 \text{ torr} = 14.7 \text{ lb/in.}^2$$

Weather reports in the United States report air pressure (barometric pressure) in inches of mercury, where 29.92 in. of Hg = 1 atm.

EXAMPLE 1   How many atmospheres are there in 900 torr?

Solution:

$$1 \text{ atm} = 760 \text{ torr}$$

$$900 \text{ torr} \times \frac{1 \text{ atm}}{760 \text{ torr}} = 1.18 \text{ atm} \quad \blacksquare$$

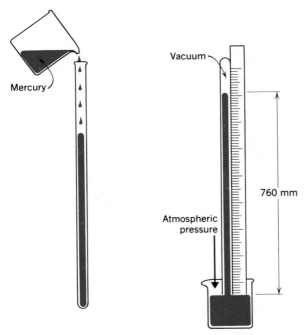

FIGURE 12-4 Making a mercury barometer. A piece of glass tubing, about 800 mm long and closed at one end, is filled with mercury. Then it is inverted and put into a beaker filled with mercury. Atmospheric pressure, acting downward on the surface of the mercury in the beaker, prevents the mercury in the tube from draining out. The mercury level drops somewhat until the weight of the mercury in the tube is equal to the air pressure pushing on the surface of the mercury. If the length of the column of mercury is 760 mm, the air pressure is exactly 1 atm. The weight of this column of mercury is 14.7 lb for each square inch of cross section of the column.

EXAMPLE 2   How many torr are there in 5.00 lb/in.^{2}?

Solution:

$$760 \text{ torr} = 14.7 \text{ lb/in.}^2$$

$$5.00 \cancel{\text{lb/in.}}^2 \times \frac{760 \text{ torr}}{14.7 \cancel{\text{lb/in.}}^2} = 259 \text{ torr} \quad \blacksquare$$

Another unit of pressure is being used more frequently because it is the pressure unit of the SI system. This is the **pascal**, named after the French scientist Blaise Pascal (1623–1662). The abbreviation for the pascal is **Pa**, and

$$1 \text{ atm} = 101,325 \text{ Pa}$$

Since the pascal is rather small, it is usually more convenient to use the kilopascal, abbreviated kPa. We have

$$1000 \text{ Pa} = 1 \text{ kPa}$$

and, to three significant figures,

$$1 \text{ atm} = 101 \text{ kPa}$$

EXAMPLE 3   How many kilopascals are there in 5.00 atm?

Solution: Since 1 atm = 101 kPa, we have

$$5.00 \text{ atm} \times \frac{101 \text{ kPa}}{1 \text{ atm}} = 505 \text{ kPa} \quad \blacksquare$$

12-3
DEGREES CELSIUS AND DEGREES FAHRENHEIT ARE TWO UNITS OF TEMPERATURE

The everyday unit of temperature that we use is the Fahrenheit degree, named after Gabriel Fahrenheit (1686–1736), a German–Dutch physicist who invented the mercury thermometer. As you know, 32 degrees Fahrenheit (abbreviated °F) is the temperature of the freezing point of water. Normal body temperature is 98.6 °F. Fahrenheit chose 212 °F as the boiling point of water so that his body temperature would come out to be 100 °F. He must have had a slight fever while he was doing his experiments.

In science, we don't use the Fahrenheit scale. We use the Celsius degree (abbreviated °C), which is named after Anders Celsius (1701–1744), a Swedish astronomer. The Celsius degree is also called the centigrade degree, but the official name is Celsius. In the Celsius system, 0 °C is the freezing point of water and 100 °C is the boiling point of water. Both the Celsius and Fahrenheit degrees are being mentioned on radio and television weather reports.

In degrees Fahrenheit, the temperature difference between the freezing point and boiling point of water is 180 °F. In the degrees Celsius system, the difference is 100 °C. Thus one degree Celsius is larger than one degree Fahrenheit, as is shown in Figure 12-5.

A temperature in degrees Celsius can be converted to the equivalent temperature in degrees Fahrenheit by the formula

$$°F = \tfrac{9}{5} °C + 32$$

FIGURE 12-5 The relative sizes of the Celsius and Fahrenheit degrees. The Celsius degree is 1.8 times larger than the Fahrenheit degree.

To derive this formula, we draw the graph of a straight line in Figure 12-6.

The equation of any straight line can be expressed as $y = mx + b$, where m is the slope of the line and b is the y intercept. For our straight line we can write the equation as $°F = m °C + b$. To find out what m and b are, we simply substitute the values of the two known points from our graph in Figure 12-6:

the equation:	$°F = m °C + b$
substituting the first point:	$32 = m(0) + b$
solving for b:	$b = 32$
substituting the second point and the value of b:	$212 = m(100) + 32$
solving for m:	$m = \dfrac{212 - 32}{100} = \dfrac{180}{100} = 1.8$
the final equation:	$°F = 1.8 °C + 32$

To get the equation to look like the one you are familiar with, notice that $1.8 = 18/10$ and that $18/10 = 9/5$. Thus $°F = \frac{9}{5} °C + 32$. Maybe the form $°F = 1.8 °C + 32$ is more convenient if you are using a calculator.

EXAMPLE 4   How many degrees Fahrenheit are equal to 20 °C?

Solution:

$$°F = 1.8 °C + 32 = 1.8(20) + 32 = 36 + 32 = 68 °F \quad \blacksquare$$

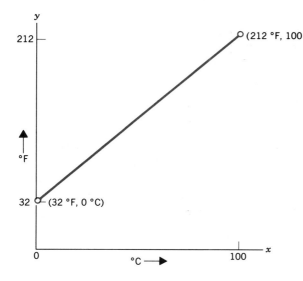

FIGURE 12-6 The x-axis is the degrees Celsius axis. The y-axis is the degrees Fahrenheit axis. The two points whose coordinates we know are 0 °C = 32 °F and 100 °C = 212 °F. These two points determine a straight line.

It is useful to solve the equation °F = 1.8 °C + 32 for °C so that we can easily convert from degrees Fahrenheit to degrees Celsius:

$$°F = 1.8 \ °C + 32$$

$$°F - 32 = 1.8 \ °C$$

$$\frac{°F - 32}{1.8} = °C$$

The form you may be familiar with is $\frac{5}{9}(°F - 32) = °C$. The two forms are the same since 5/9 = 10/18 = 1/1.8. To convert 10/18 to 1/1.8, divide numerator and denominator by 10:

$$\frac{10}{18} = \frac{\dfrac{10}{10}}{\dfrac{18}{10}} = \frac{1}{1.8}, \qquad \text{where } \frac{10}{10} = 1 \text{ and } \frac{18}{10} = 1.8$$

EXAMPLE 5   Normal body temperature is 98.6 °F. What is this in degrees Celsius?

Solution:

$$°C = \frac{°F - 32}{1.8} = \frac{98.6 - 32}{1.8} = \frac{66.6}{1.8} = 37.0 \ °C \quad ∎$$

12-4
AS THE TEMPERATURE OF A GAS DECREASES, THE VOLUME OF THE GAS DECREASES

To study the effect of temperature on the volume (abbreviated V) of a gas, we will use the piston and cylinder arrangement shown in Figure 12-7. The piston has a seal around its edge so that no gas can get in or out of the cylinder. Since the piston is very light, it easily moves up and down. This keeps the pressure inside the cylinder equal to the pressure outside. Since the outside pressure equals atmospheric pressure, we can assume that the pressure is held constant. If we place a candle under the cylinder, it will become warmer. This causes the gas inside the cylinder to expand, pushing up the piston. The volume inside the cylinder will increase.

If we now surround the cylinder with ice (Figure 12-8), the piston will go down, decreasing the volume of the gas inside the cylinder.

Now we are going to do an experiment in which we keep cooling the cylinder, well below the temperature of ice. As the gas inside the cylinder gets colder, its volume decreases. Let's plot what happens to the volume of the gas as the temperature decreases. As you can see in Figure 12-9, as the gas gets colder and colder, its volume gets smaller and smaller. But at a certain point, the gas liquefies. All gases will liquefy if they get cold enough.

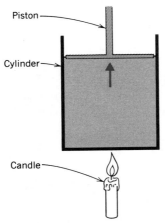

FIGURE 12-7 A cylinder with a lightweight piston that can move up and down without friction is heated by a candle. The walls of the cylinder are very strong, and no gas can escape through the piston seals. Before the candle is put under the cylinder, the pressure inside is equal to the pressure outside. When the cylinder is heated, the piston moves up because the pressure inside the cylinder is increasing. The volume of the gas inside the cylinder increases as the piston moves up. If we assume that the heating is done very slowly, the pressure inside the cylinder is only a tiny bit greater than the pressure outside. We can assume that inside and outside pressures are "equal."

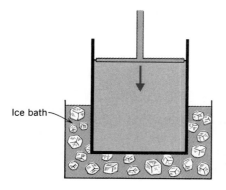

FIGURE 12-8 The cylinder is surrounded by ice, which cools it and causes the piston to go down. The volume of gas inside the cylinder decreases as the piston falls.

Since the volume change of a liquid is extremely small as the liquid is cooled, our graph shows a leveling off of the volume as the temperature decreases. If we continue drawing the original line (the dotted part of the line in Figure 12-9), however, we find that it crosses the temperature axis at −273.15 °C. This is the temperature that would be needed to reduce the gas volume to **zero** if the gas had not liquefied. However, since real gases liquefy long

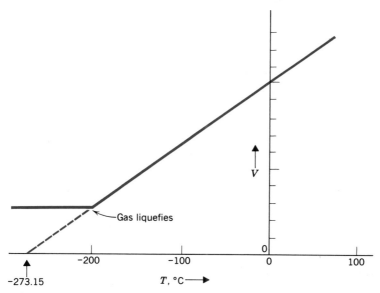

FIGURE 12-9 A graph showing the volume of a gas versus the temperature in degrees Celsius. As the gas cools, the volume decreases. When the gas becomes sufficiently cold, it liquefies; the volume remains fairly constant upon further cooling. The dashed line shows what would happen if the gas didn't liquefy but kept decreasing in volume. Its volume would become zero at −273.15 °C.

before they reach −273.15 °C, volumes never do become zero. But −273.15 °C is the lowest temperature conceivable.

12-5
THE LOWEST POSSIBLE TEMPERATURE IS ZERO KELVINS

The lowest conceivable temperature is so important that it is the basis of a temperature scale. This is the absolute temperature scale and has the units of kelvins (abbreviated K). It is named after Lord Kelvin (William Thomson, 1824–1907), the British scientist who first suggested that −273.15 °C is the lowest temperature that can exist.

The new temperature scale *starts* at −273.15 °C. This temperature is called zero kelvins or **absolute zero** and is written as 0 K. There is much evidence suggesting that absolute zero cannot be attained. The closest that scientists have gotten to absolute zero is 0.00000002 K. **NOTE:** We do not say "degrees kelvin"—just "kelvins"! The word kelvin starts with a lower-case letter, but the abbreviation is a capital K. This is the convention scientists use for units named after people. Remember our use of pascal and Pa. (The exceptions are degrees Celsius and Fahrenheit.)

Since the Kelvin scale starts at −273 °C (let's round it off a bit), we can write

$$-273 \text{ °C} = 0 \text{ K}$$

Adding 273 to each side of this equation, we have

$$0 \text{ °C} = 273 \text{ K}$$

Although kelvins are the same size as Celsius degrees, the kelvin scale starts 273 degrees below the Celsius scale. Thus we have that

$$\text{°C} + 273 = \text{K}$$

EXAMPLE 6   Convert 37 °C to kelvins.

Solution: 37 °C + 273 = 310 K. ■

EXAMPLE 7   Convert 200 K to degrees Celsius.

Solution: Since °C + 273 = K, we can write K − 273 = °C. Therefore, 200 K − 273 = −73 °C. ■

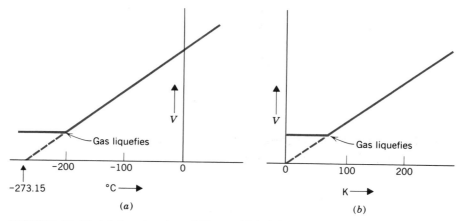

FIGURE 12-10 (*a*) A redrawing of Figure 12.9. (*b*) The vertical axis (the volume or *V* axis) has been moved over to −273.15 °C. The temperature scale is now in kelvins (K), since the vertical axis is now at zero kelvins.

If we redraw Figure 12-9 and move the vertical axis (the volume axis) left to −273.15 °C, we see that the volume of a gas is directly proportional to the absolute temperature (*T*) of the gas. This means that as *T* increases, *V* increases, and when *T* = 0, *V* = 0 (see Figure 12-10). This is why degrees Celsius is not proportional to *V*—when degrees Celsius equals zero, *V* does not equal zero.

12-6
A NOTE ABOUT PROPORTIONALITY AND PROPORTIONALITY CONSTANTS

If you go to the store to buy 10 apples, the price you pay depends on the price per apple. Thus we have

$$¢ \text{ you pay for 10 apples} = (¢ \text{ per apple})(10 \text{ apples})$$

To be a bit more mathematical, we can write (the "per" in "¢ per apple" means divided by)

$$¢ = \left(\frac{¢}{\text{apple}}\right) (\text{number of apples})$$

Suppose each apple costs 25¢. Then for the price of 10 apples we have

$$¢ = \left(\frac{25¢}{\text{apple}}\right)(10 \text{ apples}) = 250¢ = \$2.50$$

The cents you pay is **directly proportional** to the number of apples you buy. The actual price paid for your apples is determined by the price per apple—in this case 25¢. The term 25¢/apple is called the **constant of proportionality.** It converts the unit apples to the unit cents; it also sets the scale of the conversion with the number 25.

12-7
THE VOLUME OF A GAS IS DIRECTLY PROPORTIONAL TO THE TEMPERATURE IN KELVINS

Just as cents is directly proportional to apples, volume (V) is directly proportional to absolute temperature (T) in kelvin (K). We can say that (the word "directly" is usually left out for simplicity)

$$V \text{ is proportional to } T$$

The symbol for "is proportional to" is "$\propto$." Thus

$$V \propto T$$

We will not discuss the actual value of the proportionality constant relating V and T. You will see why later.

EXAMPLE 8   If x is proportional to y, and k is the proportionality constant, write an equation relating x and y.

Solution: $x = ky$.   ∎

12-8
THE PRESSURE OF A GAS IS INVERSELY PROPORTIONAL TO THE VOLUME OF THE GAS

Let's look at our piston–cylinder arrangement again (Figure 12-11). This time we will hold the temperature constant. The greater the pressure (P) on the piston, the more the piston sinks, compressing the gas in the cylinder. The volume is **inversely proportional** to the pressure. This means that as the pressure gets larger, the volume gets smaller. We can write the relationship mathematically as

$$P \propto 1/V$$

Let's see how this works. If V goes from 2 to 4, P goes from $\frac{1}{2}$ to $\frac{1}{4}$. We see that when V doubles, P becomes one-half of its original value.

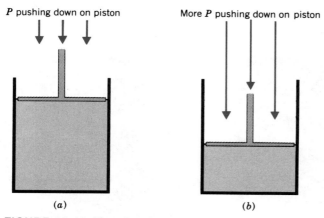

P pushing down on piston

More P pushing down on piston

(a)   (b)

FIGURE 12-11 The piston–cylinder arrangement at constant temperature. As the pressure increases from part a to part b, the piston sinks and the volume of the gas in the cylinder decreases.

When you breathe, you illustrate the pressure–volume relationship of a gas. Try taking a breath, but at the same time block your throat by putting your tongue toward the back of your mouth so no air can rush into your lungs. What have you done? You've lowered your diaphragm (the muscle at the bottom of your chest cavity) and expanded your rib cage, thus increasing the volume of your chest cavity. You have also decreased the air pressure inside your chest. Now, if you relax your throat, air rushes into your lungs. The higher air pressure outside your chest forces air into the lower-pressure region inside your chest. When you exhale, you do the opposite, decreasing chest volume and forcing air out of your lungs.

EXAMPLE 9 If x is inversely proportional to y, and c is the proportionality constant, what is the equation relating x and y?

Solution: $x = c(1/y) = c/y.$ ■

12-9
THE PRESSURE OF A GAS IS DIRECTLY PROPORTIONAL TO THE TEMPERATURE IN KELVINS

Instead of a piston–cylinder arrangement, we will now use a closed box as shown in Figure 12-12. In this system, the volume is constant because the walls of the box are rigid and cannot move. As the temperature of the gas

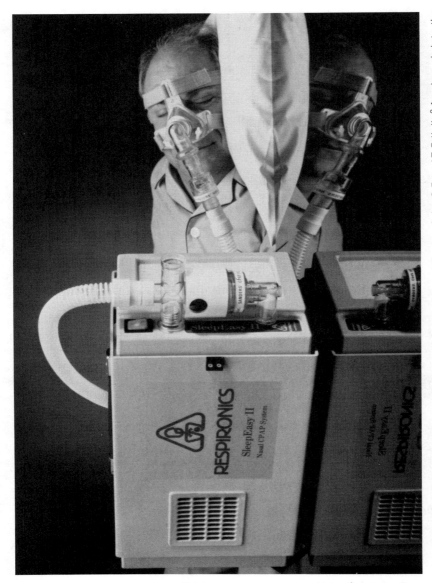

A pump in the Sleep Easy II increases the air pressure at the mask by 2.5 to 17.5 lb/in^2 forcing air into the nose and lungs. This instrument supplying constant positive air pressure is used by people with obstructive sleep apnea. It forces air past the obstructions in the throat that cause them to stop breathing while asleep.

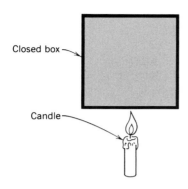

FIGURE 12-12 A closed box being heated by a candle. The walls of the box are very strong, and no gas can pass through them. As the flame heats the gas in the box, the pressure inside increases. The volume cannot change because of the strong walls—the volume is constant.

increases, its pressure increases. Experiments show that the pressure (P) is proportional to the absolute temperature (T):

$$P \propto T$$

12-10
THE PRESSURE OF A GAS IS DIRECTLY PROPORTIONAL TO THE AMOUNT OF GAS

Again we will need a closed box that is held at constant temperature. In addition, we need a gas syringe that we will use to inject gas into the box (see Figure 12-13). As gas is injected, the pressure increases. The pressure is found to be directly proportional to the amount of gas,

$$P \propto n$$

where n is the number of moles of gas in the box.

EXAMPLE 10   If the pressure is 2 atm when 3 mol of gas are in a box, what is the pressure when 6 mol of gas are in the box?

Solution: Since $P \propto n$, and n has doubled, P must double. Thus, when there are 6 mol of gas in the box, the pressure is 4 atm. ■

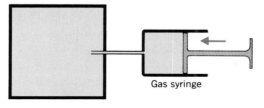

Gas syringe

FIGURE 12-13 As the gas is injected into the box, the pressure increases. The volume and temperature of the gas in the box are constant.

12-11
THE VOLUME OF A GAS IS DIRECTLY PROPORTIONAL TO THE AMOUNT OF GAS

Here we will use our piston–cylinder arrangement and a gas syringe. Both the temperature and the pressure are held constant (see Figure 12-14). As gas is injected into the cylinder, the piston rises and the volume of the gas in the cylinder increases. The volume is found to be directly proportional to the amount of gas. We can write

$$V \propto n$$

EXAMPLE 11 If the volume is 20 L when 1 mol of gas is in the piston–cylinder arrangement, what is the volume when 2 mol of gas are in the piston–cylinder arrangement? Assume that both the temperature and pressure are held constant.

Solution: Since $V \propto n$, when the number of moles doubles (from 1 to 2), the volume doubles. The final volume is 40 L. ■

12-12
THE IDEAL GAS LAW RELATES PRESSURE, VOLUME, MOLES, AND TEMPERATURE

In previous sections, we derived five proportionalities that describe the behavior of gases. We list them together here.

$$V \propto T \qquad \text{(at constant } P \text{ and } n)$$
$$P \propto 1/V \qquad \text{(at constant } n \text{ and } T)$$
$$P \propto T \qquad \text{(at constant } n \text{ and } V)$$
$$P \propto n \qquad \text{(at constant } T \text{ and } V)$$
$$V \propto n \qquad \text{(at constant } P \text{ and } T)$$

Let's rewrite the second proportionality so that V is on the left side of the proportionality sign. We get $V \propto 1/P$. The proportionalities involving V are

$$V \propto T$$
$$V \propto 1/P$$
$$V \propto n$$

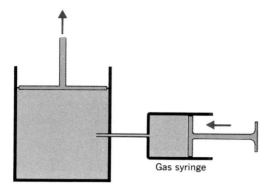

FIGURE 12-14 As gas is injected into the cylinder, the volume increases. The pressure of the gas in the cylinder is constant because the piston can move until both the inside and outside pressures are equal.

Gas syringe

Since V is proportional to T, $1/P$, and n individually, it must be proportional to all three at the same time. So we can write

$$V \propto nT/P$$

Multiplying both sides of the expression by P, we get

$$PV \propto nT$$

Now let's look at the proportionalities involving P. These are

$$P \propto 1/V$$

$$P \propto T$$

$$P \propto n$$

Since P is proportional to $1/V$, T, and n individually, it must be proportional to all three at the same time, and as we did before, we can write

$$P \propto nT/V$$

Multiplying both sides by V, we get

$$PV \propto nT$$

This is the same expression that we got above. Let's make this proportionality an equality by putting in a constant of proportionality. It is customary to use the symbol R for the constant of proportionality and to insert it between the n and the T. Doing this we get

$$PV = nRT$$

This equation is called the **ideal gas law**, and R is called the **ideal gas law constant** or the **universal gas constant**. As you study more chemistry, you will learn more about ideal gases and why real gases, such as air, do not exactly obey the ideal gas law. The ideal gas law is called *ideal* because it is a *model* of the way gases behave. It is pretty good but not perfect. Under normal conditions, air "obeys" the ideal gas law with an error of less than 1%.

To use the ideal gas law equation, we must decide on the units we will use. The following units are very common.

Pressure (P) is measured in atmospheres (atm).

Volume (V) is measured in liters (L).

Amount of gas (n) is measured in moles (mol).

Temperature (T) is measured in kelvins (K).

To determine the units of R, we will substitute these units into the ideal gas law:

$$PV = nRT$$

$$\text{atm} \times \text{L} = \text{mol} \times R \times \text{K}$$

Solving for R we get

$$R = \frac{\text{atm} \times \text{L}}{\text{mol} \times \text{K}}$$

Now we must determine the numerical value of R. To do this it is necessary to perform an experiment with a gas. In Figure 12-15, which shows our piston–cylinder arrangement, we take 1.00 mol of a gas at 1.00 atm pressure and 273 K. Putting the gas into the piston–cylinder apparatus,

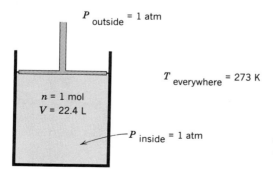

P_{outside} = 1 atm

$T_{\text{everywhere}}$ = 273 K

n = 1 mol
V = 22.4 L

P_{inside} = 1 atm

FIGURE 12-15 An experiment to determine the numerical value of R. Any values of n, T, and P could have been used; the ones in the diagram are convenient.

we allow the movable piston to come to its natural resting place and then we measure the volume. It turns out to be 22.4 L. Now we substitute the numerical values of P, V, n, and T into the ideal gas law:

$$PV = nRT$$

$$(1.00 \text{ atm})(22.4 \text{ L}) = (1.00 \text{ mol})(R)(273 \text{ K})$$

Solving for R, we get

$$R = \frac{(1.00 \text{ atm})(22.4 \text{ L})}{(1.00 \text{ mol})(273 \text{ K})} = 0.0821 \frac{\text{atm} \cdot \text{L}}{\text{mol} \cdot \text{K}}$$

Any combination of P, V, n, and T values will give the same numerical value for R if these units are used. Other units will give different numerical values for R.

Since you will use the units atmospheres, liters, moles, and kelvins in solving problems with $PV = nRT$, be sure to convert all units to atmospheres, liters, moles, and kelvins. For example, pressure might be reported in torr or kilopascals. Remember that

$$1 \text{ atm} = 760 \text{ torr} = 101 \text{ kPa}$$

Temperature is commonly measured in degrees Celsius. Remember that

$$°C + 273 = K$$

Volume is sometimes measured in milliliters (mL). Remember that 1000 mL = 1 L.

EXAMPLE 12   What pressure (in atmospheres) would be exerted by 3.00 mol of oxygen gas in a 5.00-L container at 300 K?

Solution: The given values are $V = 5.00$ L, $n = 3.00$ mol, and $T = 300$ K. Substituting these values into $PV = nRT$, we get

$$(P)(5.00 \text{ L}) = (3.00 \text{ mol})(0.0821 \text{ atm} \cdot \text{L/mol} \cdot \text{K})(300 \text{ K})$$

$$P = \frac{(3.00 \text{ mol})(0.0821 \text{ atm} \cdot \text{L/mol} \cdot \text{K})(300 \text{ K})}{5.00 \text{ L}}$$

$$P = \frac{(3.00)(0.0821)(300)}{5.00} \text{ atm} = 14.8 \text{ atm} \quad \blacksquare$$

How do the units in Example 12 divide out? Let's write the expression for P using only units:

$$\frac{(\cancel{mol})\left(\dfrac{atm \cdot L}{\cancel{mol} \cdot \cancel{K}}\right)(\cancel{K})}{L}$$

Mol divides mol and K divides K, giving

$$\frac{atm \cdot \cancel{L}}{\cancel{L}}$$

L divides L, leaving only atm.

EXAMPLE 13   What volume (in liters) will 0.500 mol of N_2 occupy at a pressure of 300 torr and a temperature of 37.0 °C?

Solution: First convert torr to atmospheres and degrees Celsius to kelvins:

$$300 \text{ torr} \times \frac{1 \text{ atm}}{760 \text{ torr}} = 0.395 \text{ atm}$$

$$37.0 \text{ °C} + 273 = 310 \text{ K}$$

Then substitute into $PV = nRT$. Since we want volume, let's solve this equation for V first and then substitute numerical values:

$$V = \frac{nRT}{P} = \frac{(0.500 \text{ mol})(0.0821 \text{ atm} \cdot \text{L/mol} \cdot \text{K})(310 \text{ K})}{0.395 \text{ atm}} = 32.2 \text{ L}$$

Be sure that you understand how the units divide out.   ■

EXAMPLE 14   How many moles of methane are there in a sample of gas that is at the following conditions: $P = 75.0$ kPa, $V = 600$ mL, $T = 25.0$ °C?

Solution: First convert kilopascals to atmospheres, milliliters to liters, and degrees Celsius to kelvins:

$$75.0 \text{ kPa} \times \frac{1 \text{ atm}}{101 \text{ kPa}} = 0.743 \text{ atm}$$

$$600 \text{ mL} \times \frac{1 \text{ L}}{1000 \text{ mL}} = 0.600 \text{ L}$$

$$25.0 \text{ °C} + 273 = 298 \text{ K}$$

Solving $PV = nRT$ for n, as well as substituting numerical values, we get

$$n = \frac{PV}{RT} = \frac{(0.743 \text{ atm})(0.600 \text{ L})}{(0.0821 \text{ atm} \cdot \text{L/mol} \cdot \text{K})(298 \text{ K})} = 0.0182 \text{ mol}$$

Again, be sure you understand how the units divide out. ■

A useful application of the ideal gas law is the determination of the molecular weight of an unknown gas.

EXAMPLE 15   A sample of gas at 27.1 °C is contained in a 1.00-L glass bulb. The pressure of the gas is 1.00 atm, and the gas weighs 0.650 g. What is the molecular weight of the gas?

Solution: First calculate the number of moles of gas in the sample. Solve $PV = nRT$ for n and substitute numerical values:

$$n = \frac{PV}{RT} = \frac{(1.00 \text{ atm})(1.00 \text{ L})}{(0.0821 \text{ atm} \cdot \text{L/mol} \cdot \text{K})(300 \text{ K})} = 0.0406 \text{ mol}$$

Since the statement of the problem gave us the mass of the gas, we have

given mass
of gas   → 0.650 g = 0.0406 mol ←   calculated moles
(from the problem)   of gas

The molecular weight is the number of grams per mole of a substance. Another way to write this is to say that

$$\text{molecular weight} = \frac{\text{number of grams}}{\text{number of moles}}$$

Substituting 0.650 g and 0.0406 mol, we have

$$\text{molecular weight} = \frac{0.650 \text{ g}}{0.0406 \text{ mol}} = \frac{16.0 \text{ g}}{1 \text{ mol}} \text{ or } 16.0 \text{ g/mol} ■$$

12-13
CHANGING P, V, AND T WHILE KEEPING n CONSTANT: THE COMBINED GAS LAW

Sometimes the pressure, temperature, and volume of a certain number of moles of gas are given. Then the values of one or two of these quantities are changed and we are asked to find the value of the third quantity. This is easily done if the ideal gas law equation is used. Let the initial values of P, V, and T be denoted by the subscript "1". These values are written as P_1, V_1, and T_1. The ideal gas law for these initial values is $P_1V_1 = nRT_1$. Notice that n and R do not have subscripts. R is always constant; n is also constant for problems of this kind, since the number of moles of gas doesn't change.

We will denote the final condition of the gas with a subscript "2". The ideal gas law for these values is $P_2V_2 = nRT_2$.

We can solve each of these two equations for nR by dividing the first equation by T_1 and the second one by T_2.

$$nR = \frac{P_1V_1}{T_1} \quad \text{and} \quad nR = \frac{P_2V_2}{T_2}$$

Since nR is equal to nR, we have

$$\frac{P_1V_1}{T_1} = \frac{P_2V_2}{T_2}$$

This equation is called the **combined gas law.**

EXAMPLE 16   You are given the following initial conditions: $P_1 = 3.00$ atm, $V_1 = 10.0$ L, and $T_1 = 298$ K. If the final conditions are $P_2 = 5.00$ atm and $T_2 = 100$ K, what is V_2 (in liters)?

Solution: Write $P_1V_1/T_1 = P_2V_2/T_2$ and solve for V_2:

$$V_2 = \frac{P_1V_1T_2}{P_2T_1}$$

It is convenient to rearrange this equation so that all the P and T terms are together:

$$V_2 = \frac{P_1}{P_2} \times \frac{T_2}{T_1} \times V_1$$

Now substitute the given numerical values:

$$V_2 = \frac{3.00 \text{ atm}}{5.00 \text{ atm}} \times \frac{100 \text{ K}}{298 \text{ K}} \times 10.0 \text{ L} = 2.01 \text{ L}$$

To check that the answer is reasonable (Should the volume really decrease?), look at the terms P_1/P_2 and T_2/T_1. As you see, the pressure increases from 3.00 to 5.00 atm. We know that an increase in pressure will decrease the volume. The temperature decreases from 298 to 100 K. This cooling will also decrease the volume. Since both the pressure and the temperature changes decrease the volume, a volume change from 10.0 to 2.01 L is certainly in the right direction. (Sometimes both changes are not in the same direction. Then you cannot tell what happens without doing the calculation.) ∎

EXAMPLE 17   You are given the following initial conditions: $P_1 = 0.500$ atm, $V_1 = 5.00$ L, and $T_1 = 300$ K. If the final conditions are $V_2 = 2.00$ L and $T_2 = 400$ K, what is P_2 (in atmospheres)?

Solution: Write $P_1V_1/T_1 = P_2V_2/T_2$ and solve for P_2:

$$P_2 = \frac{P_1V_1T_2}{T_1V_2}$$

Rearrange to get

$$P_2 = P_1 \times \frac{V_1}{V_2} \times \frac{T_2}{T_1}$$

Substituting numerical values gives

$$P_2 = 0.500 \text{ atm} \times \frac{5.00 \text{ L}}{2.00 \text{ L}} \times \frac{400 \text{ K}}{300 \text{ K}} = 1.67 \text{ atm}$$

Is it reasonable that the pressure increases? ∎

12-14
DALTON'S LAW OF PARTIAL PRESSURES SAYS THAT THE SUM OF THE PRESSURES OF THE INDIVIDUAL GASES EQUALS THE TOTAL GAS PRESSURE

If we have a mixture of three gases in a container, the total number of gas molecules in the container is equal to the sum of the number of gas molecules of each of the three gases. It follows also that the total number of moles is equal to the sum of the number of moles of each gas. Thus

$$n_T = n_1 + n_2 + n_3$$

where n_T is the total number of moles and n_1, n_2, and n_3 are the number of moles of each component gas.

In a given container at a certain temperature, different kinds of gas all have the same volume and temperature. Thus, using the ideal gas law, $PV = nRT$, we can write

$$n_T = \frac{P_T V}{RT}$$ for the entire gaseous mixture

$$n_1 = \frac{P_1 V}{RT}$$ for the first component in the gaseous mixture

$$n_2 = \frac{P_2 V}{RT}$$ for the second component in the gaseous mixture

$$n_3 = \frac{P_3 V}{RT}$$ for the third component in the gaseous mixture

Since V and T are constant, only P can change when n changes. R is always a constant.

Adding all the moles together, we get

$$n_T = n_1 + n_2 + n_3$$

Substituting the appropriate PV/RT expression for n_T, n_1, n_2, and n_3, we get

$$\frac{P_T V}{RT} = \frac{P_1 V}{RT} + \frac{P_2 V}{RT} + \frac{P_3 V}{RT}$$

Dividing both sides of the last equation by V/RT, all the V/RT terms in the equation divide out and we get

$$P_T = P_1 + P_2 + P_3$$

This is called **Dalton's law of partial pressures** (named after the British scientist John Dalton, 1766–1844), which says that the total pressure of a mixture of gases is equal to the sum of the partial pressures of all the individual kinds of gas in the mixture. The **partial pressure** of a gas is simply the pressure exerted by the gas when it is mixed with other gases. This is the same pressure the gas would have if it were in the container by itself.

EXAMPLE 18   A gas mixture consists of the following partial pressures: 20 torr O_2, 50 torr N_2, and 10 torr Ar. What is the total pressure (in torr) of the mixture of gases?

Solution: Let's use the following notation:

$$P_{O_2} = \text{partial pressure of } O_2 = 20 \text{ torr}$$
$$P_{N_2} = \text{partial pressure of } N_2 = 50 \text{ torr}$$
$$P_{Ar} = \text{partial pressure of Ar} = 10 \text{ torr}$$

Then using Dalton's law of partial pressures, we can write

$$P_T = P_1 + P_2 + P_3 = P_{O_2} + P_{N_2} + P_{Ar}$$
$$P_T = 20 \text{ torr} + 50 \text{ torr} + 10 \text{ torr} = 80 \text{ torr} \quad \blacksquare$$

EXAMPLE 19   The aquanauts who lived for many days 200 ft underwater in Sealab breathed a mixture of about 3.0% O_2 and 97% He. If the total gas pressure they breathed was 7.0 atm, what are the partial pressures of O_2 and He in torr?

A scuba diver. A pressure regulator just above the air tanks reduces the high pressure of the air in the tanks to equal the outside water pressure. Thus the diver can breathe easily.

Solution: The partial pressure of O_2 is 3.0% × 7.0 atm = 0.030 × 7.0 atm = 0.21 atm or 160 torr. The partial pressure of He is 97% × 7.0 atm = 0.97 × 7.0 atm = 6.8 atm or 5200 torr. It is interesting to note that the partial pressure of O_2 in the air we breathe is 20.9% × 760 torr = 0.209 × 760 torr = 159 torr, about the same as the partial pressure of the oxygen the aquanauts breathed. If they had breathed a 20.9% O_2–79.1% He mixture, they would have gotten 1112 torr of O_2, which certainly would have damaged their cells. The reason they breathed O_2–He and not O_2–N_2 (as in air) is that N_2 is more soluble in body tissues than He and would make decompression (coming to the surface) a very time-consuming affair. In addition, N_2 is toxic at high pressure, causing **nitrogen narcosis** (a drunklike behavior). In fact, if a diver breathes normal compressed air, each 50 ft of descent is equivalent to drinking one martini on an empty stomach. ■

PROBLEMS

KEYED PROBLEMS

1. How many atmospheres are there in 1500 torr?

2. How many torr are there in 12.00 lb/in.²?

3. How many kilopascals are there in 3 atm?

4. How many degrees Fahrenheit are equal to 25 °C?

5. How many degrees Celsius are equal to 86 °F?

6. Convert 100 °C to kelvins.

7. Convert 77 K to degrees Celsius.

8. If a is directly proportional to b, and q is the proportionality constant, write an equation relating a and b.

9. If a is inversely proportional to b, and t is the proportionality constant, what is the equation relating a and b?

10. If the pressure is 5 atm when 10 mol of gas are in a box, what is the pressure (in atmospheres) when 20 mol of gas are in the box?

11. If the volume is 15 L when 3 mol of gas are in the cylinder, what is the volume (in liters) when 6 mol of gas are in the cylinder?

12. What pressure (in atmospheres) would 2.0 mol of nitrogen gas exert if it were in a 7.0-L container at 400 K?

13. What volume (in liters) will 0.25 mol Cl_2 occupy at a pressure of 500 torr and a temperature of 80 °C?

14. How many moles of CO_2 are there in a sample of gas that is at the following conditions: $P = 80.0$ kPa, $V = 200$ mL, and $T = 30$ °C?

15. A sample of gas at 25 °C is contained in a 1.50-L glass bulb. The pressure of the gas is 1.30 atm, and the gas weighs 4.40 g. What is the molecular weight of the gas?

16. The initial conditions of a gas are $P_1 = 1.5$ atm, $V_1 = 5.0$ L, and $T_1 = 273$ K. If the final conditions are $P_2 = 4.0$ atm and $T_2 = 200$ K, what is V_2 (in liters)?

17. The initial conditions of a gas are $P_1 = 0.20$ atm, $V_1 = 5.0$ L, and $T_1 = 210$ K. If the final conditions are $V_2 = 0.70$ L and $T_2 = 310$ K, what is P_2 (in atmospheres)?

18. A gas mixture consists of 150 torr CH_4, 150 torr O_2, and 500 torr Ar. What is the total pressure of the gas (in torr)?

19. Calculate the partial pressure (in torr) of the oxygen in a 20% O_2–80% He mixture at 7 atm.

SUPPLEMENTAL PROBLEMS

20. How many atmospheres are there in 500 kPa?

21. Convert 0.50 atm to torr.

22. A high fever is 104 °F. What is this temperature in degrees Celsius?

23. A child has a fever of 41 °C. What is this temperature in degrees Fahrenheit?

24. Convert 68 °F to kelvins. (Hint: First convert °F to °C.)

25. The boiling point of sodium is 892 °C. What is this in kelvins?

26. The boiling point of nitrogen gas is -195.8 °C. What is this in kelvins?

27. The temperature of the glowing filament of an incandescent light bulb is about 2800 K. What is this in degrees Celsius? In degrees Fahrenheit?

28. At what temperature are degrees Fahrenheit and degrees Celsius equal? (Hint: In the equation °F = 1.8 °C + 32, let °F = °C = T, substitute T into the equation, and solve for T.)

29. What volume (in liters) will 8.0 g O_2 occupy at a pressure of 400 torr and a temperature of 100 °F? (Hint: Convert grams to moles, convert degrees Fahrenheit to kelvins, and then use $PV = nRT$.)

30. What pressure (in torr) would 15 g F_2 have if it were in a 2.0-L container at 25 °C?

31. How many grams of N_2O are there in a sample of gas that is at the following conditions: $P = 200$ kPa, $V = 700$ mL, and $T = 40.0$ °C?

32. What is the molecular weight of 3.00 g of gas that is in a 400-mL container at 300 torr and 22 °C?

33. The total pressure of a gas mixture is 1 atm. The gas consists of three components: CH_3CN, O_2, and H_2. The partial pressure of the CH_3CN is 50 torr, and that of the O_2 is 150 torr. What is the partial pressure of the H_2 (in torr)? (Hint: Convert atmospheres to torr and use Dalton's law of partial pressures.)

34. A sample of gas in a rigid closed 200-mL container has a P of 150 kPa and a T of 50.0 °C. The container is heated to 100 °C. What is the new pressure (in kilopascals)? (Assume that the volume is constant.)

35. A gas in a piston–cylinder arrangement has a P of 800 torr and a T of 20.0 °C. The piston is allowed to expand from a volume of 2.00 L so that the gas pressure is 1.00 atm. What is the final volume (in liters)? (Assume that T doesn't change.)

36. The given conditions of a gas are $P = 20.0$ kPa, $V = 800$ mL, and $T = 23.0$ °C. Conditions are changed so that $V = 400$ mL and $T = 30.0$ °C. What is the new pressure (in kilopascals)?

37. What pressure (in atmospheres) will be exerted by 3.0×10^{10} molecules of a gas that is in a 3.0-mL container at 15 °C? (Hint: Find the number of moles of gas by using Avogadro's number.)

38. What is the molecular weight of a gas from the following data: m (mass) $=$ 7.2 g, $P = 95$ kPa, $V = 2.5$ L, and $T = 100$ °C?

39. What pressure (in atmospheres) will be exerted by 50.0 g C_2H_4 in a 100-mL container at 35.0 °C?

40. How many grams of benzene, C_6H_6, are in a sample of gas that is in a 5.0-L container at 50 torr and 75 °C?

41. Substitute $n = m/MW$ into $PV = nRT$ and solve for MW.

(NOTE: MW is the abbreviation for molecular weight and m is a symbol for mass in grams.)

42. Starting with the combined gas law, $P_1V_1/T_1 = P_2V_2/T_2$, derive Boyle's law, $P_1V_1 = P_2V_2$ (named after the British chemist and physicist Robert Boyle, 1627–1691). This is done by holding the temperature constant and then dividing out the temperature. (In other words, $T_1 = T_2$.)

43. Starting with the combined gas law, $P_1V_1/T_1 = P_2V_2/T_2$, derive Charles's law, $V_1/T_1 = V_2/T_2$ (named after the French physicist Jacques Charles, 1746–1823). This is done by holding the pressure constant and then dividing out the pressure. (In other words, $P_1 = P_2$.)

44. Because of the combined effects of nitrogen narcosis and oxygen poisoning, divers breathing compressed air cannot safely go below 200 ft. What is the partial pressure (in torr) of N_2 a diver would breathe at 200 ft? (See Example 19 for data. Air is 78.1% N_2.)

13

ATOMIC ORBITALS

In Chapter 1 we promised to discuss the way electrons exist in the space surrounding the nucleus. In this chapter we shall try to give you an idea of how chemists "visualize" these electrons.

13-1
THE PROBLEM OF SEEING THINGS IS RELATED
TO THEIR SIZE AND HOW YOU LOOK AT THEM

Electrons have very little mass. Remember that they are almost 2000 times lighter than a proton. Because of the extremely small mass of the electron, it is impossible to clearly "see" where the electron is in the space surrounding the nucleus. It is also impossible to "see" what the electron looks like. This is because almost anything that collides with the electron, even light beams, causes it to be moved around.

It will help you to understand this moving around if you think of the following example. Suppose you wanted to know what a book looks like. The way that you see any object is to shine a light on it. The light is reflected by the object, in this case the book, and is focused into an image on the retina of your eye. The brain then interprets the image as that of a book. The light seems to have no effect on the book except to allow you to see it.

Now let's "look" at the book, not with light but with machine gun bullets. Assume that the book is made of stainless steel so that the bullets bounce off. The way that you can tell what the book looks like is to form, in some way, an image from the bullets that bounce off the book. But there is a problem—the book moves around from the force of the impact of the bullets. The image of the book that you see does not come from a stationary

(Opposite) A high-power carbon dioxide laser beam directed at a rotating ceramic rod causes molten droplets to fly out. Quantum mechanics is necessary to understand lasers.

book—it comes from a book that is moving around. The image of the book that you will get is not clear—you will see a *blurry image of a book*. In addition to not being able to see the book clearly, you cannot even be sure where the book is in space. It keeps moving around from the impact of the bullets.

It is more difficult to understand why we cannot "see" an electron than to understand why we couldn't "see" the book that was hit with machine gun bullets. But the basic idea is the same. The next section will try to give you some insight into the problem of "seeing" electrons.

13-2
THE UNCERTAINTY PRINCIPLE LIMITS OUR ABILITY TO "SEE" AN ELECTRON

The law of physics that explains why we cannot "see" an electron clearly is called the **uncertainty principle.** This principle, first stated by Werner Heisenberg (1901–1976) in 1927, is one of the most important principles in science. This law governs the behavior of electrons as far as our ability to "see" them is concerned. The uncertainty principle can be stated in the following way: *There is a limit to the accuracy of any measurement, and this limit becomes important when we try to measure (or "see") extremely small particles such as the electron.*

To help you to understand why the uncertainty principle is necessary, we must first discuss a few properties of light.

13-3
THE PROPERTIES OF LIGHT DEPEND ON ITS WAVELENGTH

Light can be considered as being both a particle and a wave. This "duality" of light is hard to understand, but you can get some feeling for it if you think in the following way. Light consists of particles called **photons.** Photons have energy but no mass, and they travel at the speed of light. There is also a **wave** associated with the photon. It is this wave that gives light many of its optical properties. For instance, light can be focused by the lenses of a camera, a telescope, or a microscope.

We can describe a light wave by its **wavelength**. The wavelength is the distance between two nearest peaks of a wave and has the symbol λ (Greek lambda), as shown in Figure 13-1. The wavelength of visible light determines its color. Red light has a long wavelength and violet light has a short one.

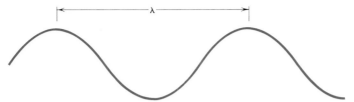

FIGURE 13-1 The wavelength λ is the distance between two closest peaks of the wave.

The other colors have wavelengths shorter than that of red light but longer than that of violet light. The colors of the rainbow are

red, orange, yellow, green, blue, indigo, violet

decreasing wavelength →

In addition to visible light, there are other waves that are in many ways similar to light waves. These have wavelengths that are longer and shorter than visible light. The following is a list of these waves in order of decreasing wavelength:

radio, radar, microwave, infrared, visible, ultraviolet, x-rays, gamma rays

decreasing wavelength →

The photon associated with the light wave has more energy when the wavelength is short and less energy when the wavelength is long:

wavelength: **short**   **long**

photon energy: **high**   **low**

The radio-wave photon has very little energy. The gamma-ray photon has a great deal of energy. The photon energy of the other kinds of light are in between these two.

You become aware of the different photon energies of different kinds of light every time you get a suntan. Ultraviolet-light photons have enough energy to cause the chemical reaction that tans the skin. Visible light will not cause a tan, no matter how bright the light is. Each visible-light photon has insufficient energy, no matter how bright it is. The brightness is related to the number of photons, not the energy of the photons. However, even a few ultraviolet-light photons can cause tanning, because their energy is high enough.

Another property of light that has to do with our ability to see an object

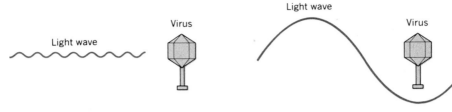

Short wavelength of light:
        Clear image of virus

Long wavelength of light:
        Clear image not possible

FIGURE 13-2 In order to see an object clearly, the wavelength of the light must be much smaller than the object.

is related to the wavelength of the light. *It is an optical principle that to be able to see an object clearly, the object must be larger than the wavelength of the light used to look at it.* This is shown in Figure 13-2.

Objects that need to be magnified more than about 2000 times to be seen are smaller than the wavelength of visible light. This is why a visible-light microscope has a maximum magnification of only about 2000 times.

13-4
WE CANNOT SEE AN ELECTRON BECAUSE OF THE UNCERTAINTY PRINCIPLE

Now to return to our discussion of electrons. Electrons are *very* small and have *very* little mass. Taking these facts into consideration, what conditions would you need in order to see an electron clearly? You would need light with very low energy photons (because of the small mass of the electron) and a very short wavelength (because of the small size of the electron). The very low energy photons won't move the electron around. The very short wavelength allows you to focus the image clearly. The wavelength must be shorter than the size of the electron.

Now we have a real dilemma. To focus on an electron clearly, we would have to use gamma rays, which have a wavelength shorter than the size of the electron. But gamma rays have *very* high energy photons. These high-energy photons would bounce the electron around all over the place. And we cannot lower the energy of the photon, because then the wavelength would be too long to get a clear image.

This is a dilemma that cannot be solved. There is no way to arrange the wavelength and photon energy so that we can see an electron. And scientists are pretty sure that there will *never* be a way to see an electron. This is what the uncertainty principle tells us.

You might ask, "Even if we cannot see an electron clearly, doesn't it really have a definite shape and isn't it really in some definite place?" The answer is that we don't know. Since we are pretty sure that we can never answer this question, it is not too useful to spend a great deal of time on speculation.

13-5
THE ELECTRON IS DESCRIBED WITH FUZZY PICTURES CALLED ORBITALS

According to the uncertainty principle, we cannot see an electron clearly. The interesting thing is that any theory describing the electron bound to a nucleus that does not take into account the uncertainty principle gives results that do not agree with experiments. Another way of putting this is to say that any theory that assumes, either implicitly or explicitly, that we can see an electron, will give results that do not agree with experiments. Niels Bohr's great theory, mentioned in Appendix 1-1 in Chapter 1, assumed that the orbit of the electron around the nucleus was completely known. His theory was great in the sense that it was the first theory that could explain *any* of the experiments involving the electron of the hydrogen atom. Even though Bohr's theory agreed with some of the experiments done on hydrogen atoms, it failed when applied to other atoms. We now know that there are things about the hydrogen atom that his theory doesn't even begin to explain.

Between 1924 and 1928, physicists developed a theory that *did* include the uncertainty principle and that *did* agree with the experimental findings for all atoms. This theory is called **quantum theory** or **quantum mechanics.** Naturally, quantum mechanics gives us a blurry picture of the electron.

The mathematics of quantum mechanics is rather involved and will not be discussed in this book. But the *results* of quantum mechanics will be discussed. It is these results that chemists use as their model to describe the electrons that are in the space surrounding the nucleus. Be warned, however, that the diagrams chemists draw are only to help them understand electrons in atoms—they are not real pictures of the electrons. Nobody, not even with these diagrams, can possibly visualize what the electron looks like or how it exists in the space around the nucleus. Only the mathematics of quantum mechanics can do that. But after a while, you will come to regard these diagrams as a good substitute for the real thing.

The kinds of diagrams that chemists draw of the electrons in the space surrounding the nucleus are called **orbitals.** *Orbitals are a description of the shape of the space in which the electrons exist.* The orbitals are our fuzzy picture of the electron.

There are many kinds of orbitals as shown in Figure 13-3. Some are shaped like spheres, some like dumbbells, and some like four-leaf clovers. Others are even more complicated.

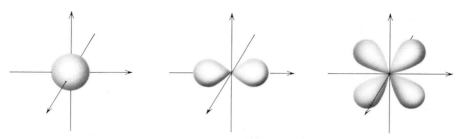

FIGURE 13-3 Examples of a few types of orbitals.

13-6
THE RULES FOR PLACING ELECTRONS IN ORBITALS DEPEND ON THE PROPERTIES OF THE ELECTRON

As you know, there are more than 105 different known elements. The number of electrons in an atom of each of these elements depends on the atomic number of the element and ranges from one electron in hydrogen to, say, 105 electrons in element 105, which is sometimes called hahnium, after Otto Hahn (1879–1968), who co-discovered nuclear fission in 1938. The following questions then arise.

1. What kinds of orbitals are there in these different elements?
2. Since the electrons are in orbitals, how many orbitals of each kind are there in each different kind of atom?

We will spend the rest of this chapter answering the first question; the second question will be answered in Chapter 14.

To begin answering these questions, we must first look at some rules that electrons follow as they are arranged in orbitals. These rules come from quantum mechanics, and it has been found from many experiments that electrons really do obey them.

RULE 1
Electrons surrounding a nucleus are arranged in orbitals.

RULE 2
An orbital can contain a maximum of two electrons.

RULE 3
Electrons like to be as far apart from one another as possible because they have negative charges and like charges repel.

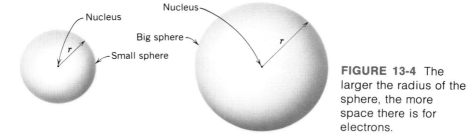

FIGURE 13-4 The larger the radius of the sphere, the more space there is for electrons.

RULE 4

Electrons are ''lazy.'' This means that they like to be as close to the nucleus as possible, because being close to the nucleus means that they need less energy. Since the negative charge of the electron attracts the positive charge of the nucleus, the electron has to do work to get farther from the nucleus, just as you have to do work in climbing a flight of stairs. Being ''lazy,'' the electron prefers not to do this work, and thus it likes to be close to the nucleus.

RULE 5

The farther away the electrons are from the nucleus, the more room there is for them. Or, the closer the electrons are to the nucleus, the less room there is for electrons. (See Figure 13-4.)

13-7
ORBITALS ARE ARRANGED IN ENERGY LEVELS

The orbitals are not spaced at equal distances from the nucleus. They tend to be bunched together in groups or levels. All the orbitals in a group or level are at about the same average distance from the nucleus and therefore have about the same energy. (This statement will be modified in Chapter 14.) The usual way to refer to orbitals having about the same energy is to say that they are in the same **energy level.**

The energy levels are numbered, beginning closest to the nucleus with energy level 1 and working outward to higher levels. Each energy level has a different number of orbitals in it, which is reasonable in light of Rule 5. The higher the energy level, the greater the number of orbitals it contains. There is a simple formula that tells us how many orbitals there are in each energy level. The formula is

$$\text{number of orbitals} = n^2$$

where n is the number of an energy level; n needs only to take on the values 1

TABLE 13-1

VALUES OF n, n^2, and $2n^2$

ENERGY LEVEL, n	NUMBER OF ORBITALS, n^2	NUMBER OF ELECTRONS, $2n^2$
1	1	2
2	4	8
3	9	18
4	16	32
5	25	50
6	36	72
7	49	98

through 7 to get all the energy levels needed to contain enough electrons for the 105 elements.

Naturally in atoms, the lower energy levels fill up with electrons first. This is a consequence of Rule 4.

We also need to know the maximum number of electrons in each energy level. Because each orbital can contain up to two electrons (Rule 2), the formula

$$\text{number of electrons} = 2n^2$$

gives the maximum number of electrons in a given energy level.

Table 13-1 lists values of n^2 and $2n^2$ for the first seven energy levels.

If we add up all the numbers in the last column of Table 13-1, we get 280. But since we only need 105 electrons to make element 105, why take 7 energy levels? The reason is that the last three energy levels, 5, 6 and 7, never get completely filled. We will see how this works in Chapter 14.

13-8
SPIN IS A PROPERTY OF ELECTRONS

Before we proceed, we should talk about one more property of electrons—their **spin**. The word "spin" may suggest that electrons spin like a top, but because of the uncertainty principle, we cannot say that they do. The spin of an electron is a property that seems to be like a top's spin in only one respect: It has what's called **spin angular momentum,** as do tops, which is why this property of the electron was called **spin**. Remember that the uncertainty principle prevents us from visualizing the spin of an electron.

However, we will talk of spin as if we actually could visualize it. We say that electrons can have a spin of either $+\frac{1}{2}$ or $-\frac{1}{2}$. Each of these two spin

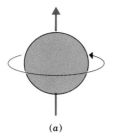

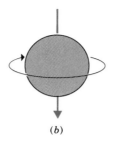

FIGURE 13-5 The two spin states of an electron: (*a*) "spin-up" electron; (*b*) "spin-down" electron.

conditions is called a **spin state.** Chemists call the $+\frac{1}{2}$ spin state "spin up" and the $-\frac{1}{2}$ spin state "spin down." The numbers $+\frac{1}{2}$ and $-\frac{1}{2}$ come out of the mathematics of quantum mechanics and are related to the spin angular momentum. But these matters need not concern you in this course. You can think of the "spin-up" condition as the electron spinning with its top "up" and the "spin-down" condition as the electron having flipped over and having its top pointing "down." "Up" and "down" are our descriptions. The electron only "knows" that it has two spin states, which are illustrated in Figure 13-5.

If it bothers you that you are continually told "We cannot possibly visualize this or that property of electrons, but here is the way to look at it," please try not to be upset. That's the way nature has made electrons and human beings—"indescribable" and "describers," respectively. Since we do have to talk about electrons, we use the next best method, a model. So you will continually be given pictures of the unpicturable and told to imagine the unimaginable.

This discussion about spin was included here to bring out one very important point. We said before that only two electrons can fit into an orbital (Rule 2). These two electrons *must* have different spin states. One electron has to have spin *up*, and the other electron has to have spin *down*. We can symbolize a spin-up electron with an up arrow, ↑, and a spin-down electron with a down arrow, ↓. So in each orbital that is filled, the two electrons are ↑ and ↓.

To summarize: **Two electrons can be in the same orbital only if their spins are different.**

13-9
THE FOUR BASIC TYPES OF ORBITALS ARE THE *s*, *p*, *d*, AND *f* ORBITALS

Earlier in this chapter, we raised the question, "What kinds of orbitals are there in the different elements?" We are now in a position to answer this question.

All the elements have their electrons arranged in four basic types of

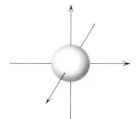

FIGURE 13-6 The s orbital in the first energy level.

orbitals. These orbitals are referred to as "s," "p," "d," and "f." Unfortunately, this system is a carryover from the old days, before scientists knew the shapes of the orbitals. The names aren't related to the shapes.

The s orbital. The s orbital has the shape of a fuzzy sphere, as shown in Figure 13-6. There is only *one* s orbital possible in each energy level. Like all orbitals, the s orbital can hold up to two electrons. In each energy level, electrons will go into an s orbital first. The lowest energy level, $n = 1$, has *only* an s orbital.

The p orbitals. The p orbital has the shape of a fuzzy dumbbell. There are *three* p orbitals in each energy level except for the first level. Those in the second energy level are shown in Figure 13-7. The three p orbitals are at right angles to one another and have names that tell us the axis they are on. The p orbital that lies on the x-axis is called p_x, the p

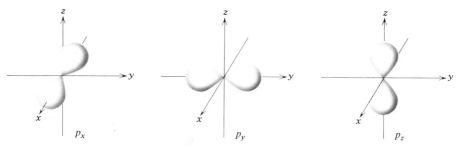

FIGURE 13-7 The three p orbitals in the second energy level.

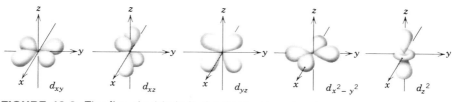

FIGURE 13-8 The five d orbitals in the third energy level.

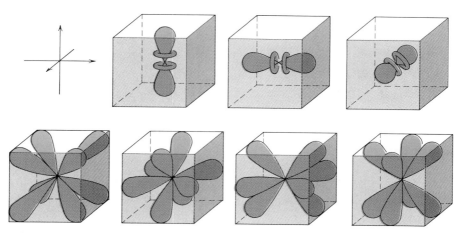

FIGURE 13-9 The seven f orbitals in the fourth energy level.

orbital that lies on the y-axis is called p_y, and the p orbital that lies on the z-axis is called p_z. Since each p orbital can hold up to two electrons, all three p orbitals can hold up to six electrons. In each energy level, electrons will go into the p orbitals only after the s orbital has been filled.

The d orbitals. The d orbitals have shapes that can best be described with the help of the pictures in Figure 13-8. There are *five* d orbitals in each energy level starting with the third level ($n = 3$). Since each d orbital can hold a maximum of two electrons, the five d orbitals can hold up to 10 electrons. Each of the d orbitals has a name (see Figure 13-8). However, it is not necessary to learn these names at this time.

The f orbitals. The f orbitals have shapes that can best be described with the help of the pictures in Figure 13-9. They are more complicated than the d orbitals. There are *seven* f orbitals in each energy level starting with the fourth level ($n = 4$). Since each f orbital can hold a

TABLE 13-2

ORBITALS AND ELECTRONS

ORBITAL	NUMBER OF ORBITALS OF A GIVEN TYPE IN AN ENERGY LEVEL	MAXIMUM NUMBER OF ELECTRONS IN ALL THE ORBITALS OF A GIVEN TYPE
s	1	2
p	3	6
d	5	10
f	7	14

maximum of two electrons, the seven f orbitals can hold up to 14 electrons. There is no need for you to learn the names, and they will not be listed in the figure.

Table 13-2 gives (a) the number of orbitals of each type in an energy level and (b) the maximum number of electrons in all the orbitals of a given type.

PROBLEMS[1]

1. Define the uncertainty principle.

2. Define wavelength.

3. What is the relationship between the photon energy and the wavelength of light?

4. Discuss the two reasons why the uncertainty principle exists.

5. Why are the effects of the uncertainty principle not noticeable in everyday living?

6. What do we mean by the term "orbital"?

7. List five rules that electrons follow as they are arranged in orbitals.

8. How many orbitals and electrons can there be in the eighth energy level?

9. What are the "quantum numbers" of the two spin states that an electron can have?

10. Complete the following: Two electrons can be in the same orbital *only* if . . . .

11. Sketch the s orbital in the first energy level.

12. Sketch the p_x, p_y, and p_z orbitals in the second energy level.

13. a. How many g orbitals can exist in one of the higher energy levels? (The g orbitals come right after the f orbitals.)

    b. How many electrons does this correspond to?

[1]There are no keyed problems in this chapter.

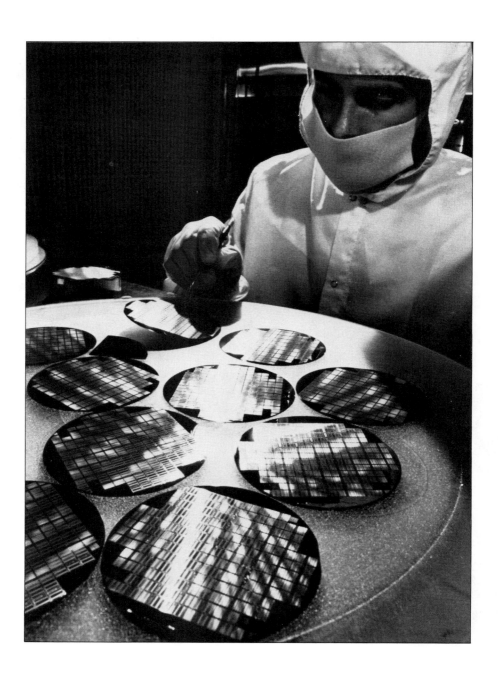

ORBITALS IN ATOMS
AND THE PERIODIC TABLE

In Chapter 13 we raised two questions: "What kinds of orbitals are in the different elements?" and "How many orbitals of each kind are there in each different kind of atom?" We answered the first question in Chapter 13. We will answer the second question in this chapter.

14-1
ORBITALS ARE ARRANGED IN ENERGY
LEVELS IN THE ORDER s, p, d, AND f

In each energy level, the order in which the orbitals are filled with electrons is as follows:

s first, p second, d third, f fourth

Thus, for the first energy level ($n = 1$), which has only one orbital, we see that this orbital must be an s orbital.

For the second energy level ($n = 2$), which has four orbitals, we can have one s orbital and three p orbitals.

For the third energy level ($n = 3$), which has nine orbitals, we can have one s orbital, three p orbitals, and five d orbitals.

For the fourth energy level ($n = 4$), which has 16 orbitals, we can have one s orbital, three p orbitals, five d orbitals, and seven f orbitals.

Higher energy levels (n greater than 4) have room for additional types of orbitals. However, these additional orbitals are not needed to discuss the

(Opposite) The manufacture of silicon chips used in computers. Quantum mechanics is necessary to understand semiconductors.

TABLE 14-1

ORBITAL TYPES IN EACH ENERGY LEVEL

ENERGY LEVEL, n	MAXIMUM NUMBER OF ORBITALS, n^2	ORBITAL TYPES IN ENERGY LEVEL	NUMBER OF ORBITALS OF EACH TYPE
1	1	s	one s
2	4	s, p	one s, three p
3	9	s, p, d	one s, three p, five d
4	16	s, p, d, f	one s, three p, five d, seven f
5	25	s, p, d, f^a	—
6	36	s, p, d^a	—
7	49	s^a	—

[a]There is room in these energy levels for more orbital types. But these orbitals do not have electrons in them. We can get all 105 known elements with only the orbitals indicated.

known elements. Table 14-1 summarizes the information just given on orbitals and energy levels.

14-2
ATOMS OF DIFFERENT ELEMENTS HAVE THEIR ORBITALS FILLED IN A SPECIFIC ORDER

Before we can discuss the order in which electrons fill up the orbitals to make the different elements, a bit of notation is necessary. An s orbital in the first energy level ($n = 1$) is called a $1s$ orbital. An s orbital in the second energy level ($n = 2$) is called a $2s$ orbital. A p orbital in the second energy level is called a $2p$ orbital. And so on. As you can see, the number of the energy level of an orbital precedes the name of the orbital.

EXAMPLE 1   What would a p orbital in the fourth energy level be called?

Solution: It would be called a $4p$ orbital.   ■

EXAMPLE 2   What would a d orbital in the fifth energy level be called?

Solution: It would be called a $5d$ orbital.   ■

Figure 14-1 shows the order in which the orbitals are filled by electrons. Each little circle represents one orbital. The shading of the circles in each

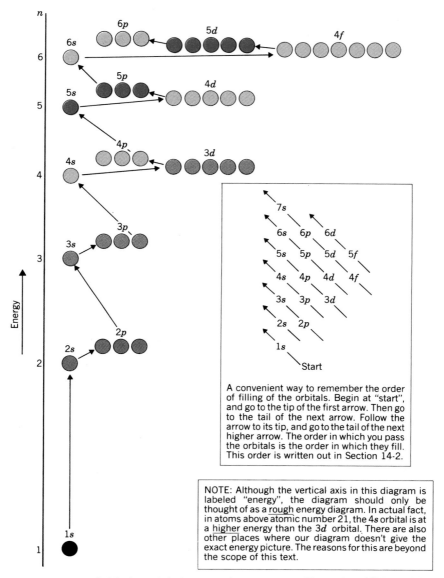

A convenient way to remember the order of filling of the orbitals. Begin at "start", and go to the tip of the first arrow. Then go to the tail of the next arrow. Follow the arrow to its tip, and go to the tail of the next higher arrow. The order in which you pass the orbitals is the order in which they fill. This order is written out in Section 14-2.

NOTE: Although the vertical axis in this diagram is labeled "energy", the diagram should only be thought of as a <u>rough</u> energy diagram. In actual fact, in atoms above atomic number 21, the 4s orbital is at a <u>higher</u> energy than the 3d orbital. There are also other places where our diagram doesn't give the exact energy picture. The reasons for this are beyond the scope of this text.

FIGURE 14-1 Orbitals and their approximate energy. The order of filling of the orbitals is indicated by the arrows. In the upper boxed area, we present a convenient way to remember the order of filling of the orbitals. In the lower boxed area, we briefly discuss the actual energy of some orbitals.

energy level is different. This is so you can easily tell which energy level a circle is in. The arrows show the order of the filling of the orbitals with electrons.

The lowest energy level is the first one ($n = 1$). There is an s orbital in it, and this orbital is called the 1s orbital. An s orbital can contain a maximum of two electrons.

The second energy level has four orbitals: one s orbital, called the 2s, and three p orbitals, known as the 2p orbitals. Notice that the three 2p orbitals have a slightly higher energy than the single 2s orbital. See Appendix 14-1 at the end of this chapter for an explanation. From Rule 4 in Chapter 13, the 2s orbital will fill up first and then the electrons will begin to fill the 2p orbitals. Again, each orbital can contain a maximum of two electrons (Rule 2 in Chapter 13).

The third energy level has nine orbitals: one s, three p, and five d. The 3s orbital will fill up first and then the electrons will go into the 3p orbitals. However, now we have a problem. You would normally think that after the 3p orbitals are filled, the 3d orbitals would fill up with electrons. But nature doesn't work this way. It has been found by experiment that it is the 4s orbital that now begins to fill with electrons. It is only after the 4s orbital is filled that the 3d orbitals begin to fill. After the 3d orbitals are filled, the 4p orbitals start to fill.

Notice that in Figure 14-1, the 3d orbitals are at about the same energy as the 4s and 4p orbitals. However, they are still called 3d orbitals, which means that they formally belong to the third energy level. The d orbitals fill *one* energy level later than you would think. This may seem strange, but it all has to do with the mathematics of quantum mechanics and many laboratory experiments. More advanced courses will cover this in more detail.

After the 4p orbitals are filled, the 5s orbital fills up. Then the 4d orbitals fill up, followed by the 5p orbitals.

After the 5p come the 6s, then the 4f, the 5d, and the 6p. Notice that the 4f orbitals are also out of place. They are on the energy level of the 6s and 6p and 5d orbitals. The 4f orbitals fill *two* energy levels later than expected.

Figure 14-1 illustrates the order of filling up to the 6p orbitals; the order of filling the remaining few orbitals is 7s, 5f, 6d, 7p. Actually, the 6d orbitals only have two electrons in them (for elements 104 and 105). The 7p orbitals, which would begin filling at element 113, do not have any electrons in them.

A summary of the order in which the orbitals fill with electrons is as follows:

1s 2s 2p 3s 3p 4s 3d 4p 5s 4d 5p 6s 4f 5d 6p 7s 5f 6d 7p

Now look at Figure 14-2. The diagram shows the order in which the elements are built up by the filling of the various orbitals. Looking at Figure 14-2, we see that if there is one electron in the 1s orbital, we have hydrogen, H. Two electrons in the 1s orbital gives us helium, He.

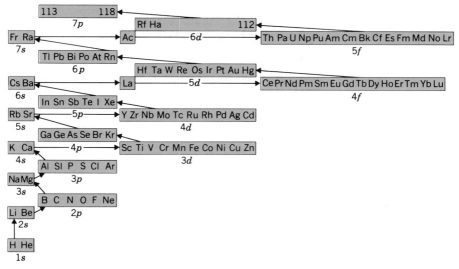

FIGURE 14-2 The order in which the elements are formed by the filling of the orbitals. Begin reading the diagram in the lower left-hand corner and follow the arrows. Compare with Figure 14.1. In each rectangle, the atomic numbers of the elements increase from left to right.

Adding an electron to the $2s$ orbital gives lithium, Li. Two electrons in the $2s$ orbital gives beryllium, Be. Now we have a total of four electrons, corresponding to the atomic number of Be, which is 4.

The three $2p$ orbitals can contain a total of 6 electrons. These orbitals are filled one electron at a time to give the elements B, C, N, O, F, and Ne.

EXAMPLE 3   Using Figure 14-2, list all the elements in which the $5s$ orbital, the $4d$ orbitals, and the $4f$ orbitals are being filled.

Solution: $5s$ orbitals: Rb, Sr.
$4d$ orbitals: Y, Zr, Nb, Mo, Tc, Ru, Rh, Pd, Ag, Cd.
$4f$ orbitals: Ce, Pr, Nb, Pm, Sm, Eu, Gd, Tb, Dy, Ho, Er, Tm, Yb, Lu. ∎

There are two things about Figure 14-2 that may seem a little strange. After the $6s$ orbital is filled, one electron goes into the $5d$ orbital to give the element La. The next electron then goes into a $4f$ orbital to give Ce. After that, all the $4f$ orbitals can be filled. Then, beginning with Hf and finishing with Hg, we return to the $5d$ orbitals.

The same thing happens in the $6d$ and $5f$ orbitals.

14-3
ORBITAL DIAGRAMS AND ELECTRON CONFIGURATIONS ARE WAYS OF REPRESENTING THE ARRANGEMENT OF ELECTRONS IN ATOMS

In Figure 14-1, we showed the order of the filling of the orbitals. If we put one electron at a time into the diagram, we will build up the **orbital diagram** (or arrangement) of the different elements.

Each electron will be represented by an arrow, either $\uparrow$ or $\downarrow$, to correspond with the two spin states, "spin up" or "spin down."

In the following discussion, Figure 14-1 has been reduced in size to save space. Notice that the circles for all the orbitals are still present, but only some of them will be filled with electrons. That's all right, since the space for the electrons is always there, even though there are no electrons in that space.

The first element is hydrogen, H, which has one electron in the $1s$ orbital. We can write this electron as $1s^1$, where the superscript "1" tells us the number of electrons that are in the $1s$ orbital. This is shown in Figure 14-3. This way of writing the electron description is called the **electron configuration** of an element.

Figure 14-4 has one arrow drawn in the circle that represents the $1s$ orbital. This is the orbital diagram of hydrogen. Below the diagram we have written the electron configuration of hydrogen.

The second electron also goes into the $1s$ orbital to give helium, He. We can write the electron configuration of He as $1s^2$, since there are two electrons in the $1s$ orbital. Figure 14-5 shows the orbital diagram and electron configuration for He.

Since each orbital can have a maximum of two electrons, the superscript on the orbital letter can only take on the values of 1 or 2.

The third electron goes into the $2s$ orbital to give the element lithium, Li. The orbital diagram and electron configuration for Li is $1s^2 2s^1$, as shown in Figure 14-6.

The fourth electron also goes into the $2s$ orbital to give beryllium, Be. The orbital diagram and electron configuration for Be is shown in Figure 14-7.

Next the $2p$ orbitals fill up to give the elements B, C, N, O, F, and Ne. The electron configurations and orbital diagrams are shown in Figure 14-8.

We will now discuss some very important points in the way electrons fill the $2p$ orbitals. Remember that there are three p orbitals—p_x, p_y, and p_z. In studying Figure 14-8, you may wonder why we put the electrons into the p orbitals in the order p_x, p_y, p_z. We do this to make our diagrams look

Energy level $\rightarrow 1s^1$ $\leftarrow$ Number of electrons in orbital
$\leftarrow$ Type of orbital

FIGURE 14-3 The electron configuration of hydrogen.

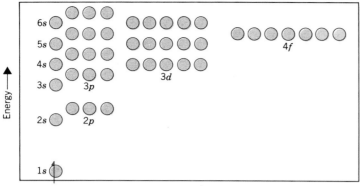

H: $1s^1$

FIGURE 14-4 The orbital diagram and electron configuration for hydrogen.

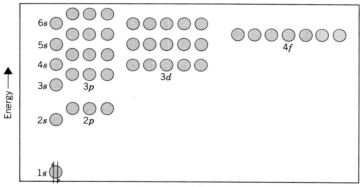

He: $1s^2$

FIGURE 14-5 The orbital diagram and electron configuration for helium.

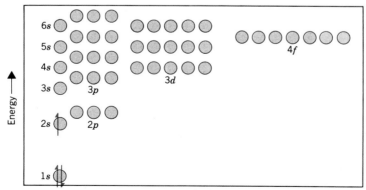

Li: $1s^2 2s^1$

FIGURE 14-6 The orbital diagram and electron configuration for lithium.

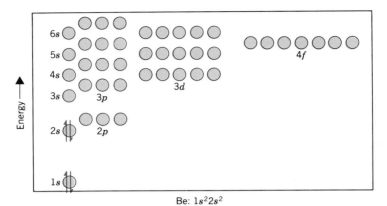

Be: $1s^2 2s^2$

FIGURE 14-7 The orbital diagram and electron configuration for beryllium.

consistent. We don't actually know which orbital is occupied first. Notice, however, that when electrons go into the p_x, p_y, and p_z orbitals, the first three always go in with their spins in the same direction. This is the lower-energy arrangement.

The orbital diagrams demonstrate a very important point. According to Rule 3 in Chapter 13, electrons like to be as far apart from each other as possible. But this must be consistent with Rule 4, which states that they are "lazy" and like to be at the lowest energy. Following these rules, look at the orbital diagram and electron configuration for carbon. The sixth electron *doesn't* join the fifth one in the $2p_x$ orbital, which would fill up the $2p_x$ orbital. The sixth electron can get farther from the p_x electron by going into the $2p_y$ orbital with *no* increase in energy. So it does just that.

The same thing happens with nitrogen: The seventh electron goes into the $2p_z$ orbital. This way the electron is as far from the other electrons as possible with no increase in energy. But after nitrogen, the electrons will start doubling up in the $2p$ orbitals, because it would take some extra energy for the electrons to go into the $3s$ and higher orbitals. And electrons are so "lazy" that they would rather "tolerate" each other than do extra work. Refer to the diagrams of oxygen, fluorine, and neon for illustrations of this phenomenon.

After neon, the eleventh electron will go into the $3s$ orbital to form sodium, Na. We could continue describing the orbital diagrams and electron configurations for all the elements, but now you can figure them out for yourself. Just use Figures 14-1 and 14-2 as guides, and remember the rules (see Chapter 13) that electrons follow when arranging themselves in orbitals.

EXAMPLE 4   Using a diagram like Figure 14-1, draw the orbital diagram for the element silicon, atomic number 14. Also write the electron configuration for silicon.

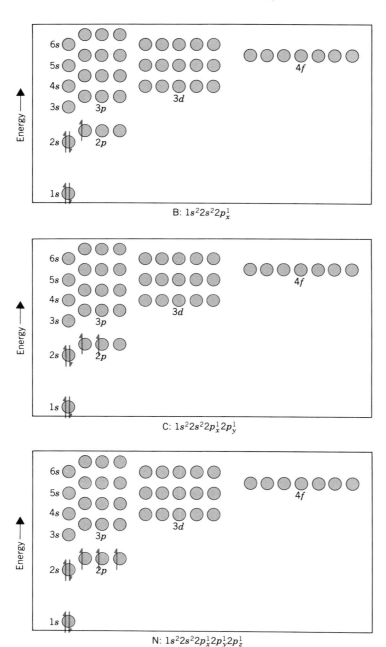

FIGURE 14-8 The orbital diagrams and electron configurations for boron, carbon, nitrogen, oxygen, fluorine, and neon.

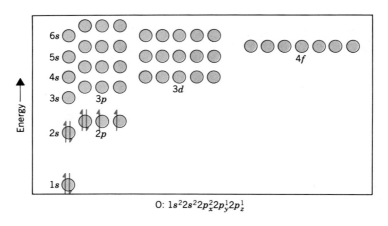

O: $1s^2 2s^2 2p_x^2 2p_y^1 2p_z^1$

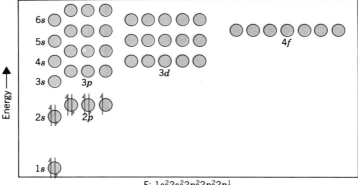

F: $1s^2 2s^2 2p_x^2 2p_y^2 2p_z^1$

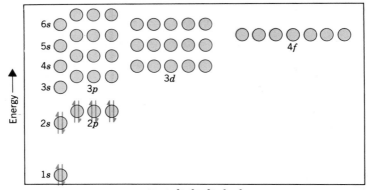

Ne: $1s^2 2s^2 2p_x^2 2p_y^2 2p_z^2$

FIGURE 14-8 continued.

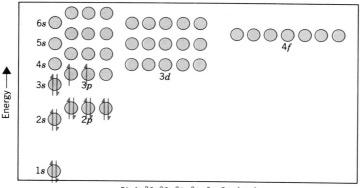

Si: $1s^2 2s^2 2p_x^2 2p_y^2 2p_z^2 3s^2 3p_x^1 3p_y^1$

FIGURE 14-9 The orbital diagram and electron configuration for silicon.

Solution: See Figure 14-9. ■

As you can see, it takes considerable space to draw the orbital diagrams as we have done. So chemists have developed a way of writing them across the page. Instead of circles to represent orbitals, they use a short line like this: ____. The electrons are drawn in as arrows. An orbital with one electron in it is drawn as ___↑___, and an orbital with two electrons in it is drawn as ___↑↓___. (Again, we draw in the "spin-up" electrons first. Of course the electron doesn't know up from down, but chemists want all their diagrams to look the same.) The names of the orbitals are put under the short lines in the same way as the electron configurations are written under the orbital diagrams. These **horizontal orbital diagrams** for the first 10 elements are written as follows:

H: $\dfrac{\uparrow}{1s^1}$

He: $\dfrac{\uparrow\downarrow}{1s^2}$

Li: $\dfrac{\uparrow\downarrow}{1s^2}$ $\dfrac{\uparrow}{2s^1}$

Be: $\dfrac{\uparrow\downarrow}{1s^2}$ $\dfrac{\uparrow\downarrow}{2s^2}$

B: $\dfrac{\uparrow\downarrow}{1s^2}$ $\dfrac{\uparrow\downarrow}{2s^2}$ $\dfrac{\uparrow}{2p_x^1}$

C: $\dfrac{\uparrow\downarrow}{1s^2}$ $\dfrac{\uparrow\downarrow}{2s^2}$ $\dfrac{\uparrow}{2p_x^1}$ $\dfrac{\uparrow}{2p_y^1}$

N: $\underset{1s^2}{\uparrow\downarrow}$ $\underset{2s^2}{\uparrow\downarrow}$ $\underset{2p_x^1}{\uparrow}$ $\underset{2p_y^1}{\uparrow}$ $\underset{2p_z^1}{\uparrow}$

O: $\underset{1s^2}{\uparrow\downarrow}$ $\underset{2s^2}{\uparrow\downarrow}$ $\underset{2p_x^2}{\uparrow\downarrow}$ $\underset{2p_y^1}{\uparrow}$ $\underset{2p_z^1}{\uparrow}$

F: $\underset{1s^2}{\uparrow\downarrow}$ $\underset{2s^2}{\uparrow\downarrow}$ $\underset{2p_x^2}{\uparrow\downarrow}$ $\underset{2p_y^2}{\uparrow\downarrow}$ $\underset{2p_z^1}{\uparrow}$

Ne: $\underset{1s^2}{\uparrow\downarrow}$ $\underset{2s^2}{\uparrow\downarrow}$ $\underset{2p_x^2}{\uparrow\downarrow}$ $\underset{2p_y^2}{\uparrow\downarrow}$ $\underset{2p_z^2}{\uparrow\downarrow}$

EXAMPLE 5   Draw a horizontal orbital diagram for phosphorus.

Solution: The atomic number of phosphorus is 15. This means that we will have to put 15 electrons into orbitals.

P: $\underset{1s^2}{\uparrow\downarrow}$ $\underset{2s^2}{\uparrow\downarrow}$ $\underset{2p_x^2}{\uparrow\downarrow}$ $\underset{2p_y^2}{\uparrow\downarrow}$ $\underset{2p_z^2}{\uparrow\downarrow}$ $\underset{3s^2}{\uparrow\downarrow}$ $\underset{3p_x^1}{\uparrow}$ $\underset{3p_y^1}{\uparrow}$ $\underset{3p_z^1}{\uparrow}$

The sum of the superscripts equals 15. Notice that the electrons in the $3p$ orbitals are as far from one another as possible with no increase in energy. Each is in a different p orbital. ■

EXAMPLE 6   Which of the following horizontal orbital diagrams is correct?

a. $\underset{1s^2}{\uparrow\downarrow}$ $\underset{2s^2}{\uparrow\downarrow}$ $\underset{2p_x^1}{\uparrow}$ $\underset{2p_y^1}{\downarrow}$ $\underset{2p_z^1}{\uparrow}$

b. $\underset{1s^2}{\uparrow\downarrow}$ $\underset{2s^2}{\uparrow\downarrow}$ $\underset{2p_x^1}{\downarrow}$ $\underset{2p_y^1}{\downarrow}$ $\underset{2p_z^1}{\downarrow}$

c. $\underset{1s^2}{\uparrow\downarrow}$ $\underset{2s^2}{\uparrow\downarrow}$ $\underset{2p_x^2}{\uparrow\downarrow}$ $\underset{2p_y^1}{\uparrow}$ $\underset{2p_z}{}$

d. $\underset{1s^2}{\uparrow\downarrow}$ $\underset{2s^2}{\uparrow\downarrow}$ $\underset{2p_x^2}{\uparrow}$ $\underset{2p_y}{}$ $\underset{2p_z^1}{\uparrow}$

Solution:

   a. The orbital diagram is incorrect because the three unpaired p electrons do not have the same spin.

   b. The orbital diagram is correct because the unpaired p electrons all have the same spin. However, we should always draw this diagram with the spins in the up direction so that all our diagrams will look alike.

   c. The orbital diagram is incorrect. One of the $2p_x^2$ electrons should be in the $2p_z$ orbital.

d. The orbital diagram is correct. However, we should always draw this diagram with the $2p_y$ orbital, not the $2p_z$ orbital, containing the second unpaired electron so that all our diagrams will look alike. ∎

14-4
THE PERIODIC TABLE ARRANGES THE ELEMENTS ACCORDING TO THEIR ORBITAL CONFIGURATION AND CHEMICAL PROPERTIES

Chemists have developed a table of the elements that is organized according to the way orbitals are filled; it is called the **periodic table**.[1] Figure 14-10 is an example of a periodic table.

The periodic table is divided into groups and periods. The **groups** (also called **chemical families**) are in vertical columns. Each group or chemical family is identified by a Roman numeral at the top of the column. The elements in each group have similar chemical properties, which is why they are also called a chemical family. The **periods** are in horizontal rows that correspond to the filling of the energy levels. The numbers at the left of each period show the energy level that is being filled for the A group elements.

Even though hydrogen, H, is listed in group IA, it has different properties than the other elements in group IA, which are all silvery metals. Hydrogen is a gas and really belongs in a group by itself.

In the element boxes in Figure 14-10, the number above the symbol for each element is the atomic number; the number below the symbol is the atomic weight.

The two groups on the left side of the table (columns IA and IIA) represent elements that have electrons in the s orbitals of their highest occupied energy level. Helium, which is listed in group VIIIA, also has an s orbital in its highest energy level.

The six groups to the right side of the table (columns IIIA, IVA, VA, VIA, VIIA, and VIIIA) represent elements whose p orbitals are being filled. (Remember that helium, which is in group VIIIA, is the exception.)

The eight groups toward the center represent elements whose d orbitals are being filled (columns IIIB, IVB, VB, VIB, VIIB, VIII, IB, and IIB). These elements are called the **transition metals**. The numbering system of these groups is related to the chemical properties of the elements and won't be discussed further in this book.

The rows without column numbers, which contain elements 58 through 71 and 90 through 103, represent elements whose f orbitals are being filled.

---

[1] Before orbitals were known, chemists had devised the periodic table based on the chemical and physical properties of the elements. Dmitri Mendeleev (1834–1907) is considered the founder of the modern periodic table. Element 101 is named in his honor.

Periodic Table of the Elements

s orbitals

p orbitals

d orbitals

f orbitals

1	IA	IIA	IIIB	IVB	VB	VIB	VIIB	VIII			IB	IIB	IIIA	IVA	VA	VIA	VIIA	VIIIA
1	1 H 1.0079																	2 He 4.00260
2	3 Li 6.941	4 Be 9.01218											5 B 10.81	6 C 12.011	7 N 14.0067	8 O 15.9994	9 F 18.99840	10 Ne 20.179
3	11 Na 22.98977	12 Mg 24.305											13 Al 26.98154	14 Si 28.0855	15 P 30.97376	16 S 32.06	17 Cl 35.453	18 Ar 39.948
4	19 K 39.0983	20 Ca 40.08	21 Sc 44.9559	22 Ti 47.88	23 V 50.9415	24 Cr 51.996	25 Mn 54.9380	26 Fe 55.847	27 Co 58.9332	28 Ni 58.69	29 Cu 63.546	30 Zn 65.38	31 Ga 69.72	32 Ge 72.59	33 As 74.9216	34 Se 78.96	35 Br 79.904	36 Kr 83.80
5	37 Rb 85.4678	38 Sr 87.62	39 Y 88.9059	40 Zr 91.22	41 Nb 92.9064	42 Mo 95.94	43 Tc (98)	44 Ru 101.07	45 Rh 102.9055	46 Pd 106.42	47 Ag 107.868	48 Cd 112.41	49 In 114.82	50 Sn 118.69	51 Sb 121.75	52 Te 127.60	53 I 126.9045	54 Xe 131.29
6	55 Cs 132.9054	56 Ba 137.3	57* La 138.9055	72 Hf 178.49	73 Ta 180.9479	74 W 183.85	75 Re 186.207	76 Os 190.2	77 Ir 192.22	78 Pt 195.08	79 Au 196.9665	80 Hg 200.59	81 Tl 204.383	82 Pb 207.2	83 Bi 208.9804	84 Po (209)	85 At (210)	86 Rn (222)
7	87 Fr (223)	88 Ra 226.0254	89** Ac 227.0278	104 Rf (261)	105 Ha (262)													

Atomic number and atomic weight key:
1 H 1.0079
← atomic number
← atomic weight

*Lanthanoid series
**Actinoid series

*	58 Ce 140.12	59 Pr 140.9077	60 Nd 144.24	61 Pm (145)	62 Sm 150.36	63 Eu 151.96	64 Gd 157.25	65 Tb 158.9254	66 Dy 162.50	67 Ho 164.9304	68 Er 167.26	69 Tm 168.9342	70 Yb 173.04	71 Lu 174.967
**	90 Th 232.0381	91 Pa 231.0359	92 U 238.02	93 Np 237.0482	94 Pu (244)	95 Am (243)	96 Cm (247)	97 Bk (247)	98 Cf (251)	99 Es (252)	100 Fm (257)	101 Md (258)	102 No (259)	103 Lr (260)

These rows have special names: The row containing elements 58 through 71 is called the **lanthanoid** or **rare earth series**, and the row containing elements 90 through 103 is called the **actinoid series**. The *lanthanoid* elements appear right after the element *lanthanum,* and the *actinoid* elements appear right after the element *actinium.*

(**NOTE:** Older, but still widely used, names for lanthanoid and actinoid are lanthanide and actinide.)

EXAMPLE 7   List the elements that are in the following groups in the periodic table: Groups IIA, VIIB, VIA.

Solution: From Figure 14-3, we have, reading down each group:

group IIA: Be, Mg, Ca, Sr, Ba, Ra
group VIIB: Mn, Tc, Re
group VIA: O, S, Se, Te, Po   ■

The first row (period 1) across the table lists elements in which the highest-energy orbital that has electrons in it is the first energy level ($n = 1$). The second row (period 2) lists elements in which the highest-energy orbitals that have electrons in them are the second energy level ($n = 2$). Similarly for the third row (period 3).

EXAMPLE 8   List the elements that are in periods 1, 2, and 3 of the periodic table.

Solution: From Figure 14-3, we have, reading across each period:

period 1:   H, He
period 2:   Li, Be, B, C, N, O, F, Ne
period 3:   Na, Mg, Al, Si, P, S, Cl, Ar   ■

The fourth row (period 4) introduces some complications. Elements 19, 20, and 31 through 36 all have their highest-energy occupied orbitals in the fourth energy level ($n = 4$), but elements 21 through 30 have their highest-energy occupied orbitals in the third energy level.[2] The *d* orbitals in the

---

[2]It is actually more complicated than this. The highest-energy occupied orbital for some of the transition elements (elements 21–30) is not the 3*d* orbitals but the 4*s* orbital. A complete discussion is beyond the scope of this book.

fourth row are $3d$ orbitals. Remember that after the $4s$ orbital, the $3d$ orbitals begin to fill (refer to Figures 14-1 and 14-2).

The d orbitals in periods 5 through 7 are also in one lower energy level than that of the period they are listed in.

The lanthanoid series (Ce through Lu) is in row 6 of the periodic table. For the 14 elements in this series, the $4f$ orbitals begin to fill after the $6s$ orbital has been filled and one $5d$ electron (for the element La) has been added.

So we see that **the d orbitals are one energy level behind** and **the f orbitals are two energy levels behind** the rows that they are in.

14-5
CHEMICAL FAMILIES HAVE SIMILAR CHEMICAL PROPERTIES

In addition to the transition metals, the lanthanoid series, and the actinoid series, there are other areas of the periodic table that have special names. Some of them are shown in Figure 14-11. As we have said, these areas are called chemical families because they have similar chemical properties.

1. The **alkali metals** (group IA): Li, Na, K, Rb, Cs, and Fr all have one electron in their last or highest-energy occupied orbital, which is an s orbital.

2. The **alkaline earth metals** (group IIA): Be, Mg, Ca, Sr, Ba, and Ra all have two electrons in their last or highest-energy occupied orbital, which is an s orbital.

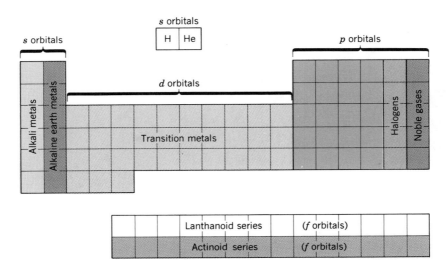

FIGURE 14-11   Chemical families and regions in the periodic table. H and He are placed in the center. These two elements are special because they are filling the s orbital in the first energy level. He is a noble gas.

3. The **halogens** (group VIIA): F, Cl, Br, I, and At all have seven electrons (two *s* and five *p*) in their last or highest occupied energy level.
   We can say that they all have one electron in their last p_z orbital. (As
   we said in section 14-3, we really don't know which *p* orbital the lone
   electron is in.)

4. The **noble gases** (group VIIIA): Helium has two *s* electrons in its highest occupied energy level, which is the first energy level. Ne, Ar, Kr,
   Xe, and Ra all have eight electrons (two *s* and six *p*) in their highest
   occupied energy level. All their *p* orbitals are filled.

EXAMPLE 9   Using the periodic table (Figure 14-10), list all the elements in
which (a) the 3*d* orbitals are being filled and (b) the 4*d* orbitals are being
filled.

Solution: The elements in which the 3*d* orbitals are filling are found in period
4, and those in which the 4*d* orbitals are filling are in period 5:

   a. 3*d*:   Sc, Ti, V, Cr, Mn, Fe, Co, Ni, Cu, Zn
   b. 4*d*:   Y, Zr, Nb, Mo, Tc, Ru, Rh, Pd, Ag, Cd   ∎

EXAMPLE 10   Using the periodic table, list all the elements in which the 4*f*
orbitals are being filled.

Solution: These are the lanthanoid series of elements:

   4*f*:   Ce, Pr, Nd, Pm, Sm, Eu, Gd, Tb, Dy, Ho, Er, Tm, Yb, Lu   ∎

The periodic table is very useful to chemists because it allows them to
correlate the properties of the elements. A simple example of this correlation
is shown in Figure 14-12.

In Figure 14-12, the periodic table is divided into three areas—metals,
nonmetals, and semimetals. **Metals** have a characteristic luster (they shine
when polished), conduct electricity very well, and aren't brittle (they can be
bent somewhat without breaking). **Nonmetals** usually don't conduct electricity very well, do not shine like metals, and are generally rather brittle as
solids. Many nonmetals are gases. **Semimetals** have properties that are intermediate between those of metals and nonmetals.

In Figure 14-12, the heavy zigzag black line going down through the
dark blue boxes is the formal separation of the periodic table into metals and
nonmetals. The dark blue boxes represent the semimetals.

Another useful feature of the periodic table is that we can tell at a
glance the number of electrons in the highest occupied energy level for the

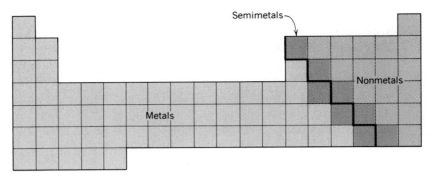

FIGURE 14-12 Metals, nonmetals, and semimetals.

"A" group elements (IA and IIA on the left, IIIA through VIIIA on the right). The Roman numeral of the group tells us the number of electrons in the highest occupied energy level. Group VIIIA elements have eight electrons in their highest energy level; the only exception is helium, which has two electrons in its highest energy level.

Information about the number of electrons in the highest occupied energy level is useful because it is these "outer" electrons that are responsible for chemical bonding. These "outer" electrons are called **valence electrons,** and the highest occupied energy level is usually called the the **valence shell**.

EXAMPLE 11   How many valence electrons are there in the valence shell of the following groups in the periodic table: IA, IVA, VIIA, VIIIA?

Solution:

IA   one valence electron
IVA   four valence electrons
VIIA   seven valence electrons
VIIIA   eight valence electrons (except for helium, which has two valence electrons)

NOTE: The elements in group VIIIA all have filled p orbitals in their valence shell (except for helium, which has a filled s orbital in its valence shell, the first energy level). These filled orbitals, with all electrons paired, are very stable; elements in group VIIIA are particularly unreactive chemically. However, in 1962 Niel Bartlett (born 1932), then working at the University of British Columbia, prepared the first noble gas compound, which contained xenon. Since then, many compounds of krypton, xenon, and radon have been prepared. No compounds of helium, neon, and argon have ever been made. ▪

The number of valence electrons in the transition metals and the lanthanoid and actinoid series cannot always be determined by looking at the periodic table and will not be discussed in this book.

APPENDIX 14-1
ORBITAL SPACING IN MULTIELECTRON ATOMS

THE ENERGY RELATIONSHIP OF ORBITALS IN A ONE-ELECTRON ATOM IS DIFFERENT FROM THAT IN MULTIELECTRON ATOMS

To understand better the arrangement of the orbitals in a many-electron atom, we must first describe the orbital arrangement in a one-electron atom such as the hydrogen atom. The hydrogen atom has only one electron, but the space for the orbitals is still around the nucleus. It is like an apartment house with only one occupied apartment on the first floor.

The orbital arrangement in the hydrogen atom is shown in Figure 14-13. All the orbitals in each energy level have the same energy. In contrast, as we saw Figure 14-1, in a multielectron atom, different types of orbitals (i.e., the s, p, d, and f orbitals) in an energy level have different energies. To explain this difference, we will use an analogy with gravity.

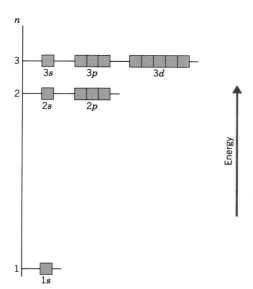

FIGURE 14-13 The orbital diagram of the hydrogen atom.

AS YOU LIFT AN OBJECT IN A GRAVITATIONAL FIELD, IT GAINS ENERGY

If you lift an object above the surface of the earth, it gains energy. The higher you lift it, the more energy it has. How can you tell that it has more energy? Just drop it. The higher it has been lifted, the more damage it can do when it hits the ground. So we have our first energy principle: *The farther away from the surface an object is lifted, the more energy it has.*

Now let's shift to the moon. The gravitational field on the moon is one-sixth that on earth. (This means that if you weigh 180 lb on earth, you weigh only 30 lb on the moon.) Now lift the same object we used on earth above the surface of the moon. Again, the higher you lift the object, the more energy it has. But since the moon has only one-sixth the gravity force of the earth, the object lifted 60 ft on the moon will have the same energy that it would have if lifted only 10 ft on the earth. So we have our second energy principle: *The stronger the gravitational field, the more energy an object has when lifted in that gravitational field.*

HOW DO GRAVITATIONAL FIELDS APPLY TO ELECTRONS IN ATOMS?

In an atom, we have a positive nucleus and negative electrons. The gravitational field of the nucleus is extremely small and can be ignored. But the positive electrical charge of the nucleus is large and attracts the negative electrons. So, even though gravity isn't important in the atom, we can use the *analogy* of gravity on both the earth and the moon to help us understand energy relationships in an atom.

THE ENERGY OF AN ELECTRON DEPENDS ON THE NUCLEAR CHARGE AND HOW FAR THE ELECTRON IS FROM THE NUCLEUS

Since the positive nucleus attracts the negative electrons, our two energy principles listed above apply. We will restate them in terms of electrical attraction.

1. The farther away an electron is from the nucleus, the higher its energy.
2. The stronger the positive nuclear charge that an electron "feels," the higher the energy of that electron around the nucleus.

We can use these principles to help explain the fact that in a multielectron atom, $2p$ orbitals have a higher energy than $2s$ orbitals. But first we must introduce the concept of shielding.

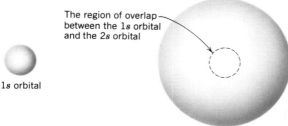

The region of overlap between the 1s orbital and the 2s orbital

1s orbital

2s orbital

FIGURE 14-14 The overlap between the 1s orbital and the 2s orbital.

INNER ELECTRONS CAN SHIELD OUTER ELECTRONS FROM THE NUCLEUS

Suppose that you are sitting on a 2s electron in a boron atom. Remember that a boron atom has the following electron configuration: $1s^2 2s^2 2p^1$. The 2s electrons are at a higher energy than the 1s electrons because they are, on the average, farther away from the nucleus than the 1s electrons. Therefore, the 2s electrons don't feel the full positive charge of the boron nucleus. The two 1s electrons get in the way and neutralize some of the positive nuclear charge. They partially *shield* the 2s electrons from the nucleus. The 2s electrons don't "feel" the full +5 nuclear charge of boron; they feel a somewhat smaller nuclear charge. But they *don't* "feel" only a +3 charge as you might expect if the two 1s electrons neutralized two positive nuclear charges. The reason is that the 2s electrons aren't always outside the 1s orbital. They penetrate the 1s orbital and spend some of their time close to the nucleus. There is a region of *overlap* between the 1s and the 2s orbitals. But on the *average* the 2s electrons are farther away from the nucleus than the 1s electrons. This is shown in Figure 14-14. So the shielding by the two 1s electrons is not complete—it is only a partial shielding.

The one 2p electron in boron is also only partially shielded from the nucleus by the two 1s electrons. The 2p electron can also penetrate the 1s orbital and spend some of its time close to the nucleus. This is shown in Figure 14-15.

However, the average distance between the nucleus and the 2p electron is greater than the average distance between the nucleus and the 2s

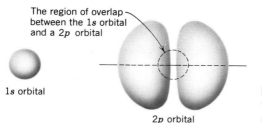

The region of overlap between the 1s orbital and a 2p orbital

1s orbital

2p orbital

FIGURE 14-15 The overlap between the 1s orbital and a 2p orbital.

electrons, and so the penetration of the $2p$ electron is less than the penetration of the $2s$ electrons. This is reasonable because their shapes are so different. Thus the nuclear charge that the $2p$ electron "feels" is less than the charge that the $2s$ electrons "feel." So, according to our energy principles, the energy of the two kinds of electrons is different.

THERE ARE ALSO ELECTRON—ELECTRON INTERACTIONS THAT AFFECT THE ENERGY OF THE ELECTRONS

There is an additional effect that can change the energy of the electrons. It is the repulsion that the electrons have for each other. Remember that electrons are negative, and negative charges repel each other. The situation is similar to the effect that a planet in our solar system has on the planet nearest to it. Their gravitational attraction for each other actually disturbs their orbits and changes their energy. With electrons it is not an attraction, of course, but a repulsion. The $2s$ and the $2p$ electrons repel one another, and both are repelled by the $1s$ electrons.

THE $2p$ ELECTRON HAS A HIGHER ENERGY THAN THE $2s$ ELECTRON

It turns out that when all these effects are taken into account, the $2p$ electron in boron has a slightly higher energy than the $2s$ electrons. We can generalize this and say that the three $2p$ orbitals (which all have the same energy because their shapes are the same) have a slightly higher energy than the $2s$ orbital.

The same kinds of arguments also apply to higher energy levels and different kinds of orbitals.

PROBLEMS

KEYED PROBLEMS

1. What would a p orbital in the fifth energy level be called?

2. What would a d orbital in the sixth energy level be called?

3. Using Figure 14.2, list all the elements that are formed by the filling of the $4s$ orbital, the $3d$ orbitals, and the $5f$ orbitals.

4. Using an orbital diagram like Figure 14-1, draw in the electrons for the element phosphorus, P, atomic number 15. Also, write the electron configuration for phosphorus.

5. Draw a horizontal orbital diagram for silicon.

6. Which of the following horizontal orbital diagrams is correct? Explain what is wrong with the incorrect ones.

a. $\frac{\uparrow\downarrow}{1s^2}$ $\frac{\uparrow\downarrow}{2s^2}$ $\frac{\uparrow}{2p_x^1}$ $\frac{\downarrow}{2p_y^1}$ $\frac{\downarrow}{2p_z^1}$

b. $\frac{\uparrow\downarrow}{1s^2}$ $\frac{\uparrow\downarrow}{2s^2}$ $\frac{\uparrow}{2p_x^1}$ $\frac{\uparrow}{2p_y^1}$ $\frac{\uparrow}{2p_z^1}$

c. $\frac{\uparrow\downarrow}{1s^2}$ $\frac{\uparrow\downarrow}{2s^2}$ $\frac{\uparrow}{2p_x^1}$ $\frac{\uparrow\downarrow}{2p_y^2}$ $\frac{}{2p_z}$

d. $\frac{\uparrow\downarrow}{1s^2}$ $\frac{\uparrow\downarrow}{2s^2}$ $\frac{}{2p_x}$ $\frac{\uparrow}{2p_y^1}$ $\frac{\uparrow}{2p_z^1}$

7. List the elements that are in the following groups in the periodic table: groups IIIA, VB, and VIA.

8. List the elements that are in periods 4 and 5 of the periodic table.

9. Using the periodic table, Figure 14-10, list all the elements that are formed by having their $5d$ orbitals filled.

10. Using the periodic table, Figure 14-10, list all the elements that are formed by having their $5f$ orbitals filled.

11. How many valence electrons are there in the elements in the following groups in the periodic table: IIA, VA, and VIA?

SUPPLEMENTAL PROBLEMS

12. Explain what the following symbols mean.

   a. $3p_x^2$ c. $2s^1$
   b. $4f^7$ d. $4d^{10}$

13. Name the elements that correspond to the following electron configurations.

   a. $1s^2 2s^1$ d. $1s^2 2s^2 2p^6 3s^2 3p_x^2 3p_y^1 3p_z^1$
   b. $1s^2 2s^2 2p_x^1 2p_y^1$ e. $1s^2 2s^2 2p^6 3s^2 3p^6 4s^1$
   c. $1s^2 2s^2 2p^6 3s^1$ f. $1s^2 2s^2 2p^6 3s^2 3p^6 4s^2 3d^7$

14. Write a complete horizontal orbital diagram for each of the following elements.

   a. C   c. Mg   e. Ar
   b. O   d. P   f. V

15. List the chemical symbols of all the elements in the following chemical families or regions of the periodic table.

    a. alkali metals   c. halogens   e. lanthanoid series
    b. alkaline earth metals   d. noble gases   f. actinoid series

16. Using Figures 14-12 and 14-10, list the names and chemical symbols of all the elements indicated as semimetals.

17. Write out the names and chemical symbols of all the transition elements from atomic number 21 to atomic number 30.

18. Write out the names and chemical symbols of all the elements in group IVA of the periodic table.

19. The following are some horizontal orbital diagrams. Which ones are correct and which are incorrect? Explain the incorrect ones.

a. $\dfrac{\uparrow\downarrow}{1s^2}\quad\dfrac{\uparrow}{2s^1}\quad\dfrac{\uparrow}{2p_x^1}\quad\dfrac{\uparrow}{2p_y^1}\quad\dfrac{\uparrow}{2p_z^1}$

b. $\dfrac{\uparrow\downarrow}{1s^2}\quad\dfrac{\uparrow\downarrow}{2s^2}\quad\dfrac{\uparrow\downarrow}{2p_x^2}\quad\dfrac{}{2p_y}\quad\dfrac{}{2p_z}$

c. $\dfrac{\uparrow\downarrow}{1s^2}\quad\dfrac{\uparrow\downarrow}{2s^2}\quad\dfrac{\uparrow\downarrow\uparrow}{2p_x^3}\quad\dfrac{\uparrow}{2p_y^1}\quad\dfrac{}{2p_z}$

d. $\dfrac{\uparrow\downarrow}{1s^2}\quad\dfrac{\uparrow\downarrow}{2s^2}\quad\dfrac{\downarrow}{2p_x^1}\quad\dfrac{\downarrow}{2p_y^1}\quad\dfrac{\downarrow}{2p_z^1}$

e. $\dfrac{\uparrow\downarrow}{1s^2}\quad\dfrac{\uparrow\downarrow}{2s^2}\quad\dfrac{}{2p_x}\quad\dfrac{\uparrow}{2p_y^1}\quad\dfrac{}{2p_z}$

15

THE CHEMICAL BOND

In Chapters 13 and 14 we described atomic orbitals as the space around the atom in which electrons exist. In this chapter we will briefly discuss the space around a molecule in which electrons exist. We will also discuss chemical bonds. The discussion will be limited to atoms of the first three periods of the periodic table because the bonding of these atoms is easier to explain. This chapter, then, will try to answer the question "How and why do atoms combine using chemical bonds to form molecules?"

15-1
THE COVALENT BOND OF THE HYDROGEN MOLECULE
IS THE SIMPLEST COVALENT BOND

The hydrogen molecule is the simplest molecule, and thus we will use it to introduce the basic concepts of chemical bonding. A hydrogen molecule is formed, as we mentioned in Chapter 3, when two hydrogen atoms combine to form a hydrogen molecule according to the equation

$$H + H \rightarrow H_2$$

We also said that this reaction gives off a lot of heat, so the H_2 molecule has less energy than the two separate H atoms. (Remember from Chapter 2 that all substances in nature like to be at the lowest possible energy. This is true of electrons, as we saw in Chapters 13 and 14, and it is also true of atoms and molecules.) Since the energy of the newly formed H_2 molecule is less than the energy of the two separate H atoms, the H atoms "want" to combine. They are at a lower energy state as a molecule.

(Opposite) A large-scale integrated circuit made from a silicon chip. These devices have made modern computers and calculators possible.

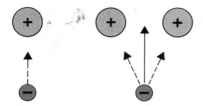

FIGURE 15-1 The dashed arrows represent the force of attraction of the electron to one nucleus. The full arrow represents the total force of attraction of the electron to two nuclei. The longer the arrow, the greater the force.

Why should the H_2 molecule be at a lower energy than the two separate H atoms? In Chapter 13, we said that electrons like to be as close as possible to the nucleus. After all, the nucleus is positive and it attracts the negative electrons.

Now think of the situation whereby an electron is attracted to two nuclei at the same time. It seems reasonable that the two nuclei will have a greater attraction for an electron than will one nucleus. This is shown in Figure 15-1.

Thus the electrons would rather be in the space between the two nuclei of a molecule than surrounding only one nucleus of an atom. The greater attraction of two nuclei causes the electron to be at a lower energy than it would be were it surrounding only one nucleus. The energy that is lost by the electron appears as heat in a chemical reaction.

Each H atom has one electron, so the H_2 molecule has two electrons. What about the second electron? Where does it go? The second electron also goes into the space between the two nuclei, also getting its energy lowered. Figure 15-2 will give you some idea of what this looks like.

Notice that the electrons aren't found between the nuclei all of the time, just *most* of the time. This arrangement, in which two electrons are found between two nuclei most of the time, is called a **molecular orbital.**

The reaction $H + H \rightarrow H_2$ gives off heat because the electrons have lost energy in going from the individual atoms to the molecule.

You might wonder, "If two H atoms combine so readily, why not combine three or even more H atoms? Then the electrons would have even more nuclei to go around, and their energy would be lowered even further." To answer this, we have to go back to the discussion of *spin* in Chapter 13. We said there that "Two electrons can be in the same atomic orbital only if their spins are different." This rule also applies to molecular orbitals

FIGURE 15-2 Formation of the H_2 molecule. The relative darkness of the shading gives an idea of where the "electron cloud" is densest—between the two nuclei. This is where the electrons spend most of their time.

because electrons are in the same region of space in a molecular orbital. **Two electrons can be in the same molecular orbital only if their spins are different.** Now you can see why more H atoms don't add onto the H_2 molecule—the additional electrons could *not* join the first two in their molecular orbital because there is already a spin-up and a spin-down electron in that molecular orbital. The additional electrons would have to go into other molecular orbitals that are farther from the nucleus and higher in energy. There is no room for more molecular orbitals at the same energy as the first one. These higher-energy molecular orbitals do *not* lower the energy of the electrons, and so there is no reason for more H atoms to add to H_2.[1]

The kind of chemical bond that we have been discussing, whereby the electrons spend most of their time between two nuclei, is called a covalent chemical bond, or simply a **covalent bond.** *In a covalent bond, two electrons are shared between two nuclei.* Before we can discuss the covalent bond in more detail, however, we must discuss the Lewis formulas of atoms and molecules.

15-2
LEWIS FORMULAS OF ATOMS IN THE FIRST AND SECOND PERIODS SHOW THE VALENCE ELECTRONS

It is time-consuming to draw the fuzzy cloud picture of a covalent bond each time we want to discuss the bonding in a molecule. So chemists use a symbolic notation to show the electrons. This notation is called the Lewis electron-dot formula, or simply the **Lewis formula.** It was first developed in 1916 by Gilbert Newton Lewis (1875–1946), a professor of chemistry at the Berkeley campus of the University of California.

In Lewis formulas, each valence electron is represented by a dot. (Remember that valence electrons are electrons in the valence shell, which is the highest occupied energy level.) A covalent bond would be represented by two dots. For a hydrogen atom, we would write H·. The hydrogen atom has one valence electron. For a hydrogen molecule, we would write H:H. The two dots drawn between the atoms represent the covalent bond. They are drawn between the atoms because that is where the electrons spend most of their time.

<table>
<tr><td>H·</td><td>H:H</td></tr>
<tr><td>↑</td><td>↑</td></tr>
<tr><td>This dot represents
the 1s electron.</td><td>These dots represent the
covalent bond formed with
two electrons.</td></tr>
</table>

[1]Actually, the H_3 molecule does exist. The additional electron does go into a higher-energy molecular orbital. But because this higher-energy molecular orbital doesn't lower the energy of the electron, the H_3 molecule is very unstable. It exists only for a fraction of a second in some chemical reactions.

Gilbert Newton Lewis received his Ph.D. degree from Harvard University. In 1912, he became professor of physical chemistry at the University of California, Berkeley. His work on the relation of the electron pair to chemical bonding laid the foundation for modern theories of the chemical bond.

The reaction between two H atoms could be written with dots.

$$H\cdot\ +\ H\cdot \rightarrow H\mathbin{:}H$$

EXAMPLE 1   Write the Lewis formula for helium, He.

Solution: Helium has the electron configuration $1s^2$. So we draw two dots, that is, He:. The helium atom has two valence electrons.   ■

In the Lewis formula for helium in Example 1, we drew the dots on the same side of the symbol He. Why didn't we draw the dots separated like ·He·? The reason is that the two electrons in He are in the same atomic orbital (the $1s$ orbital), and by drawing the Lewis formula as He:, we emphasize this fact. **In a Lewis formula, two dots drawn together always represent two electrons in either an atomic orbital or a molecular orbital. The spins of these electrons are always different.**

Let's draw Lewis formulas of some other atoms. For example, lithium, Li, has three electrons, $1s^2 2s^1$. What is the Lewis formula for Li? It is simply Li· because we have drawn in only the $2s^1$ electron. The reason for this is that the $1s^2$ electrons do not take part in chemical bonding for Li. They are already in a filled energy level (the first). There is no way that they could interact with the electrons of another atom unless one of them was raised to a higher-energy atomic orbital. This would take so much energy that it just doesn't happen. Only the $2s^1$ electron of Li is available for chemical bonding. So, electrons that are available for chemical bonding are the electrons in the highest-energy occupied orbitals or valence shell. These are called, as we have said, the valence electrons. Li has one valence electron in its valence shell.

We repeat our rule for drawing Lewis formulas. **Only valence electrons are drawn as dots. Electrons in lower, filled energy levels do not take part in chemical bonding and are not indicated by the use of dots.**

EXAMPLE 2   Draw the Lewis formula for beryllium, Be.

Solution: The electron configuration for Be is $1s^2 2s^2$. Only the $2s^2$ electrons, the valence electrons, are available for bonding. The Lewis formula for Be is Be:.   ∎

In Table 15-1, we have drawn the Lewis formulas for all the atoms in period 2 of the periodic table. We will also indicate the electron configurations in Table 15-1 so you can compare them with the Lewis formulas.

TABLE 15-1

LEWIS FORMULAS OF SECOND-PERIOD ELEMENTS

ELEMENT	ELECTRON CONFIGURATION	LEWIS FORMULA[a]	GROUP NUMBER[b]
Li	$1s^2 2s^1$	Li·	IA
Be	$1s^2 2s^2$	Be:	IIA
B	$1s^2 2s^2 2p_x^1$	:B·	IIIA
C	$1s^2 2s^2 2p_x^1 2p_y^1$	:C·	IVA
N	$1s^2 2s^2 2p_x^1 2p_y^1 2p_z^1$	:N·	VA
O	$1s^2 2s^2 2p_x^2 2p_y^1 2p_z^1$	:O·	VIA
F	$1s^2 2s^2 2p_x^2 2p_y^2 2p_z^1$	:F·	VIIA
Ne	$1s^2 2s^2 2p_x^2 2p_y^2 2p_z^2$	:Ne:	VIIIA

[a]The number of dots is the number of electrons of all the orbitals in the second energy level.

[b]Note that the group number is the same as the number of dots (valence electrons) in the Lewis formula.

EXAMPLE 3   Draw the Lewis formulas for Na and P.

Solution: The electron configurations are

$$\text{Na:} \quad 1s^2 2s^2 2p_x^2 2p_y^2 2p_z^2 3s^1$$
$$\text{P:} \quad 1s^2 2s^2 2p_x^2 2p_y^2 2p_z^2 3s^2 3p_x^1 3p_y^1 3p_z^1 .$$

Since the first and second energy levels are filled and do not take part in chemical bonding for Na and P, only the third-energy-level electrons make up the valence shell; only these electrons will be shown as dots.

$$\text{Na} \cdot \quad : \dot{\text{P}} \cdot$$

Note the similarity between these Lewis formulas and those of Li and N. Compare Li with Na and compare N with P. Find these elements in the periodic table (Figure 14-10). Elements in the same *group* of the periodic table have the same Lewis formulas.   ■

15-3
COVALENT BONDING FOR SOME MOLECULES IN THE FIRST AND SECOND PERIODS: THE ELEMENTS C, N, O, AND F OBEY THE OCTET RULE

In Section 15-2 we said that the Lewis formula for the molecule H_2 was H:H, where the two dots represent the covalent bond formed from two electrons. Helium, on the other hand, cannot form bonds because the atom already has two electrons in the $1s$ orbital and there is no room for any more electrons in that orbital. Helium could form bonds only if one of the electrons went into an orbital in the second energy level. This would take so much energy that it doesn't happen.

Most atoms in the periodic table do form compounds (He, Ne, and Ar are the exceptions), and we would like to be able to draw Lewis formulas of these compounds in order to understand bonding better. In this section we will develop the rules to do just that. Most of the ideas we will develop were first stated in the 1930s by Linus Pauling, John C. Slater, and others.

We have already seen that each hydrogen atom can only have two electrons around it. But atoms in the second period of the periodic table, from Li to Ne, have electrons in the second energy level. There is enough room for up to eight electrons around each of these atoms.

Since electrons lower their energy by being around more than one nucleus, it seems reasonable that an atom would form as many bonds as possible. Atoms in the second period can form *up to four* covalent bonds (each with two electrons) with other atoms. This gives each atom eight

Linus Carl Pauling received his Ph.D. degree from the California Institute of Technology and was a professor there until his retirement. His book *The Nature of the Chemical Bond,* first published in 1939, had a profound effect on a generation of chemists. Pauling received the Nobel Prize in Chemistry in 1954 for his discovery of the alpha helix structure of protein and the Nobel Peace Prize in 1962 for his work against nuclear tests in the atmosphere.

electrons around it, the maximum number for which there is room. Some of these atoms form fewer than four bonds. We shall see why.

The first element in the second period of the periodic table is lithium. Li has one valence electron, and its Lewis formula is Li·. Lithium can react with a hydrogen atom to form lithium hydride, LiH, according to the equation

$$\text{Li·} + \text{H·} \rightarrow \text{Li:H}$$

Lithium doesn't form more than one bond. Apparently the positive charge on the Li nucleus is strong enough to attract only one additional electron.

The second element in the second period is beryllium. Beryllium has two valence electrons, and its Lewis diagram is Be:. You might think that Be wouldn't form a compound with H atoms because both of the valence electrons in Be are paired in the $2s$ orbital. But Be actually does react with two H atoms to form beryllium hydride, BeH_2:

$$\text{Be:} + 2\text{H·} \rightarrow \text{H:Be:H}$$

How do we explain this? Well, the valence electrons in beryllium's $2s$ orbital can *separate* while Be is reacting, giving two unpaired electrons. The Lewis formula would then be ·Be·. If you want to, you can think of one of these electrons as going into one of the vacant $2p$ orbitals. You can see that there is room for the electron in unoccupied p orbitals of the second energy level. These vacant p orbitals are only slightly higher in energy than the $2s$ orbital. Notice that now the electrons in the Be atom are as far apart from one another as possible, on opposite sides of the atom. Since electrons repel one another, it is only natural for them to try to get as far apart from one another as possible. Thus the BeH_2 molecule is a **linear** molecule, which means that all the atoms are in a straight line.

Why do the electrons in Be separate in this way? The reason must have to do with energy. If the electrons separate, then Be can share two electrons from two H atoms. And since the $2p$ orbitals are only a little higher in energy than the $2s$ orbital, the overall process lowers the energy so much that it pays for the electrons to separate. (Electrons may be "lazy," but they "know" a good situation when they "see" one.)

We can now write the reaction between Be and two H atoms as

$$\cdot Be \cdot \ + \ 2H \cdot \ \rightarrow \ H:Be:H$$

Beryllium forms only two bonds. Apparently, the positive charge on the nucleus of the Be atom is strong enough to attract only two additional electrons.

The third element in the second period is boron. The Lewis formula for boron is $:B\cdot$ because boron has three valence electrons. You might think that boron would combine with one H atom to form BH according to the following equation:

$$:B \cdot \ + \ H \cdot \ \rightarrow \ :B:H$$

Alternatively, using the ideas just developed above for BeH_2, you might think that boron would combine with *three* H atoms to form BH_3:

$$\dot{B}. \ + \ 3H \cdot \ \rightarrow \ \begin{matrix} H \\ \cdot \ddot{B} \cdot \\ H \quad H \end{matrix}$$

This is actually what happens, although the BH_3 molecule, called borane, is unstable and exists only for short times during certain chemical reactions. The explanation for the formation of BH_3 is similar to the case of the BeH_2 molecule discussed above. The valence electrons in boron's $2s$ orbital (the paired dots) can separate while the boron atom is reacting, giving three unpaired electrons. The Lewis formula would then be $\dot{B}.$. Again, you can think of a $2s$ electron as going into one of the vacant $2p$ orbitals. Notice that now the electrons in boron are as far apart from one another as possible, at the corners of an equilateral triangle.

Since boron now has three unpaired electrons, it is easy to see why three H atoms can bond to the boron atom, forming three covalent bonds:

$$\overset{\displaystyle .}{\underset{\displaystyle .}{B}} \ + \ 3H\cdot \ \rightarrow \ \ \overset{\displaystyle H}{\underset{\displaystyle H \quad H}{\cdot \overset{\displaystyle ..}{B}\cdot}}$$

The BH_3 molecule is a **triangular planar** molecule, which means that all the atoms are in the same plane and that the H atoms are at the corners of an equilateral triangle. A plane is a two-dimensional surface, such as a table top.

The boron atom in BH_3 has six electrons around it. It must be that the positive charge on the boron nucleus is strong enough to attract only three extra electrons. Boron *can* form compounds in which it has eight valence electrons. In these compounds the boron atom gets two electrons from another atom or molecule. We will not discuss these compounds in this book.

Carbon is the fourth element in the second period. Carbon has four electrons in its valence shell, and its Lewis formula is $:\overset{\displaystyle .}{C}\cdot$. You might think that carbon would form a species like CH_2 according to the equation[2]

$$:\overset{\displaystyle .}{C}\cdot \ + \ 2H\cdot \ \rightarrow \ \overset{\displaystyle H}{:\overset{\displaystyle ..}{C}:H}$$

Carbon actually forms stable molecules like methane, CH_4. The explanation for this is similar to that mentioned for Be and B. The paired electrons in the $2s$ orbital of carbon split up to give

$$\cdot\overset{\displaystyle .}{\underset{\displaystyle .}{C}}\cdot$$

Now four H atoms can easily combine with a C atom to give four covalent bonds:

$$\cdot\overset{\displaystyle .}{\underset{\displaystyle .}{C}}\cdot \ + \ 4H\cdot \ \rightarrow \ \overset{\displaystyle H}{\underset{\displaystyle H}{H:\overset{\displaystyle ..}{C}:H}}$$

Carbon now has eight electrons around it, the maximum number for which there is room in the second energy level.

Actually, the four separated electrons in carbon are not planar, that is, in the same plane. They are as far apart as they can be from one another in three dimensions. The geometric arrangement of $\cdot\overset{.}{\underset{.}{C}}\cdot$ and CH_4 is a **regular**

---

[2]The species CH_2, called methylene, does exist. It is *very* reactive and exists only for fractions of a second during certain chemical reactions. This type of reactive species is called a **radical.**

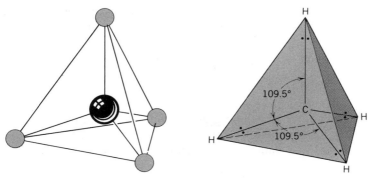

FIGURE 15-3 Two diagrams of the methane (CH_4) molecule. The geometric structure shown is a **regular tetrahedron**. All the edges and faces of a regular tetrahedron are equal. The angle between the H atoms is 109.5°, which is called the **tetrahedral angle**.

tetrahedron, as shown in Figure 15-3. The corners of a regular tetrahedron are the farthest apart in space that four points can get from one another at a given distance from a central point. Since electrons repel one another, it is reasonable that they arrange themselves as far from one another as they can.

At this point, we should mention that equations like

$$\cdot\overset{\cdot}{\underset{\cdot}{C}}\cdot + 4H\cdot \rightarrow H\overset{H}{\underset{H}{:\overset{\cdot\cdot}{C}:}}H$$

are *not* the chemical reactions that chemists use to make CH_4, BH_3, and so on. These equations are written here to help you see how the electrons act during the formation of a chemical bond. These equations are more like ''thought'' equations than real ones.

Nitrogen is the fifth element in the second period. Nitrogen has five electrons in its valence shell, and its Lewis formula is $:\overset{\cdot}{N}\cdot$. Nitrogen forms ammonia, NH_3, by combining with three H atoms:

$$:\overset{\cdot}{N}\cdot + 3H\cdot \rightarrow \overset{H}{\underset{\overset{\cdot\cdot}{H}}{:N:}}H$$

There are eight electrons around the N atom in NH_3. Six electrons are in three covalent bonds, and two electrons are in a **nonbonded pair:**

$$\text{Nonbonded pair} \rightarrow \overset{H}{\underset{\overset{\cdot\cdot}{H}}{:N:}}H$$

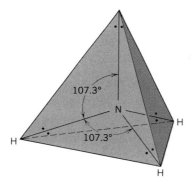

FIGURE 15-4 A diagram of the ammonia (NH₃) molecule. The nonbonded pair of electrons is at the top of the pyramid. The geometric figure formed by just the N and H atoms is called a **trigonal pyramid**. The angle between the H atoms is 107.3°. This angle is smaller than in CH₄ because the nonbonded pair of electrons pushes away the H atoms more than the bonded pair in CH₄ does.

This nonbonded pair does *not* split up, unlike the electron pairs in beryllium, boron, and carbon. There is room for only eight electrons in the second energy level of the nitrogen atom. The only place an electron could go if the pair split up is into a vacant orbital in the third energy level. This would require so much energy that it doesn't happen.

The structure of NH_3 is a **trigonal pyramid** (a pyramid whose faces are triangles), as shown in Figure 15-4. This structure is similar to a tetrahedron if we include the nonbonded pair of electrons at the top of the pyramid. As usual, the electrons are as far away from one another as possible. Actually, the nonbonded electron pair "wants" to be a little farther away from the rest, and it pushes the three H atoms down a bit. This is because the nonbonded electrons are only influenced by one positive nucleus, unlike bonded electrons, which are influenced by two positive nuclei. Thus, nonbonded electrons repel bonded electrons to a greater extent than bonded electrons repel other bonded electrons. That's why the angle between the H atoms in NH_3 is 107.3° instead of the tetrahedral angle of 109.5° in CH_4.

Oxygen has six valence electrons, and its Lewis formula is $:\ddot{O}\cdot$. Oxygen can combine with two H atoms to form water:

$$:\ddot{O}\cdot \, + \, 2H\cdot \, \rightarrow \, :\ddot{O}:H$$
$$\phantom{:\ddot{O}\cdot \, + \, 2H\cdot \, \rightarrow \, :\ddot{O}:}\dot{H}$$

In water, oxygen has eight electrons around it, four in two nonbonded pairs and four in two covalent bonds with two H atoms. The structure of the water molecule is shown in Figure 15-5.

Fluorine, the next-to-last element in period 2, has seven valence electrons, and its Lewis formula is $:\ddot{F}\cdot$. Fluorine can combine with one H atom to form hydrogen fluoride, HF:

$$H\cdot \, + \, :\ddot{F}\cdot \, \rightarrow \, H:\ddot{F}:$$

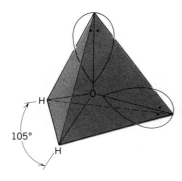

FIGURE 15-5 A diagram of the water (H_2O) molecule. The two pairs of nonbonded electrons point to the corners of a tetrahedron. The angle between the H atoms is 105°. This angle is smaller than in NH_3 because the two pairs of nonbonded electrons push away the H atoms more than the one pair in NH_3 does. Water is an **angular** or **bent** molecule.

In HF the fluorine atom has eight electrons around it, six in three nonbonded pairs and two in one covalent bond with the H atom. HF is a **linear** molecule.

Neon, the last element in period 2, has eight valence electrons, and its Lewis formula is :Ṇe:. Since Ne already has eight electrons in the second energy level, there is no room for any more. Neon has never been observed to form molecules.

If we study the previous examples, we can find a rule for determining the number of covalent bonds that an atom in the second period of the periodic table can form. The rule is: **The number of covalent bonds an atom will form is equal to the maximum number of half-filled orbitals the atom can have.** Let's study Table 15-2 to see whether our rule is consistent with our previous discussion. As you can see from Table 15-2, this rule *is* consistent with everything that we said in our previous discussion.

Notice in Table 15-2 that the period-2 elements, C, N, O, and F have eight electrons around them when they form compounds. Their valence shell is now similar to that of neon, a noble gas that is unreactive chemically. The valence shell configuration of a noble gas is a low-energy configuration. This is one of the reasons why certain period-2 elements attain this configuration when they form compounds. It is so common for these atoms to have eight electrons around them in compounds that this tendency is referred to as the **octet rule** ("oct" means eight).

Students sometimes ask whether it matters where the dots go in a Lewis formula. Are :Ċ·, ·Ċ:, and Ċ· the same? Yes, all these are equivalent. So what should you draw? The best idea is to try to make your drawings like those of either the book or your instructor. But note that :Ċ· is not the same as ·Ċ·. You should keep the correct number of lone electrons and paired electrons in your Lewis formulas.

TABLE 15-2

THE NUMBER OF COVALENT BONDS IN SECOND-PERIOD ELEMENTS

ATOM	NUMBER OF VALENCE ELECTRONS	LEWIS FORMULA	REARRANGED LEWIS FORMULA	MAXIMUM NUMBER OF HALF-FILLED ORBITALS[a]	NUMBER OF COVALENT BONDS	TYPICAL COMPOUND WITH H ATOMS	GEOMETRY
Li	1	Li·	Li·	1	1	Li:H	linear
Be	2	Be:	·Be·	2	2	H:Be:H	linear
B	3	:B·	·Ḃ·	3	3	H:B̈:H (with H above)	triangular planar
C	4	·C̈·	·Ċ·	4	4	H:C̈:H (with H above)	tetrahedral
N	5	·N̈·	·N̈·	3	3	:N̈:H (with H above)	trigonal pyramid
O	6	·Ö·	·Ö·	2	2	:Ö:H	angular or bent
F	7	:F̈·	:F̈·	1	1	H:F̈:	linear
Ne	8	:N̈e:	:N̈e:	0	0	—	—

[a]The maximum number of half-filled orbitals is the same as the number of lone dots in the column "rearranged Lewis formula."

15-4
COVALENT BONDING IN OTHER MOLECULES OF THE SECOND-PERIOD ELEMENTS: THE ELEMENTS C, N, O, AND F OBEY THE OCTET RULE

Elements in the second period of the periodic table can form many different molecules. We will look at a few simple ones as examples of covalent bonding.

Two fluorine atoms can combine to form a fluorine molecule, F_2, as follows:

$$:\ddot{F}\cdot + :\ddot{F}\cdot \rightarrow :\ddot{F}:\ddot{F}:$$

The unpaired electrons in each F atom join to form a covalent bond. This is shown in the following diagram, where we have drawn arrows to indicate the electrons that combine to form the covalent bond. One covalent bond connecting two atoms is called a **single bond:**

$$:\ddot{F}\cdot \rightarrow \leftarrow \cdot\ddot{F}:$$

Notice that one F atom has been turned around so that the two unpaired electrons face each other. This was done so you could easily see how they combine.

Two nitrogen atoms can combine to form a nitrogen molecule, N_2, as follows:

$$:\dot{N}\cdot + :\dot{N}\cdot \rightarrow :N:::N:$$

Here we have *three* covalent bonds connecting the two N atoms. Three covalent bonds connecting two atoms is called a **triple bond:**

$$:\dot{N}\cdot \overset{\rightarrow}{\underset{\rightarrow}{\rightrightarrows}} \overset{\leftarrow}{\underset{\leftarrow}{\leftleftarrows}} \cdot\dot{N}:$$

Again, one N atom has been turned around so that the unpaired electrons face each other.

Carbon can combine with two oxygen atoms to form carbon dioxide, CO_2, as follows:

$$\cdot\dot{C}\cdot + 2 :\ddot{O}\cdot \rightarrow :\ddot{O}::C::\ddot{O}:$$

There are two covalent bonds to each oxygen atom in CO_2. Two covalent bonds connecting two atoms is called a **double bond:**

$$:\ddot{O}\cdot \overset{\rightarrow}{\rightrightarrows} \Leftarrow \cdot\dot{C}\cdot \overset{\rightarrow}{\rightrightarrows} \Leftarrow \cdot\dot{O}:$$

Two oxygen atoms can combine to form an oxygen molecule, O_2. You might think that you could draw its Lewis formula as $\ddot{:}O::\ddot{O}:$, but this is really wrong. Experiments show that the oxygen molecule has *two unpaired electrons*. So you might try drawing the Lewis formula as $:\ddot{O}:\ddot{O}:$. Now there are two unpaired electrons, seven electrons around each oxygen atom, and a single bond connecting the two atoms. But experiments and theory suggest that the bond between the oxygen atoms is a double bond. O_2 is one of the few molecules that cannot be represented by a Lewis formula. Lewis formulas work *most* of the time, but *not* all the time.

We have not discussed the molecules Li_2, B_2, and C_2. These molecules are unstable and exist only as gases at very high temperatures. In Li_2 there is a single bond joining the two Li atoms, as you might expect. The bonding in B_2 and C_2 cannot be described by Lewis formulas. But in case you are curious, in B_2 there is a single bond, and in C_2 there is a double bond. If you tried to use Lewis formulas and followed the rules, you would probably end up with a triple bond for B_2 and a quadruple bond for C_2. The predictions from the Lewis formulas are wrong.

The molecule Be_2 has never been found to exist. As you study more chemistry, you may learn a theory of bonding called **molecular orbital theory.** This theory is more involved than the one we have been discussing, but it does explain the bonding in molecules like O_2, B_2, and C_2 as well as explaining why Be_2 does not exist.

We can summarize the rules for forming covalent bonds for period-2 elements.

1. Each covalent bond contains two electrons whose spins are different.
2. The number of covalent bonds an atom will form is equal to the maximum number of half-filled orbitals that the atom can have.
3. The maximum number of electrons that a period-2 atom can have in its second energy level is eight. Period-2 atoms will form covalent bonds with other atoms in order to get eight electrons in their second energy level. Exceptions: Li and Be can have only two and four electrons, respectively, in their valence shell. Boron can form molecules with both six electrons and eight electrons in its valence shell.
4. The statement that many atoms tend to have eight valence electrons in their valence shell when they form compounds is called the **octet rule.**

EXAMPLE 4   Draw Lewis formulas for the following molecules: NF_3, CF_4, and OF_2.

Solution: We will write the "thought" equations to make things clearer:

$$NF_3 \quad :\overset{\cdot}{\underset{\cdot}{N}}\cdot \; + \; 3 \; :\overset{\cdot\cdot}{\underset{\cdot\cdot}{F}}\cdot \; \rightarrow \; \begin{array}{c} :\overset{\cdot\cdot}{\underset{\cdot\cdot}{F}}: \\ :N:\overset{\cdot\cdot}{\underset{\cdot\cdot}{F}}: \\ :\overset{\cdot\cdot}{\underset{\cdot\cdot}{F}}: \end{array}$$

$$CF_4 \quad \cdot\overset{\cdot}{\underset{\cdot}{C}}\cdot \; + \; 4 \; :\overset{\cdot\cdot}{\underset{\cdot\cdot}{F}}\cdot \; \rightarrow \; \begin{array}{c} :\overset{\cdot\cdot}{\underset{\cdot\cdot}{F}}: \\ :\overset{\cdot\cdot}{\underset{\cdot\cdot}{F}}:C:\overset{\cdot\cdot}{\underset{\cdot\cdot}{F}}: \\ :\overset{\cdot\cdot}{\underset{\cdot\cdot}{F}}: \end{array}$$

$$OF_2 \quad :\overset{\cdot\cdot}{\underset{\cdot\cdot}{O}}\cdot \; + \; 2 \; :\overset{\cdot\cdot}{\underset{\cdot\cdot}{F}}\cdot \; \rightarrow \; \begin{array}{c} :\overset{\cdot\cdot}{\underset{\cdot\cdot}{O}}:\overset{\cdot\cdot}{\underset{\cdot\cdot}{F}}: \\ :\overset{\cdot\cdot}{\underset{\cdot\cdot}{F}}: \end{array} \quad \blacksquare$$

15-5
LEWIS FORMULAS USING LINES INSTEAD OF DOTS IS COMMON

Chemists find that drawing all the dots in Lewis formulas is rather tedious. So sometimes they draw lines in for the dots according to the following rules.

1. A covalent bond consisting of two electrons is replaced by a line.
2. The dots representing nonbonded electrons are sometimes left out.

For example, let's draw all the molecules discussed in the previous section with lines instead of dots (see Table 15-3).

EXAMPLE 5 Draw the Lewis formulas, as was done in Table 15-3, for the following molecules: NF_3, CF_4, and OF_2.

Solution:

LEWIS FORMULA USING LINES	LEWIS FORMULA USING DOTS AND LINES
$\begin{array}{c} F \\ \| \\ N-F \\ \| \\ F \end{array}$	$\begin{array}{c} :\overset{\cdot\cdot}{F}: \\ \| \\ :N-\overset{\cdot\cdot}{F}: \\ \| \\ :F: \end{array}$
$\begin{array}{c} F \\ \| \\ F-C-F \\ \| \\ F \end{array}$	$\begin{array}{c} :\overset{\cdot\cdot}{F}: \\ \| \\ :\overset{\cdot\cdot}{F}-C-\overset{\cdot\cdot}{F}: \\ \| \\ :F: \end{array}$
$\begin{array}{c} O-F \\ \| \\ F \end{array}$	$\begin{array}{c} :\overset{\cdot\cdot}{O}-\overset{\cdot\cdot}{F}: \\ \| \\ :F: \end{array}$

$\blacksquare$

TABLE 15-3

LEWIS FORMULAS OF SOME MOLECULES USING BOTH LINES AND DOTS

MOLECULE	LEWIS FORMULA USING LINES	LEWIS FORMULA USING DOTS AND LINES[a]
H_2	H—H	H—H
BH_3	H—B(H)(H)	H—B(H)(H)
CH_4	H—C—H (with H above and below)	H—C—H (with H above and below)
NH_3	H—N—H (with H above)	:N—H (with H above and below)
H_2O	O—H (with H below)	:Ö—H (with H below)
HF	H—F	H—F̈:
CO_2	O=C=O	:Ö=C=Ö:
F_2	F—F	:F̈—F̈:
N_2	N≡N	:N≡N:

[a] Bonding electrons are drawn with lines. Nonbonding electrons are drawn as dots.

15-6
COVALENT BONDING IN THE THIRD-PERIOD ELEMENTS: THE ELEMENTS P, S AND Cl MAY OR MAY NOT OBEY THE OCTET RULE

The third period of the periodic table consists of the elements from Na to Ar. Much, but not all, of their bonding is similar to the second-period elements.

The group IA and IIA elements Na and Mg, respectively, generally form ionic compounds, as will be discussed in Section 15-7.

The group IIIA element aluminum can form compounds, some with ionic bonds and some with covalent bonds.

The group IVA through group VIIA elements Si to Cl form covalent bonds. Chlorine and sulfur can also form ionic compounds. These will be discussed in Section 15-7. Examples of the kinds of compounds with covalent bonds these elements can form are listed below. Notice the similarity with the corresponding period-2 elements. In many compounds, these period-3 elements obey the octet rule:

$$\cdot \overset{\displaystyle \cdot}{\underset{\displaystyle \cdot}{Si}} \cdot \; + \; 4 \; H\cdot \; \rightarrow \; H\!:\!\overset{\displaystyle H}{\underset{\displaystyle \ddot{H}}{\overset{\displaystyle \cdot\cdot}{Si}}}\!:\!H \quad \text{(silane)}$$

$$:\!\overset{\displaystyle \cdot}{P}\cdot \; + \; 3 \; H\cdot \; \rightarrow \; :\!\overset{\displaystyle H}{\underset{\displaystyle \ddot{H}}{\overset{\displaystyle \cdot\cdot}{P}}}\!:\!H \quad \text{(phosphine)}$$

$$:\!\overset{\displaystyle \cdot\cdot}{S}\cdot \; + \; 2 \; H\cdot \; \rightarrow \; :\!\overset{\displaystyle \cdot\cdot}{\underset{\displaystyle \dot{H}}{S}}\!:\!H \quad \text{(hydrogen sulfide)}$$

$$:\!\overset{\displaystyle \cdot\cdot}{\underset{\displaystyle \cdot\cdot}{Cl}}\cdot \; + \; H\cdot \; \rightarrow \; H\!:\!\overset{\displaystyle \cdot\cdot}{\underset{\displaystyle \cdot\cdot}{Cl}}\!: \quad \text{(hydrogen chloride)}$$

However, there are some differences between the bonding of the period-2 elements and that of the period-3 elements. The period-3 elements in the A groups of the periodic table have up to eight electrons (two s and six p electrons) in their valence shell ($n = 3$). But some compounds formed by these elements cannot be explained using just eight electrons and four bonds. In other words, the octet rule does not apply to these compounds.

For example, phosphorus can form a compound with chlorine; this compound is called phosphorus pentachloride, PCl_5. In this molecule, the phosphorus atom has *five* covalent bonds and *ten* electrons in its valence shell. A diagram of this molecule is shown in Figure 15-6.

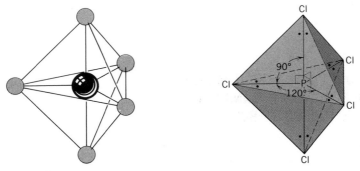

FIGURE 15-6 Two diagrams of the phosphorous pentachloride (PCl_5) molecule. The structure can be thought of as two trigonal pyramids "back to back." The geometric figure is called a **trigonal bipyramid**.

How can the phosphorus atom have five covalent bonds and ten electrons around it? Let's look at the Lewis formula of phosphorus, $:\dot{\text{P}}\cdot$. If the pair of electrons separated, five bonds could easily form, and the result would be a molecule with a lower energy:

$$:\dot{\text{P}}\cdot + 5 :\ddot{\text{Cl}}\cdot \rightarrow \begin{matrix} & :\ddot{\text{Cl}}: & :\ddot{\text{Cl}}: & \\ & :\ddot{\text{Cl}}: & \text{P} :\ddot{\text{Cl}}: & \\ & & :\ddot{\text{Cl}}: & \end{matrix}$$

In nitrogen, $:\dot{\text{N}}\cdot$, as we discussed, this separation doesn't happen because only eight electrons can be around second-period atoms. But third-period atoms can have *up to* 18 electrons around them *if they can use the 3d orbitals*. So you can think of one of the paired phosphorus electrons as going into a vacant 3d orbital, thereby allowing bonding with five Cl atoms.

Sulfur can form a compound with six fluorine atoms; this compound is called sulfur hexafluoride, SF_6. The Lewis formula of sulfur is $:\dot{\text{S}}\cdot$. If *both* pairs of electrons separate, then we have $:\dot{\text{S}}:$ and we can see how sulfur can bond to six F atoms. The structure of SF_6 is shown in Figure 15-7.

Chlorine can form a compound with five fluorine atoms; this compound is called chlorine pentafluoride, ClF_5. The Lewis formula of chlorine is $:\ddot{\text{Cl}}\cdot$. If *two* of the paired electrons separate, then we get $:\dot{\text{Cl}}\cdot$ and we can see how the Cl atom could combine with five F atoms. The structure of ClF_5 is shown in Figure 15-8. As far as we know, Cl does not form a compound with more than five bonds.

Since there is room for up to 18 electrons around third-period atoms, you might ask why six atoms (and 12 electrons) are the maximum number we find around these atoms. Although there may be room for 18 electrons, there isn't room for nine atoms (with nine covalent bonds). Atoms are much bigger

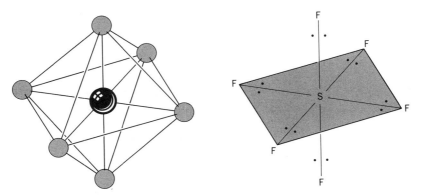

FIGURE 15-7 Two diagrams of the sulfur hexafluoride (SF_6) molecule. The sulfur atom has 12 electrons around it in six covalent bonds. The geometric figure is called a **regular octahedron**.

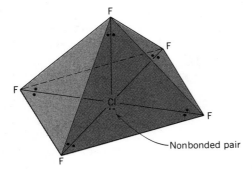

FIGURE 15-8 A diagram of the chlorine pentafluoride (ClF_5) molecule. The Cl atom is slightly below the plane of the four F atoms because the nonbonded pair of electrons pushes the four F atoms away from itself. The geometric figure is called a **square pyramid**.

than electrons, and at the particular distance allowed by the chemical bonds, only six atoms can fit. In elements of higher periods, we do find compounds with more than six atoms bonded to the central atom. In these higher-period atoms, there is much more room for such bonding. An example is iodine heptafluoride, IF_7, in which the iodine atom is surrounded by *seven* F atoms. All of the seven valence electrons of the iodine atom have separated to allow seven covalent bonds to be formed.

Now a word about using the *d* orbitals in the bonding of third-period elements. As you know from Chapter 14, the $3d$ orbitals have a higher energy than the $3p$ orbitals. You might wonder whether it is energy-wise for the electrons to go into the $3d$ orbitals. It must be, because these molecules *do* exist.

15-7
THE STRONG ELECTRICAL ATTRACTION BETWEEN OPPOSITELY CHARGED IONS FORMS THE IONIC BOND

Elements in groups IA and IIA (the alkali metals and the alkaline earth metals) of the periodic table generally form **ionic bonds** with hydrogen and with elements in groups VIA and VIIA. Note that elements in groups VIA and VIIA can form *both* ionic and covalent compounds, depending on what element they are reacting with.

Ionic bonds are bonds in which one or more electrons actually transfer from one atom to another. There is no sharing of electrons as there is in covalent bonds.

For example, in sodium chloride, NaCl, which is common table salt, the bonding takes place this way:

$$Na\cdot \ + \ :\ddot{\underset{..}{C}l}\cdot \ \rightarrow \ Na^+ \ + \ :\ddot{\underset{..}{C}l}:^-$$

The sodium atom has lost its $3s^1$ electron, and the chlorine atom has gained that electron; thus the chlorine now has a complete set of eight electrons in its valence shell. The $3s^1$ electron in Na comes off fairly easily, leaving sodium with a filled second energy level.

We have just written NaCl as

$$Na^+ \quad and \quad :\ddot{C}l:^-$$

to emphasize the transfer of the electron from the Na atom to the Cl atom. Since the Na atom now has one less electron than it has protons, the Na atom will have a charge of $+1$ (it is common to leave off the "1" in writing Na^+). We can write an equation to represent the sodium atom giving up an electron:

$$\begin{bmatrix} \text{this Na} \\ \text{atom has} \\ \text{11 } e^- \end{bmatrix} \rightarrow Na\cdot \rightarrow \underset{\underset{\begin{bmatrix} \text{this } Na^+ \text{ has} \\ \text{10 } e^- \end{bmatrix}}{\uparrow}}{Na^+} + e^- \leftarrow \begin{bmatrix} \text{this } e^- \\ \text{came off the} \\ \text{Na atom} \end{bmatrix}$$

The chlorine atom, on the other hand, has taken one electron from the sodium atom. Therefore, it has one more electron than it has protons and so has a charge of -1 (again, it is common to leave off the "1" in writing $:\ddot{C}l:^-$). We can write an equation to represent this taking of an electron:

$$\begin{bmatrix} \text{this Cl atom} \\ \text{has 17 } e^- \end{bmatrix} \rightarrow :\ddot{C}l\cdot + \underset{\underset{\begin{bmatrix} \text{this electron} \\ \text{adds to the} \\ \text{Cl atom} \end{bmatrix}}{\uparrow}}{e^-} \rightarrow :\ddot{C}l:^- \leftarrow \begin{bmatrix} \text{this } Cl^- \text{ has} \\ \text{18 } e^- \end{bmatrix}$$

It is customary to leave off the dots when writing such reactions:

$$Na \rightarrow Na^+ + e^-$$

$$Cl + e^- \rightarrow Cl^-$$

$$Na + Cl \rightarrow Na^+ + Cl^-$$

Charged atoms like Na^+ and Cl^- are called **ions**. Na^+ is called the "sodium ion," and Cl^- is called the "chloride ion." Compounds that have ionic bonds are called **ionic compounds**. *It is the strong electrical attraction between positive and negative ions that causes the bonding force of an ionic compound.*

EXAMPLE 6   Write the equation for removing an electron from an Li atom.

Solution: The Lewis formula for lithium is Li·. Li can easily lose the lone $2s$ electron:

$$\text{Li·} \rightarrow \text{Li}^+ + e^- \quad \blacksquare$$

EXAMPLE 7   Write the equation for removing two electrons from an Mg atom.

Solution: The Lewis formula for magnesium is Mg:. Mg can easily lose its two $3s$ electrons:

$$\text{Mg:} \rightarrow \text{Mg}^{2+} + 2e^-$$

The charge on the Mg ion is +2 because the Mg atom has lost two electrons. $\blacksquare$

At this point we should say a word about writing ionic formulas. Notice that in writing Mg^{2+} in Example 7, we wrote the "2" before the "+." This is the approved way of writing the ionic charge of an ion in an ionic formula. However, when we said "the charge on the Mg ion is +2 . . .," we wrote the "+" before the "2." We did so because we weren't writing an ionic formula—we were just referring to a "plus two" charge.

EXAMPLE 8   Write the equation for the addition of an electron to an F atom.

Solution: The Lewis formula for fluorine is :F̈·. Fluorine can take one electron to fill up its valence shell:

$$\text{:F̈·} + e^- \rightarrow \text{:F̈:}^- \quad \blacksquare$$

EXAMPLE 9   Write the equation for the addition of two electrons to an oxygen atom.

Solution: The Lewis formula for oxygen is :Ö·. Oxygen can acquire two electrons to fill up its valence shell:

$$\text{:Ö·} + 2e^- \rightarrow \text{:Ö:}^{2-}$$

The charge on the oxygen ion is −2 because it has gained two electrons. $\blacksquare$

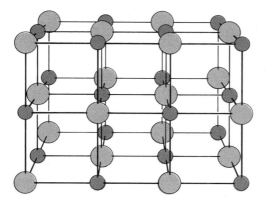

FIGURE 15-9 A diagram of the crystal structure of NaCl. The smaller Na^+ ions are shaded for clarity. Each Cl^- ion is surrounded by six Na^+ ions. Each Na^+ ion is surrounded by six Cl^- ions. In the actual crystal, there are billions of ions arranged in the pattern shown.

The electrical attraction between the positively charged Na^+ ion and the negatively charged Cl^- ion is what holds the NaCl crystal together. In the solid, the NaCl crystal is a giant array (see Chapter 4) of Na^+ ions and Cl^- ions. A diagram of the crystal structure of NaCl is shown in Figure 15-9.

How do we know that there are separate Na^+ ions and Cl^- ions in the NaCl crystal? Certain experiments have shown that this is the case; for example, careful x-ray studies of NaCl crystals have shown that the Na^+ ion has 10 electrons around it and that the Cl^- ion has 18 electrons around it. This is very strong evidence that separate Na^+ and Cl^- ions exist.

Why do ionic bonds form? After all, it must require some energy to pull an electron out of an Na atom and put it onto a Cl atom. It does indeed require some energy. But when sodium and chlorine react to form NaCl, a *great deal* of heat energy is given off. This lowering of the total energy as a result of the attraction of the Na^+ ion and the Cl^- ion in the crystal is very large. It easily compensates for the energy needed to pull an electron out of an Na atom and add it to a Cl atom. The NaCl crystal is very stable.

Why don't atoms like sodium and chlorine form covalent bonds with one another? It must be that the total energy is lowest when an ionic bond is formed.

Elements in group IA of the periodic table have one *s* electron in their highest energy level. This lone electron can be easily removed, forming a +1 ion. Elements in group IIA have two *s* electrons in their highest energy level and can easily lose both of them, forming +2 ions. Elements in group IIIA have three electrons, namely, two *s* electrons and one *p* electron. Boron forms three covalent bonds. Aluminum can form the +3 ion, Al^{3+}, or three covalent bonds, depending on what it is bonding to. The rest of the group IIIA elements also form both covalent and ionic bonds.

As you go down a group in the periodic table, it gets easier to remove the valence electrons. This is because they are farther from the nucleus and aren't attracted as much as electrons that are closer to the nucleus.

EXAMPLE 10   Write the equation for the formation of the ionic compound containing lithium and fluorine.

Solution:

$$\text{Li}\cdot + :\ddot{\text{F}}\cdot \rightarrow \text{Li}^+ + :\ddot{\text{F}}:^-$$

Without the dots we have

$$\text{Li} + \text{F} \rightarrow \text{Li}^+ + \text{F}^-$$

Under normal conditions, the fluorine gas occurs as F_2 molecules, and we should write the reaction as

$$2\text{Li} + \text{F}_2 \rightarrow 2\text{Li}^+ + 2\text{F}^-$$

This would be more realistic. ■

EXAMPLE 11   Write the equation for the formation of the ionic compound containing magnesium and fluorine.

Solution: The Mg atom can give up two electrons to become the Mg^{2+} ion. Two F atoms are needed to take these two electrons, since each F atom takes only one electron:

$$\text{Mg}: + 2 :\ddot{\text{F}}\cdot \rightarrow \text{Mg}^{2+} + 2 :\ddot{\text{F}}:^-$$

Without dots, we have

$$\text{Mg} + 2\text{F} \rightarrow \text{Mg}^{2+} + 2\text{F}^-$$

Again, since fluorine gas occurs as the F_2 molecule, we should write the reaction as

$$\text{Mg} + \text{F}_2 \rightarrow \text{Mg}^{2+} + 2\text{F}^- ■$$

In closing, we should say that most chemical bonds are not purely ionic (transfer of electrons) or purely covalent (sharing of electrons). They are usually a mixture of the two kinds of chemical bonds.

In this chapter we have only touched on the subject of chemical bonding. We have not discussed metallic bonding, bonding in the transition metals, and many other things. As you study more chemistry, you will learn more about the nature of the chemical bond and perhaps realize what a vast subject it is.

PROBLEMS

KEYED PROBLEMS

1. Write the Lewis formula for lithium, Li. Only the electrons in the second energy level should be shown since the electrons in the first energy level don't take part in the chemical reactions of lithium.

2. Draw the Lewis formula for magnesium, Mg.

3. Draw the Lewis formula for K and As.

4. Draw Lewis formulas for the following molecules: NCl_3, BCl_3, and Cl_2O.

   (NOTE: Cl_2O has a structure similar to OF_2. The reason we write it Cl_2O and not OCl_2 is indicated in Chapter 16.)

5. Draw the Lewis formulas with lines only, as well as with lines and dots, for the following molecules: NCl_3, BCl_3, and Cl_2O.

6. Write the equation for removing an electron from a K atom.

7. Write the equation for removing two electrons from a Ca atom.

8. Write the equation for the addition of an electron to a Br atom.

9. Write the equation for the addition of two electrons to an S atom.

10. Write the equation for the formation of the ionic compound containing potassium and bromine.

11. Write the equation for the formation of the ionic compound containing calcium and chlorine.

SUPPLEMENTAL PROBLEMS

12. What do we mean when we say that electrons are "lazy"?

13. Define *molecular orbital*.

14. Why can't more than two electrons be in the same molecular orbital?

15. Define *covalent bond*.

16. Define *valence electrons*.

17. In the Lewis formula of an atom, which electrons are represented as dots and which electrons are not?

18. Draw Lewis formulas for all the third-period A-group elements.

19. Draw Lewis formulas with lines only, as well as with lines and dots, for the following molecules.

a. CCl₄   c. H₂S   e. H₂CO (C in the center)

b. NBr₃   d. HBr   f. I₂

20. Draw Lewis formulas with lines only for the following molecules.

    a. C_2H_6 (carbon–carbon single bond)

    b. C_2H_4 (carbon–carbon double bond)

    c. C_2H_2 (carbon–carbon triple bond)

21. Draw the Lewis formula for PCl_3.

22. Draw the Lewis formula for SF_4.

    (**NOTE:** S has 10 electrons around it. Two electrons form a nonbonded pair.)

23. Define an *ionic bond*.

24. Write the equation for the loss of three electrons by Al.

25. Write the equation for the gain of two electrons by Se.

26. Write the equations for the formation of the ionic compounds containing the following pairs of elements.

    a. Al and Cl_2 c. Be and Br_2 e. Li and S_8 (elemental sulfur con-
sists of molecules with eight atoms)

    b. K and O_2 d. Ba and I_2

    f. Al and H_2

27. Give an example of a molecule with each of the following geometric structures.

    a. linear   d. trigonal pyramid

    b. angular or bent   e. regular tetrahedron

    c. triangular planar

28. Give an example of a molecule with each of the following geometric structures.

    a. trigonal bipyramid   c. square pyramid

    b. octahedron

16

NOMENCLATURE
OF SIMPLE INORGANIC COMPOUNDS

In previous chapters we have mentioned many compounds. In this chapter we will explain the rules for naming various inorganic compounds and the rules for writing their formulas. By the end of the chapter, you should be able to write the name of a compound if you are given the formula and to write the formula if you are given the name.

This discussion will be limited to inorganic compounds, that is, compounds that do not involve carbon atoms (there are a few exceptions, such as CO_2, H_2CO_3, and HCN, which chemists normally regard as inorganic compounds). Most carbon compounds (methane, CH_4, is an example) are called organic compounds and are named according to a system that we will not discuss in this book.[1]

There are three classes of inorganic compounds that we will discuss here: (1) ionic compounds (formed between a metal and a nonmetal), (2) covalent compounds (formed between two nonmetals), and (3) covalent compounds called acids and oxyacids.

[1] Naming organic compounds is sometimes briefly discussed in general chemistry courses. But a complete discussion must wait for a course in organic chemistry. The International Union of Pure and Applied Chemistry (IUPAC) is the group that sets nomenclature rules in both inorganic and organic chemistry.

(Opposite) A helicopter fertilizing a stand of Douglas fir trees. Typical chemicals used in fertilizers are ammonium nitrate, ammonium phosphate, and potassium sulfate.

16-1
NAMING IONIC COMPOUNDS FORMED BETWEEN A METAL AND A NONMETAL

In this section we will explain the rules for naming ionic compounds and writing the formulas of ionic compounds. The basic rules are

1. **The formulas of ionic compounds are written without showing charges on the ions.** For example, sodium chloride is written NaCl, not Na^+Cl^-.

2. **In writing formulas, we put the positive ion first and the negative ion second.** Thus sodium chloride is written NaCl, never ClNa.

3. **The same is true in writing the name of a compound: The name of the positive ion is written first, and the name of the negative ion is written second.** Thus NaCl is always called sodium chloride.

4. **The formula of the compound must be electrically neutral; that is, the total number of positive charges on the positive ions must be equal to the total number of negative charges on the negative ions.** For example, in NaCl, the Na^+ has one positive charge and the Cl^- has one negative charge. Thus the total number of positive charges equals the total number of negative charges in the formula NaCl.

5. **The smallest whole number of atoms possible to satisfy rule 4 is used to write the formula.** Thus we write NaCl, not Na_2Cl_2.

6. **To decide which elements are considered metals and which are considered nonmetals in this naming system, look at the metals and the nonmetals in a periodic table** (Figures 14-10 and 14-12). Semimetals are considered nonmetals for naming purposes.

TABLE 16-1

CHARGES ON IONS OF GROUPS IA, IIA, AND IIIA ELEMENTS AND NH_4^+

GROUP IA	GROUP IIA	GROUP IIIA	ANOTHER POSITIVE ION
H^+	Be^{2+}	B^a	NH_4^+
Li^+	Mg^{2+}	Al^{3+} [b]	(ammonium)
Na^+	Ca^{2+}	Ga^{3+}	
K^+	Sr^{2+}	In^{3+} [c]	
Rb^+	Ba^{2+}	Tl^{3+} [c]	
Cs^+	Ra^{2+}		

[a]Considered a semimetal—only forms covalent bonds. Boron compounds will not be considered in this section. Boron does not form the +3 ion.
[b]Definitely a metal, but forms both ionic and covalent bonds.
[c]Indium and thallium also form the +1 ions In^+ and Tl^+, respectively.

As you can see, it is important to know the ionic charge that an atom can have if you want to be able to write the correct formula. So we must now discuss how to determine the ionic charge that an atom can have in order to understand rules 4 and 5 above.

For group IA, IIA, and IIIA metals, the positive charge that the ion can have is the same as the group number. Table 16-1 lists the positive charge formed in these three groups and another important positive ion. For the nonmetals in groups VA, VIA, and VIIA, the negative ionic charge is found by the following formula:

$$\text{negative ionic charge} = \text{group number} - 8$$

For example, oxygen is in group VIA. Thus the ionic charge is $6 - 8 = -2$. So oxygen forms O^{2-} ions. Notice that this formula simply tells us the number of electrons that must be added to the valence shell to fill it to eight electrons:

$$:\ddot{O}\cdot + 2e^- \rightarrow :\ddot{O}:^{2-}$$

In compounds, the names of the negative ions of groups IVA, VA, VIA, and VIIA end in *ide*. Table 16-2 lists the name and charge of these ions.

In Table 16-2, only carbon is listed in group IVA and only nitrogen and phosphorus are listed in group VA because only these elements in these groups sometimes form negative ions. As we mentioned in Chapter 15, carbon, nitrogen, and phosphorus usually form covalent bonds. All the other elements listed in Table 16-2 can also form covalent bonds under the appropriate conditions. Notice that hydrogen is mentioned in both Table 16-1 and Table 16-2. Hydrogen can either take an electron to form the hydride ion, $H:^-$, or lose an electron to form the H^+ ion.

We now give a number of examples that illustrate the use of the basic rules of writing formulas and names. All the elements used are from Tables 16-1 and 16-2.

TABLE 16-2

NAME AND CHARGE ON IONS OF GROUPS IVA, VA, VIA, AND VIIA[a]

GROUP IV	GROUP VA	GROUP VIA	GROUP VIIA
C^{4-} (carbide)[b]	N^{3-} (nitride)	O^{2-} (oxide)	F^- (fluoride)
	P^{3-} (phosphide)	S^{2-} (sulfide)	Cl^- (chloride)
		Se^{2-} (selenide)	Br^- (bromide)
		Te^{2-} (telluride)	I^- (iodide)

[a]Hydrogen can also form the negative ion H^-, which is called the hydride ion.
[b]Carbon also forms the negative ion C_2^{2-}, which is called the carbide ion.

EXAMPLE 1   Write the name and formula of the compound made from potassium and bromine.

Solution: The potassium ion is K^+, and the bromide ion is Br^-. One K^+ combines with one Br^- as follows:

$$K^+ + Br^- \rightarrow KBr$$

KBr is called potassium bromide.   ∎

EXAMPLE 2   Write the name and formula of the compound made from calcium and iodine.

Solution: The calcium ion is Ca^{2+}, and the iodide ion is I^-. Two iodide ions combine with one calcium ion to give electrically neutral calcium iodide:

$$Ca^{2+} + 2I^- \rightarrow CaI_2$$ ∎

EXAMPLE 3   Write the name and formula of the compound made from Al and F.

Solution: The ions are Al^{3+} and F^-. Thus we need three F^- ions for each Al^{3+} ion to form electrically neutral aluminum fluoride:

$$Al^{3+} + 3F^- \rightarrow AlF_3$$ ∎

EXAMPLE 4   Write the name and formula of the compound made from Mg and N.

Solution: The ions are Mg^{2+} and N^{3-}. Thus we need three Mg^{2+} ions to two N^{3-} ions, for a total of six positive charges and six negative charges. The name of the compound is magnesium nitride:

$$3Mg^{2+} + 2N^{3-} \rightarrow Mg_3N_2$$ ∎

EXAMPLE 5   Write the name and formula of the compound made from Al and O.

Solution: The ions are Al^{3+} and O^{2-}. We need two Al^{3+} ions to three O^{2-} ions, for a total of six positive charges and six negative charges. The name of the compound is aluminum oxide:

$$2Al^{3+} + 3O^{2-} \rightarrow Al_2O_3$$ ∎

EXAMPLE 6   Write the correct names of the following compounds: K_2S, $CaBr_2$, LiH, BaS, Al_2Se_3, and NH_4Cl.

Solution:

K_2S	potassium sulfide	BaS	barium sulfide
$CaBr_2$	calcium bromide	Al_2Se_3	aluminum selenide
LiH	lithium hydride	NH_4Cl	ammonium chloride ∎

EXAMPLE 7   Write the correct formulas for the following names: cesium iodide, beryllium oxide, gallium sulfide, strontium bromide, potassium oxide, calcium hydride, radium iodide, and aluminum sulfide.

Solution:

cesium iodide	CsI	potassium oxide	K_2O
beryllium oxide	BeO	calcium hydride	CaH_2
gallium sulfide	Ga_2S_3	radium iodide	RaI_2
strontium bromide	$SrBr_2$	aluminum sulfide	Al_2S_3 ∎

In addition to the simple monatomic negative ions listed in Table 16-2, there are many common **polyatomic negative ions** ("poly" means many). These are listed in Table 16-3. The bonding within each polyatomic ion is covalent, but the whole negative ion can form either ionic or covalent bonds. Since there is no way to figure out their names or formulas, you should learn at least some of the ions listed in Table 16-3. The asterisked (*) ions are the most common.

In Table 16-3, we have grouped ions together in such a way that you can see some patterns. When nitrogen or sulfur form negative ions with oxygen, the names have a pattern. We list these ions as follows so that you can compare their names; we also list the nitride and sulfide ions for comparison:

N^{3-}	nitride	S^{2-}	sulfide
NO_2^-	nitrite	SO_3^{2-}	sulfite
NO_3^-	nitrate	SO_4^{2-}	sulfate

If an ion has no oxygen, its ending is *ide* (N^{3-} and S^{2-}). With the larger number of oxygen atoms, the ending is *ate* (NO_3^- and SO_4^{2-}). With the smaller number of oxygen atoms, the ending is *ite* (NO_2^- and SO_3^{2-}). In the case of chlorine, which forms more than two negative ions with oxygen, there are slight complications, as the following list indicates:

Cl^-	chloride	ClO_3^-	chlorate
ClO^-	hypochlorite	ClO_4^-	perchlorate
ClO_2^-	chlorite		

TABLE 16-3

SOME POLYATOMIC NEGATIVE IONS

NO_2^-	nitrite* ^{a}	PO_4^{3-}	phosphate*
NO_3^-	nitrate*	HPO_4^{2-}	hydrogen phosphate
SO_3^{2-}	sulfite*	$H_2PO_4^-$	dihydrogen phosphate
SO_4^{2-}	sulfate*	CrO_4^{2-}	chromate*
HSO_3^-	hydrogen sulfite	$Cr_2O_7^{2-}$	dichromate*
HSO_4^-	hydrogen sulfate	ClO^-	hypochlorite
CO_3^{2-}	carbonate*	ClO_2^-	chlorite
HCO_3^-	hydrogen	ClO_3^-	chlorate*
	carbonate*	ClO_4^-	perchlorate*
$C_2H_3O_2^-$	acetate*	$C_2O_4^{2-}$	oxalate
AsO_4^{3-}	arsenate	O_2^{2-}	peroxide*
BO_3^{3-}	borate	MnO_4^-	permanganate*
OCN^-	cyanate	SiO_4^{4-}	silicate
CN^-	cyanide*	$S_4O_6^{2-}$	tetrathionate
OH^-	hydroxide*	SCN^-	thiocyanate
		$S_2O_3^{2-}$	thiosulfate

^{a}The asterisked (*) ions are the most common.

The prefix *hypo* (meaning "below") is used for the smallest number of oxygen atoms, and the prefix *per* is used for the largest number of oxygen atoms. So, in order of increasing number of oxygen atoms, we have *ide, hypo---ite, ite, ate,* and *per---ate.*

If hydrogen is included in the formula of the negative ion, this is usually stated in the name. For example,

CO_3^{2-}	carbonate	PO_4^{3-}	phosphate
HCO_3^-	hydrogen carbonate	HPO_4^{2-}	hydrogen phosphate
		$H_2PO_4^-$	dihydrogen phosphate

HCO_3^- is sometimes called the *bicarbonate* ion, where *bi* is an old-fashioned term that stands for hydrogen. You may have heard of sodium bicarbonate ($NaHCO_3$, bicarbonate of soda, baking soda). It is more properly called sodium hydrogen carbonate.

The peroxide ion, O_2^{2-}, may be a little puzzling at first glance. To form the peroxide ion, two oxygen atoms are bonded together. A typical compound with the peroxide ion is hydrogen peroxide, H_2O_2. The structure is shown at the top of page 293. The H atoms are not in the plane of the page—one is in front of the page and one is behind the page:

$$\begin{array}{c} H \\ \diagdown \\ O - O \\ \diagdown \\ H \end{array}$$

structural formula
of hydrogen peroxide

Oxygen also forms a superoxide ion, O_2^-, which we will not discuss except to offer Problem 27 at the end of Chapter 17.

The following examples all use ions taken from Tables 16-1 and 16-3.

EXAMPLE 8   Write the correct formula for the compounds that have the following names: sodium nitrate, calcium hypochlorite, potassium permanganate, aluminum acetate, magnesium hydroxide, rubidium peroxide, barium perchlorate, potassium dichromate, ammonium phosphate, and aluminum hydrogen sulfite.

Solution:

sodium nitrate	$NaNO_3$	rubidium peroxide	Rb_2O_2
calcium		barium perchlorate	$Ba(ClO_4)_2$
hypochlorite	$Ca(ClO)_2$	potassium	
potassium		dichromate	$K_2Cr_2O_7$
permanganate	$KMnO_4$	ammonium	
aluminum acetate	$Al(C_2H_3O_2)_3$	phosphate	$(NH_4)_3PO_4$
magnesium		aluminum hydrogen	
hydroxide	$Mg(OH)_2$	sulfite	$Al(HSO_3)_3$ ∎

EXAMPLE 9   Write the correct names of the compounds that have the following formulas: $NaNO_2$, K_2O_2, $Be(OH)_2$, $CaCO_3$, $Sr(ClO_2)_2$, $AlPO_4$, $Na_2S_2O_3$, and MgC_2O_4.

Solution:

$NaNO_2$	sodium nitrite	$Sr(ClO_2)_2$	strontium chlorite
K_2O_2	potassium peroxide	$AlPO_4$	aluminum phosphate
$Be(OH)_2$	beryllium hydroxide	$Na_2S_2O_3$	sodium thiosulfate
$CaCO_3$	calcium carbonate	MgC_2O_4	magnesium oxalate ∎

Now we must consider the problem of naming compounds containing the transition metals and some other metals of the periodic table. These present a slight problem because *many of them can form more than one positive ion.* We will consider only the most common ions, and fortunately there are no more than two of these for a metal. Consider iron, which can form the $+2$ and the $+3$ ions:

$$Fe^{2+} \quad and \quad Fe^{3+}$$

There are two ways of naming these ions, the modern system and the old system. Both systems are in use today and you should learn both.
In the modern system, the ionic charge is indicated after the name of the metal by a Roman numeral in parentheses. Thus for iron we have

Fe^{2+} iron(II)
Fe^{3+} iron(III)

Notice that there is no space between the name and the parentheses. Thus $FeCl_2$ is called iron(II) chloride, and $FeCl_3$ is called iron(III) chloride. Read them as "iron two chloride" and "iron three chloride." As another example, FeS is called iron(II) sulfide, and Fe_2S_3 is called iron(III) sulfide.
In the old system, the Latin name of the metal is sometimes used together with the suffixes "ous" and "ic." Ous stands for the smaller positive charge and *ic* stands for the larger positive charge. Thus for iron we have

Fe^{2+} ferrous
Fe^{3+} ferric

(**NOTE:** The Latin name for iron is *ferrum.*)
So in the old system, $FeCl_2$ is called ferrous chloride and $FeCl_3$ is called ferric chloride. For our second example, FeS is called ferrous sulfide and Fe_2S_3 is called ferric sulfide.

Many transition metals can form more than two positive ions. For these, the old system can be used to name only the most common positive ions. These are the ones listed in Table 16-4. The other not-so-common ions are named by the modern system only.

Table 16-4 lists some of the transition metals that have two common positive ions. Table 16-5 lists the transition metals that have only one positive ion. These metals are named just like the ones in Table 16-1, that is, with no Roman numerals and no *ous* or *ic* forms.

Examples 10 and 11 use ions from Tables 16-2 through 16-5.

EXAMPLE 10   Write the correct formulas for the following names: gold(I) chloride, nickel(III) hydroxide, stannous fluoride, silver thiocyanate, lead(II) sulfate, chromic carbonate, copper(I) phosphate, mercuric nitrate, zinc hydrogen carbonate, and mercurous chloride.

TABLE 16-4

METALS THAT CAN FORM TWO COMMON POSITIVE IONS

ION[a]	MODERN NAME	OLD NAME	LATIN NAME OF METAL
Au^+	gold(I)	aurous	aurum
Au^{3+}	gold(III)	auric	aurum
Co^{2+}	cobalt(II)	cobaltous	
Co^{3+}	cobalt(III)	cobaltic	
Cr^{2+}	chromium(II)	chromous	
Cr^{3+}	chromium(III)	chromic	
Cu^+	copper(I)	cuprous	cuprum
Cu^{2+}	copper(II)	cupric	cuprum
Fe^{2+}	iron(II)	ferrous	ferrum
Fe^{3+}	iron(III)	ferric	ferrum
Hg_2^{2+}	mercury(I)[b]	mercurous	hydrargyrum
Hg^{2+}	mercury(II)	mercuric	hydrargyrum
Mn^{2+}	manganese(II)	manganous	magnes
Mn^{3+}	manganese(III)	manganic	magnes
Ni^{2+}	nickel(II)	nickelous	
Ni^{3+}	nickel(III)	nickelic	
Pb^{2+}	lead(II)	plumbous	plumbum
Pb^{4+}	lead(IV)	plumbic	plumbum
Sn^{2+}	tin(II)	stannous	stannum
Sn^{4+}	tin(IV)	stannic	stannum

[a] The most common ones, namely, Cr, Cu, Fe, Hg, Pb, and Sn, should be memorized.

[b] The Hg_2^{2+} ion is unusual. Each Hg atom has a charge of $+1$ in the Hg_2^{2+} ion, but two Hg^+ ions always join together. A typical compound is mercury(I) chloride, Hg_2Cl_2.

TABLE 16-5

TRANSITION METALS WITH ONE COMMON POSITIVE ION

ION	NAME	LATIN NAME
Ag^+	silver	argentum
Cd^{2+}	cadmium	cadmia
Zn^{2+}	zinc	

Solution:

gold(I) chloride	AuCl	chromic carbonate	$Cr_2(CO_3)_3$
nickel(III) hydroxide	$Ni(OH)_3$	copper(I) phosphate	Cu_3PO_4
stannous fluoride	SnF_2	mercuric nitrate	$Hg(NO_3)_2$
silver thiocyanate	AgSCN	zinc hydrogen carbonate	$Zn(HCO_3)_2$
lead(II) sulfate	$PbSO_4$	mercurous chloride	Hg_2Cl_2 ∎

In Example 11, use the known charge on the negative ion to figure out the charge on the positive ion. Once you know the charge on the positive ion, you can use Table 16.4 to look up its name. Thus, when given the formula Co_2S_3, you know that each sulfide ion has a -2 charge. You can then write an equation breaking up the chemical formula into its ions to help you figure out the charge on the cobalt ion:

$$Co_2S_3 \rightarrow \quad 2Co^? \quad + \quad 3S^{2-}$$

$$\uparrow \qquad\qquad\qquad \uparrow$$

+6 total charge   **−6 total charge**

$$+6/2 = +3$$

As you can see, three S^{2-} ions have a total -6 charge. Both cobalt ions must then have a total $+6$ charge, and each cobalt ion has a $+3$ charge. The name of Co_2S_3 is therefore cobalt(III) sulfide or cobaltic sulfide.

EXAMPLE 11   Write the correct names for the following formulas: Co_2S_3, $FeSO_4$, PbO_2 (oxide, not peroxide), $Ni(ClO_4)_2$, $CdBr_2$, CuO, $Mn(NO_3)_3$, Hg_2SO_4, $CrAsO_4$, and AuH_2PO_4.

Solution: We shall list the modern name and the old name if commonly used.

Co_2S_3	cobalt(III) sulfide	cobaltic sulfide
$FeSO_4$	iron(II) sulfate	ferrous sulfate
PbO_2	lead(IV) oxide	lead dioxide[2]

---

[2] The name lead dioxide for PbO_2 is a common name and doesn't follow any of the rules we have mentioned. There are a number of compounds that have common names—we will mention some others in Section 16-2.

$Ni(ClO_4)_2$	nickel(II) perchlorate	nickelous perchlorate
$CdBr_2$	cadmium bromide	
CuO	copper(II) oxide	cupric oxide
$Mn(NO_3)_3$	manganese(III) nitrate	manganic nitrate
Hg_2SO_4	mercury(I) sulfate	mercurous sulfate
$CrAsO_4$	chromium(III) arsenate	chromic arsenate
AuH_2PO_4	gold(I) dihydrogen phosphate	aurous dihydrogen phosphate ■

16-2
NAMING COVALENT COMPOUNDS FORMED BETWEEN NONMETALS

The second class of inorganic compounds whose names and formulas we will discuss are the covalent compounds formed from two nonmetals. To decide which elements are nonmetals, look at a periodic table (Figures 14-10 and 14-12). Semimetals and hydrogen are considered nonmetals for naming purposes. **Since very often more than one compound can form from two elements, prefixes are used to tell how many of each atom there are.** The prefixes used are

mono	1	**hexa**	6
di	2	**hepta**	7
tri	3	**octa**	8
tetra	4	**nona**	9
penta	5	**deca**	10

If the two nonmetals form only one compound, no prefixes are used. Thus H_2S is called hydrogen sulfide, *not* dihydrogen monosulfide.

The second element in the compound always has the *ide* ending, and the prefix "mono" is usually left out if it would normally precede the first element. Another deviation from the rules is that many compounds have common names that are used almost all the time. Examples are

water	H_2O	phosphine	PH_3
hydrogen peroxide	H_2O_2	arsine	AsH_3
ammonia	NH_3	nitric oxide	NO
hydrazine	N_2H_4	nitrous oxide	N_2O

As you must be aware, nobody ever calls water "dihydrogen monoxide." The six oxides of nitrogen illustrate the rules for naming nonmetal–nonmetal compounds. Two of these are always known by their common names.

FORMULA	MODERN NAME	COMMON NAME
N_2O	dinitrogen monoxide	nitrous oxide ("laughing gas")
NO	nitrogen monoxide	nitric oxide
NO_2	nitrogen dioxide	
N_2O_3	dinitrogen trioxide	
N_2O_4	dinitrogen tetraoxide	
N_2O_5	dinitrogen pentaoxide	

There is a preferred order of the nonmetals used for naming nonmetal–nonmetal compounds. The order is

Rn, Xe, Kr, B, Si, C, Sb, As, P, N, H, Te, Se, S, I, Br, Cl, O, F

That is why we wrote BF_3 for boron trifluoride, HF for hydrogen fluoride, NH_3 for ammonia, OF_2 for oxygen difluoride, and H_2O for water in Chapter 15. We were following the order of this list, which is arranged by the groups (or columns) of the nonmetals and semimetals in the periodic table. It starts with radon in group VIIIA and moves from the *bottom up* of group VIIIA. Boron in group IIIA comes next, and then we again move from the *bottom up* of groups IVA through VIIA. The exceptions are (1) H, which is put between N (group VA) and Te (group VIA), and (2) O, which is put between Cl and F. This was so that the ordering in compounds that were known before the list was made up wouldn't have to be changed.

Germanium (Ge) is a semimetal that is not included in the list just given. However, most authors treat Ge as if it belonged in the list between B and Si. So the name of $GeCl_4$ is germanium tetrachloride.

EXAMPLE 12   Write the formulas of the following names: carbon dioxide, carbon monoxide, hydrogen sulfide*, chlorine dioxide, silicon carbide*, boron nitride*, phosphorus pentachloride, silicon tetrafluoride, diboron hexahydride (the common name is diborane). Compounds with an asterisk form only one compound from the two elements.

Solution:

carbon dioxide	CO_2	boron nitride	BN
carbon monoxide	CO	phosphorus pentachloride	PCl_5
hydrogen sulfide	H_2S	silicon tetrafluoride	SiF_4
chlorine dioxide	ClO_2	diborane	B_2H_6
silicon carbide	SiC		

■

EXAMPLE 13   Write the names for the following formulas: PCl_3, SO_2, SO_3, NO_2, BP*, P_4O_{10}, BrO_3, I_2O_7, BrF_5, and As_4O_6. Compounds with an asterisk form only one compound from the two elements.

Solution:

PCl_3	phosphorus trichloride	P_4O_{10}	tetraphosphorus decaoxide
SO_2	sulfur dioxide	BrO_3	bromine trioxide
SO_3	sulfur trioxide	I_2O_7	diiodine heptaoxide
NO_2	nitrogen dioxide	BrF_5	bromine pentafluoride
BP	boron phosphide	As_4O_6	tetrarsenic hexaoxide ■

In Example 12 the name "monoxide" is spelled with an "o" left out. It is never spelled "monooxide." In Example 13 the name "tetrarsenic" is spelled with an "a" left out. It is never spelled "tetraarsenic."

16-3
NAMING ACIDS AND OXYACIDS

The third class of compounds we will discuss are the covalent compounds known as **acids** and **oxyacids.** Oxyacids are just acids that have oxygen in the molecule. Acids are compounds that contain hydrogen and give up a hydrogen ion (H^+) rather easily.

The name of an acid often depends on whether the acid is a solid, liquid, or gas that can be dissolved in water. For example, HCl is called hydrogen chloride or hydrochloric acid. Hydrogen chloride is pure gaseous HCl, and hydrochloric acid is a water solution of HCl gas. On the other hand, H_2SO_4, which is a liquid, is always called sulfuric acid, never hydrogen sulfate. And H_3BO_3, which is a solid, is always called boric acid, never hydrogen borate.

The following is a list of some gaseous acids and their two names.

FORMULA	NAME OF PURE ACID	NAME OF ACID DISSOLVED IN WATER
HF	hydrogen fluoride	hydrofluoric acid
HCl	hydrogen chloride	hydrochloric acid
HBr	hydrogen bromide	hydrobromic acid*
HI	hydrogen iodide	hydroiodic acid*
H_2S	hydrogen sulfide	hydrosulfuric acid*
HCN	hydrogen cyanide	hydrocyanic acid*

TABLE 16-6

OXYACIDS AND THEIR CORRESPONDING NEGATIVE IONS

FORMULA	NAME	NEGATIVE ION	NAME OF NEGATIVE ION
HNO_2	nitrous acid	NO_2^-	nitrite
HNO_3	nitric acid	NO_3^-	nitrate
H_2SO_3	sulfurous acid	SO_3^{2-}	sulfite
H_2SO_4	sulfuric acid	SO_4^{2-}	sulfate
H_2CO_3	carbonic acid	CO_3^{2-}	carbonate
H_3PO_4	phosphoric acid	PO_4^{3-}	phosphate
$HClO$	hypochlorous acid	ClO^-	hypochlorite
$HClO_2$	chlorous acid	ClO_2^-	chlorite
$HClO_3$	chloric acid	ClO_3^-	chlorate
$HClO_4$	perchloric acid	ClO_4^-	perchlorate
H_3BO_3	boric acid	BO_3^{3-}	borate

The names with an asterisk after them are much less commonly used than the pure acid name.

The oxyacids listed in Table 16-6 have only one name. This name refers to both the pure acid and the acid dissolved in water. Notice the patterns in the names of some of the series, especially that the *ic* of the acid becomes the *ate* of the negative ion, and the *ous* becomes an *ite*.

16-4
CHEMICAL NAMES SHOULD BE AS SIMPLE AS POSSIBLE BUT COMPLETELY CLEAR

Chemists name compounds with the least amount of information needed to be completely clear. For example, Al_2O_3 is called aluminum oxide, *never* aluminum(III) oxide. There is no need to specify anything more than aluminum oxide, because there is only one possible formula, namely, Al_2O_3. On the other hand, Fe_2O_3 is called iron(III) oxide or ferric oxide. If you only called it iron oxide, nobody would know whether you meant Fe_2O_3, FeO, or Fe_3O_4.

As for common names, they specify a definite compound. The names are very old but are still commonly used.

PROBLEMS

KEYED PROBLEMS

1. Write the name and formula of the compound made from cesium and iodine.

2. Write the name and formula of the compound made from magnesium and bromine.

3. Write the name and formula of the compound made from Al and Cl.

4. Write the name and formula of the compound made from Ca and N.

5. Write the name and formula of the compound made from Al and S.

6. Write the names of the following compounds: Na_2S, MgI_2, NaH, RaO, Al_2S_3, NH_4F.

7. Write the correct formulas for the following names: rubidium nitrate, magnesium oxide, calcium chloride, sodium oxide, beryllium hydride, barium bromide, aluminum oxide.

8. Write the correct formulas for the following names: potassium nitrate, magnesium hypochlorite, lithium permanganate, barium acetate, calcium hydroxide, cesium peroxide, beryllium perchlorate, sodium dichromate, ammonium carbonate, potassium hydrogen sulfite.

9. Write the correct names for the following formulas: KNO_2, Na_2O_2, $Mg(OH)_2$, $Ca(ClO_2)_2$, K_3PO_4, $NaCN$, $Rb_2S_2O_3$, BaC_2O_4, Li_2CrO_4.

10. Write the correct formulas for the following names: gold(III) chloride, nickel(II) hydroxide, stannic fluoride, cadmium thiocyanate, lead(IV) sulfate, chromous carbonate, copper(II) phosphate, mercurous nitrate, silver hydrogen carbonate.

11. Write the correct names for the following formulas: CoS, $Fe_2(SO_4)_3$, PbO, $Ni(ClO_4)_3$, $CdBr_2$, Cu_2O, $Mn(NO_3)_2$, $HgSO_4$, $Cr_3(AsO_4)_2$, $Au(H_2PO_4)_3$.

12. Write the correct formulas for the following names: sulfur dioxide, sulfur trioxide, hydrogen selenide, dichlorine monoxide, phosphorus trichloride, carbon tetrafluoride, tetraboron decahydride (commonly called tetraborane).

13. Write the correct names for the following formulas: PCl_5, CO, CO_2, N_2O_4, P_4O_6 (the common name is phosphorus oxide), I_2O_5, BrF_3, Sb_2O_5.

SUPPLEMENTAL PROBLEMS

14. Name the following compounds.

   a. CaF_2 b. $AlCl_3$ c. MgO d. $SrCl_2$ e. $CoCl_2$ f. $CoCl_3$

15. Name the following compounds.

   a. NO   b. N_2O_3 c. N_2O_5 d. SF_4 e. SF_2 f. SF_6

16. Name the following compounds.

   a. SO_3 b. CO   c. SiF_4 d. N_2O_5 e. P_4S_{10} f. XeF_4

17. Name the following compounds.

   a. PCl_5 b. IF_7 c. P_4O_6 d. As_4O_{10} e. Cl_2O_3 f. Cl_2O_7

18. Name the following oxyacids (assume they are dissolved in water).

   a. H_3BO_3 b. HNO_2 c. H_2CO_3 d. $HClO_3$ e. HIO_3 f. HNO_3
   g. H_3PO_4

19. Name the following acids (assume they are dissolved in water).

   a. HBr   b. HBrO   c. $HBrO_2$ d. $HBrO_3$ e. $HBrO_4$

20. Name the following compounds.

   a. $PbSO_3$ e. LiBrO   i. $CaCO_3$
   b. Mg_2SiO_4 f. NH_4ClO_4 j. $Ba(NO_2)_2$
   c. Na_3PO_4 g. $Fe_2(SO_4)_3$
   d. $Ca(NO_3)_2$ h. $FeSO_4$

21. Write the correct formulas for the following names.

   a. iron(III) sulfate   e. tellurium hexafluoride
   b. chromium(II) sulfite   f. mercury(I) acetate
   c. cupric sulfide   g. disulfur dichloride
   d. silver dihydrogen phosphate   h. ammonium nitrate

22. Write the correct formulas for the following names.

   a. cobalt(III) sulfide   e. calcium bicarbonate
   b. bromine pentafluoride   f. tetraphosphorus trisulfide
   c. ferrous carbonate   g. cuprous cyanide
   d. gold(I) sulfate   h. lanthanum(III) phosphate

23. Write the correct formulas for the following names.

   a. iron(II) oxide   c. magnesium nitride
   b. potassium oxide   d. magnesium nitrite

e. pentasulfur dinitride   h. lead(II) chromate

f. calcium permanganate   i. silver dichromate

g. ferric sulfite

24. Sodium azide is used to generate the nitrogen gas that inflates automobile air bags in the event of a crash. The formula of the azide ion is N_3^-. Write the formula of sodium azide.

25. Many noble gas compounds have been prepared since 1962.

   a. Name the following: XeF_6, XeF_4, XeF_2.

   b. Write the correct formulas for the following names: krypton difluoride, xenon trioxide, xenon tetraoxide.

26. A compound of nitrogen and phosphorus has been discovered in a large gas cloud in the constellation Orion by the astronomer Lucy Ziurys. Since phosphorus is in DNA, this means that the basic building blocks for life as we know it are out there. Write the name and formula for the simplest compound of nitrogen and phosphorus.

17

OXIDATION NUMBERS

Many chemical reactions involve the transfer of electrons from one atom to another. For example, in the mercury battery in your calculator or electric watch, electricity is made when two electrons from a zinc atom are transferred to mercury(II) oxide. This chapter will show you how to calculate oxidation numbers, an important concept that will allow you to keep track of where electrons go in a chemical reaction.

17-1
OXIDATION AND THE GAIN AND LOSS OF ELECTRONS

Let's write down the chemical equations that describe what happens in the mercury battery just mentioned. The reactions are as follows:

$$Zn^0 + 2OH^- \rightarrow ZnO + H_2O + 2e^-$$ zinc loses two electrons to make zinc oxide

$$HgO + H_2O + 2e^- \rightarrow Hg^0 + 2OH^-$$ mercury(II) oxide gains two electrons to make mercury

The zinc goes from a charge of zero in the zinc metal to a charge of $+2$ in the ZnO. The mercury in the HgO has a charge of $+2$ and goes to a charge of zero in the elemental mercury. In the battery, the electrons do not go directly from the Zn to the HgO. They are forced to go through your calculator or watch first, thus supplying the electricity to run these devices.

In a chemical reaction, chemists need to know how many electrons each element has lost or gained. To do this, chemists use the concept of oxidation numbers. **Oxidation** is defined as the losing of electrons. The

(Opposite) The Apollo 11 lunar module during rendezvous in lunar orbit. The chemical reactions that power its engines cause changes in oxidation numbers.

Mercury batteries come in many sizes and shapes.

oxidation number of an atom represents how many electrons an atom has lost or gained, assuming that there is no sharing of electrons. For example, in the reactions of a mercury battery, the zinc atoms, Zn, went from a charge of zero to a charge of +2. The elemental zinc atoms (plain, unreacted zinc) have an oxidation number of zero. The zinc in the zinc oxide, ZnO, has an oxidation number of +2. This means that the zinc has lost two electrons in going from Zn to ZnO. The oxidation number of Zn in ZnO is +2.

Before we proceed with a complete discussion of oxidation numbers, we will introduce the concept of electronegativity.

17-2
ELECTRONEGATIVITY IS A MEASURE OF THE ABILITY OF AN ATOM TO ATTRACT ELECTRONS TOWARD ITSELF IN A COVALENT BOND

The best way to decide which element has gained electrons and which has lost them is to use a table of electronegativities. **Electronegativity** is a measure of the ability of an atom to attract electrons toward itself in a covalent bond. Figure 17-1 lists the electronegativities of the elements arranged according to the periodic table. The concept of electronegativity we will use was defined in 1932 by Linus Pauling (born 1901), the great American chemist who is currently doing research on orthomolecular nutrition and preventive medical diagnosis. The exact details of how Pauling derived the electronegativity numbers won't be discussed in this book. The abbreviation for

	IA	IIA	IIIB	IVB	VB	VIB	VIIB	VIII			IB	IIB	IIIA	IVA	VA	VIA	VIIA	VIIIA
1	H 2.1																	He —
2	Li 1.0	Be 1.5											B 2.0	C 2.5	N 3.0	O 3.5	F 4.0	Ne —
3	Na 0.9	Mg 1.2											Al 1.5	Si 1.8	P 2.1	S 2.5	Cl 3.0	Ar —
4	K 0.8	Ca 1.0	Sc 1.3	Ti 1.5	V 1.6	Cr 1.6	Mn 1.5	Fe 1.8	Co 1.8	Ni 1.8	Cu 1.9	Zn 1.6	Ga 1.6	Ge 1.8	As 2.0	Se 2.4	Br 2.8	Kr —
5	Rb 0.8	Sr 1.0	Y 1.2	Zr 1.4	Nb 1.6	Mo 1.8	Tc 1.9	Ru 2.2	Rh 2.2	Pd 2.2	Ag 1.9	Cd 1.7	In 1.7	Sn 1.8	Sb 1.9	Te 2.1	I 2.5	Xe —
6	Cs 0.7	Ba 0.9	La 1.1	Hf 1.3	Ta 1.5	W 1.7	Re 1.9	Os 2.2	Ir 2.2	Pt 2.2	Au 2.4	Hg 1.9	Tl 1.8	Pb 1.8	Bi 1.9	Po 2.0	At 2.2	Rn —

FIGURE 17-1 A table of electronegativities arranged according to the periodic table. The number below the symbol of an element is its electronegativity. Notice that as you go up and to the right in the periodic table, the electronegativity values increase. Fluorine, in the upper right-hand part of the periodic table, has the largest electronegativity value of 4.0. Electronegativity values for the noble gases (group VIIIA) are not listed.

electronegativity (or the plural, electronegativities) that we will use in this book is EN.

Referring to Figure 17-1, notice that the EN numbers range from 4.0 for fluorine (F) to 0.7 for cesium (Cs). The higher the EN number, the greater the tendency of an atom to attract electrons toward itself in a covalent bond. Thus fluorine has the greatest tendency to attract electrons since it has the largest EN.

EXAMPLE 1   Which element has a greater tendency to attract electrons toward itself, nitrogen (N) or hydrogen (H)?

Solution: From Figure 17-1, the EN of nitrogen is 3.0. The EN of hydrogen is 2.1. Nitrogen has the greater tendency to attract electrons because its EN is greater than hydrogen's.   ■

EXAMPLE 2   Of all the elements in the periodic table, which element has the greatest tendency to attract electrons to itself?

Solution: It is the element with the largest EN, namely, fluorine (F).   ■

Since EN is a measure of the ability of an atom to attract electrons toward itself in a covalent bond, and the oxidation number represents how many electrons an atom has lost, we can say the following: In a compound, the atom with the lower EN will have the more positive oxidation number, and the atom with the higher EN will have a less positive (or more negative) oxidation number.

Remember the following:

lower electronegativity → more positive oxidation number

higher electronegativity → more negative oxidation number

Let's look at the molecule ammonia, NH_3. Which element has the greater tendency to take electrons? From Example 1, the EN of nitrogen is 3.0 and the EN of hydrogen is 2.1. The nitrogen has the greater EN; thus it has the greater tendency to take electrons. The hydrogen has the smaller EN; thus it has a lesser tendency to take electrons. The nitrogen will have the more negative oxidation number, and the hydrogen will have the more positive oxidation number.

At this point, a word of clarification is in order. In a covalent molecule like NH_3, there really isn't a "loss" and a "gain" of electrons. There is an unequal sharing. The nitrogen, being more electronegative than the hydrogen, tends to pull the electrons toward itself in the molecule. This "pulling" is noticeable and affects the properties of the molecule. With this in mind, and to simplify things in this chapter, we will say that in NH_3 the N has "gained" electrons and the H has "lost" electrons. In ionic compounds like NaCl, the Na (EN = 0.9) really does lose an electron to the Cl (EN = 3.0).

EXAMPLE 3   In methane, CH_4, which element has "lost" electrons and which has "gained" them? Which element has the more positive oxidation number?

Solution: The EN are C = 2.5 and H = 2.1. Therefore, since C has the greater EN, C has "gained" electrons and H has "lost" them. The hydrogen has the more positive oxidation number.   ■

You might have noticed that the EN difference of the elements in covalent molecules like CH_4 is small (2.5 − 2.1 = 0.4), whereas the EN difference in ionic compounds like NaCl is large (3.0 − 0.9 = 2.1). However, when we use EN to help us determine oxidation numbers, the distinction between ionic and covalent compounds is not important. In fact, you may want to think of all bonds as being ionic when you are determining oxidation numbers. Just remember in the back of your mind that many bonds are covalent.

In calculating oxidation numbers, we always assume a complete gain or loss of electrons. As we have said, this is not true for most bonds. Therefore, the oxidation numbers are somewhat artificial. Nevertheless, they are useful in keeping track of the *changes* in oxidation numbers that occur to atoms during a chemical reaction.

17-3
CALCULATING OXIDATION NUMBERS IS BASED ON THE ELECTRONEGATIVITIES OF THE ELEMENTS

There are a few rules that will help you to calculate oxidation numbers. A listing and discussions of these rules follow.

RULE 1
Atoms of elements in their elemental form have an oxidation number of zero. Thus the atoms in H_2, O_2, Cl_2, N_2, S_8, Fe, Na, and Ne all have an oxidation number of zero. Their oxidation number is zero because there is complete and equal sharing of the valence electrons in H_2, O_2, Cl_2, N_2, and S_8. Nothing has lost electrons and nothing has gained them. In Fe and Na the

valence electrons are shared throughout the metal in a special kind of bond called a **metallic bond.** In Ne the atoms exist individually, and no sharing takes place.

RULE 2
The oxidation number of a monatomic (one atom) ion is equal to the charge of the ion. In Cl^-, the oxidation number of chlorine is -1. In Al^{3+}, the oxidation number of aluminum is $+3$.

RULE 3
In a neutral molecule, the oxidation numbers of all the individual atoms add up to zero. In an ion, the oxidation numbers add up to the charge on that ion. Thus in CH_4, the sum of the oxidation numbers of the one C atom and the four H atoms equals zero. In SO_4^{2-}, the sum of the numbers of each type of atom equals -2.

RULE 4
In a molecule or ion consisting of two atoms, the atom with the larger EN has the more negative oxidation number.

RULE 5
Fluorine, F, which is the most electronegative element, always has an oxidation number of -1 in compounds. Notice that F is in group VIIA and can attract one electron to complete its valence shell.

RULE 6
Hydrogen, H, usually has an oxidation number of $+1$ in compounds. However, when hydrogen combines with metals to form compounds called hydrides, it has an oxidation number of -1. Most metals have an EN that is less than hydrogen's, so hydrogen would be negative with respect to these metals. Remember that hydrogen has one electron which it can lose. Hydrogen can also gain one electron to complete its s orbital.

RULE 7
Oxygen usually has an oxidation number of -2 in compounds. When oxygen combines with fluorine, it has an oxidation number of $+2$. In peroxides (such as H_2O_2), oxygen has an oxidation number of -1. Oxygen is in group VIA and needs two electrons to complete its valence shell. The EN of oxygen is 3.5, which is the second highest EN. Thus oxygen has a positive oxidation number *only* with respect to fluorine.

RULE 8

Metals in group IA have an oxidation number of +1 in compounds. Metals in group IIA have an oxidation number of +2 in compounds.

A few examples will illustrate the application of these rules.

EXAMPLE 4   Calculate the oxidation number of each atom in the following: He, P_4, F_2, Au, K, Kr, Br_2.

Solution: Since these are all elements, the oxidation number of each atom is zero (Rule 1).   ■

EXAMPLE 5   Calculate the sum of the oxidation numbers in each of the following: CO_2, Ar, PCl_5, H_2SO_4, $MgCl_2$, CO_3^{2-}, NH_4^+, Fe^{3+}, S^{2-}.

Solution: Using Rules 2 and 3, we have

FORMULA	SUM OF OXIDATION NUMBERS
CO_2	0
Ar	0
PCl_5	0
H_2SO_4	0
$MgCl_2$	0
HCO_3^-	-1
CO_3^{2-}	-2
NH_4^+	$+1$
Fe^{3+}	$+3$
S^{2-}	-2

Now we must learn how to calculate the actual numerical value of the oxidation number for each atom in a molecule or polyatomic ion. To do this, we will refer to the periodic table and the table of electronegativities (Figure 17-1). We will also make use of the eight rules just discussed.

EXAMPLE 6   What is the oxidation number of each atom in HCl?

Solution: The EN are H = 2.1 and Cl = 3.0. Thus H has an oxidation number of +1 (Rule 6). Cl must have an oxidation number of -1 (Rule 3). Notice that

Cl, like the other halogens, needs one electron to complete its valence shell. Thus in compounds without oxygen or fluorine, Cl will have an oxidation number of -1. Oxygen and fluorine have a larger EN than Cl; thus in compounds with oxygen and fluorine, Cl will have a positive oxidation number. ■

EXAMPLE 7   What is the oxidation number of each atom in sodium hydride, NaH?

Solution: The EN are Na $= 0.9$ and H $= 2.1$. Therefore, H has an oxidation number of -1 (Rule 6), and Na has an oxidation number of $+1$ (Rules 3 and 8). ■

EXAMPLE 8   What is the oxidation number of each atom in boron nitride, BN?

Solution: The EN are B $= 2.0$ and N $= 3.0$. Thus N is negative with respect to B. To determine the numerical values of the oxidation numbers, we see that N is in group VA of the periodic table and needs three electrons to complete its valence shell. Thus the oxidation numbers are N $= -3$ and B $= +3$. ■

EXAMPLE 9   What is the oxidation number of each element in water, H_2O?

Solution: The EN values are H $= 2.1$ and O $= 3.5$. Thus the oxidation number of O is -2 (Rule 7), and that of each H is $+1$ (Rule 6). To see whether Rule 3 is obeyed, add up all the oxidation numbers of each atom: $+1 + 1 - 2 = 0$. ■

A good way to keep track of oxidation numbers is to write them in the following way.

1. Write the chemical formula.
2. Above each atom, write its oxidation number if you know it.
3. Then use Rule 3 to calculate the oxidation numbers of the remaining atoms. Usually there will be only one remaining atom.
4. On the first line write the oxidation number of each single atom.
5. Above these numbers write the total oxidation number for all the atoms of that element.

Let's see how this would look for water:

$$+2 \ -2 \leftarrow \text{sum of the oxidation numbers of all the atoms of an element}$$

$$+1 \ -2 \leftarrow \text{oxidation number of each single atom}$$

$$H_2O$$

The top line of numbers should add up to zero for a neutral molecule. If you have an ion, the top line will add up to the charge on the ion.

EXAMPLE 10   What is the oxidation number of sulfur in sulfur trioxide, SO_3?

Solution: The EN are S = 2.5 and O = 3.5. Therefore, S is positive with respect to oxygen. The oxidation number of oxygen is -2 (Rule 7). Therefore, we have

$$+6 \ -6 \leftarrow \text{sum of oxidation numbers}$$

$$+6 \ -2 \leftarrow \text{individual oxidation numbers}$$

$$S \ O_3$$

Thus the oxidation number of S is $+6$, the number directly above the S.   ■

EXAMPLE 11   What is the oxidation number of sulfur in sulfur dioxide, SO_2?

Solution: The EN are S = 2.5 and O = 3.5. Sulfur is positive with respect to oxygen; thus we have

$$+4 \ -4$$

$$+4 \ -2$$

$$S \ O_2$$

The oxidation number of S in SO_2 is $+4$, which is the number directly above the S.   ■

Examples 10 and 11 illustrate an important idea. The oxidation number of a given element may vary, depending on what compound it is in. In fact, the oxidation number of sulfur can range from $+6$ to -2 (see Problem 28).

EXAMPLE 12   What is the oxidation number of sulfur in hydrogen sulfide, H_2S?

Solution: The EN are H = 2.1 and S = 2.5. Thus S is negative with respect to H. The oxidation number of H is +1 (Rule 6). Therefore, we have

$$\begin{array}{cc} +2 & -2 \\ +1 & -2 \end{array}$$

$$H_2\,S$$

The oxidation number of S in H_2S is −2, the number directly above the S. ∎

EXAMPLE 13   What is the oxidation number of sulfur in the sulfate ion, SO_4^{2-}?

Solution: The EN are S = 2.5 and O = 3.5. The S is positive with respect to oxygen. SO_4^{2-} is an ion with a charge of −2, so the sum of all the oxidation numbers must add up to −2 (Rule 3):

$$+6 \; -8 \; = \; -2 \; \leftarrow \; \begin{array}{l}\text{this is the charge} \\ \text{on the ion}\end{array}$$

$$\begin{array}{cc} +6 & -2 \end{array}$$

$$S \; O_4$$

The sum of the oxidation numbers is +6 − 8 = −2. The oxidation number of S = +6, which is the number directly above the S. ∎

EXAMPLE 14   What is the oxidation number of each atom in the thiosulfate ion, $S_2O_3^{2-}$?

Solution:

$$+4 \; -6 \; = \; -2 \; \leftarrow \; \begin{array}{l}\text{this is the charge} \\ \text{on the ion}\end{array}$$

$$\begin{array}{cc} +2 & -2 \end{array}$$

$$S_2O_3$$

The sum of the oxidation numbers of the three oxygen atoms is −6. Thus the two S atoms must share a +4 total oxidation number. Each S atom has a +4/2 = +2 oxidation number, the number directly above the S. ∎

EXAMPLE 15   What is the oxidation number of each element in OF_2?

Solution: The EN are O = 3.5 and F = 4.0. F is negative with respect to oxygen. The oxidation number of F is always -1 (Rule 5).

$$+2 \quad -2$$
$$+2 \quad -1$$
$$O \ F_2$$

The oxidation numbers are O = $+2$ and F = -1. You might check Rule 7 at this point. ■

EXAMPLE 16   What is the oxidation number of each element in $NaClO_4$?

Solution: The EN are Na = 0.9, Cl = 3.0, and O = 3.5. The Na and Cl are positive with respect to the oxygen. The oxidation number of Na is $+1$ (Rule 8), and that of O is -2 (Rule 7):

$$+1 \quad +7 \quad -8$$
$$+1 \quad +7 \quad -2$$
$$Na \ Cl \ O_4$$

The oxidation number of Cl is $+7$, since $+1 + 7 - 8 = 0$. ■

In Rule 7 we mentioned peroxides. In peroxides the oxygen atoms are bound to each other. The structure of hydrogen peroxide, H_2O_2, is

You can see that the oxygen atoms are bonded to each other. This increases the oxidation number of each oxygen atom (makes the oxidation number less negative) as the following example shows.

EXAMPLE 17   What is the oxidation number of each element in H_2O_2?

Solution: The EN are H = 2.1 and O = 3.5. Hydrogen is positive with respect to oxygen. Hydrogen has an oxidation number of $+1$ (Rule 6).

$$+2\ -2$$

$$+1\ -1$$

$$H_2O_2$$

The oxidation number of each oxygen atom is -1, since $-2/2 = -1$. ∎

17-4
CALCULATING THE OXIDATION NUMBER OF ATOMS IN CARBON COMPOUNDS USES THE ELECTRONEGATIVITIES OF THE ATOMS

We will discuss one more interesting point about oxidation numbers. This is how to calculate the oxidation number of carbon in certain organic compounds. For many organic compounds, we have already learned the proper technique. For instance, in CH_4 the oxidation numbers are $C = -4$ and $H = +1$. In CO_2 the oxidation numbers are $C = +4$ and $O = -2$. Notice that carbon can have oxidation numbers ranging from $+4$ to -4. This is because carbon is in group IVA in the periodic table. It has four valence electrons. Thus a carbon atom can either "gain" four electrons to get eight electrons in its valence shell or "lose" four electrons to have none in its valence shell. Carbon can also "gain" or "lose" any number of electrons between $+4$ and -4.

The problems we will discuss here is calculating oxidation numbers in compounds like acetic acid ($C_2H_4O_2$) and acetone (C_3H_6O). For acetic acid, we can write (EN are $C = 2.5$, $H = 2.1$, and $O = 3.5$)

$$0\ +4\ -4$$

$$0\ +1\ -2$$

$$C_2H_4O_2$$

Since the oxidation numbers of the H and O atoms add up to zero, the total oxidation number of both carbon atoms appears to be zero. But since there are two carbon atoms, it is possible that the apparent zero oxidation number is simply the average of equal positive and negative oxidation numbers. As we shall see later, this is actually the case. One carbon atom has an oxidation number of $+3$, and the other has an oxidation number of -3. The average of $+3$ and -3 is zero. We will discuss acetic acid in detail after we make a few more points.

Now let's look at acetone, C_3H_6O:

$$-4\ +6\ -2$$

$$-1.33\ +1\ -2$$

$$C_3H_6O$$

The oxidation number of O is -2, and that of H is $+1$. Thus the sum of the oxidation numbers of hydrogen and oxygen is $6(+1) - 2 = +4$. The total oxidation number of the three carbon atoms must equal -4. If we divide -4 equally among three carbon atoms, we get $-4/3 = -1.33$. It appears that each carbon atom has an oxidation number of -1.33. How can this be? Oxidation numbers are usually whole numbers. (However, see Problem 27 for an exception.)

The answer is that the oxidation numbers in acetone *are* whole numbers, but the oxidation numbers on different carbon atoms are not necessarily the same. To see this, we must resort to structural formulas.

Let's start with methane (CH_4) and carbon dioxide (CO_2). The structural formulas are

$$
\begin{array}{c}
\text{H} \\
| \\
\text{H}-\text{C}-\text{H} \qquad \text{O}=\text{C}=\text{O} \\
| \\
\text{H}
\end{array}
$$

A good way to figure out oxidation numbers if you have structural formulas is to put a "+" sign on each atom's bond that has the lower EN and to put a "−" sign on each atom's bond that has the higher EN.

Doing this for methane and carbon dioxide, we have (the EN are $C = 2.5$, $H = 2.1$, and $O = 3.5$)

$$
\begin{array}{c}
\overset{+}{\text{H}} \\
\overset{+}{|} \\
\text{H}\overset{+}{\underset{-}{=}}\text{C}\overset{-}{\underset{+}{=}}\text{H} \qquad \text{O}\overset{-}{\underset{=+}{=}}\text{C}\overset{+}{\underset{+}{=}}\text{O} \\
\overset{-}{\underset{+}{|}} \\
\text{H}
\end{array}
$$

Add up the "+" signs and "−" signs around each atom. The result is the oxidation number of each atom. In methane the C atom has four "−" signs around it; therefore, the oxidation number of C is -4. The oxidation number of each H is $+1$. For CO_2 the oxidation number for C is $+4$, and that of each O is -2.

Now back to acetic acid. The structural formula with "+" signs and "−" signs is

$$
\begin{array}{c}
\text{H} \qquad\qquad \text{O} \\
| \qquad\qquad \diagup\diagup \\
\text{H}-\text{C}-\text{C} \\
| \qquad\qquad \diagdown \\
\text{H} \qquad\qquad \text{O}-\text{H}
\end{array}
$$

NOTE: If a bond connects two identical atoms, such as C—C, there is no "gaining" or "losing" of electrons in this bond. There is a complete sharing, and we put a zero at each end of the bond.

Now add up all the "+" signs and "−" signs around each atom. All the H atoms have +1. All the O atoms have −2. These are the oxidation numbers of H and O atoms, respectively. Now look at the carbon atoms. The C atom on the left has a total charge of −3. Its oxidation number is −3. The C atom on the right has a total charge of +3, so its oxidation number is +3. The average of +3 and −3 is zero, just what we found before as the average oxidation number of both carbon atoms. (If you work Problem 29, you will see a carbon atom that actually has a zero oxidation number.)

For acetone the structural formula with "+" signs and "−" signs is

$$
\begin{array}{ccc}
\text{H} & \text{O} & \text{H} \\[2pt]
\overset{\scriptstyle +}{|} & \overset{\scriptstyle -\,||\,-}{} & \overset{\scriptstyle +}{|} \\
\text{H}\overset{+}{=}\overset{-}{\text{C}}\,{}^{0}\!\!-\!\!{}^{0}\,\overset{+}{\text{C}}\,{}^{0}\!\!-\!\!{}^{0}\,\overset{-}{\text{C}}\overset{+}{=}\text{H} \\
\underset{\scriptstyle +}{|} & & \underset{\scriptstyle +}{|} \\
\text{H} & & \text{H}
\end{array}
$$

The oxidation number of each H atom is +1, and that of the O atom is −2. The oxidation number of the leftmost carbon is −3 + 0 = −3. The oxidation number of the rightmost carbon atom is the same, −3. However, the oxidation number of the central carbon atom is +2 + 0 = +2. The average oxidation number of all three carbon atoms is

$$
\frac{-3 + 2 - 3}{3} = \frac{-4}{3} = -1.33
$$

which is the same as we found before.

If you study the structural formulas of acetic acid and acetone, you can see that the carbon atoms bonded to oxygen atoms (which are very electronegative) are more positive than carbon atoms bonded to hydrogen atoms (which are much less electronegative than oxygen). Carbon atoms with different oxidation numbers undergo very different kinds of chemical reactions, as you may learn in your further study of chemistry.

EXAMPLE 18 What is the oxidation number of each atom in a molecule of chloroacetic acid, $CH_2ClCOOH$? The structural formula is

$$
\begin{array}{c}
\text{H} \\
| \\
\text{Cl}-\text{C}-\text{C}\!\!\begin{array}{c} \nearrow \text{O} \\ \searrow \text{O}-\text{H} \end{array} \\
| \\
\text{H}
\end{array}
$$

Solution: The difference between chloroacetic acid and acetic acid is the Cl atom replacing an H atom. The EN are Cl = 3.0, C = 2.5, and O = 3.5. Putting in "+" signs and "−" signs we have

$$
\begin{array}{c}
\quad\quad \text{H} \quad\quad\quad\quad \text{O} \\
\quad\quad |^{+} \quad\quad\quad\quad \nearrow^{-} \\
\text{Cl}{=}^{+}\text{C}^{0}{-}^{0}\text{C}^{+} \\
\quad\quad |^{+} \quad\quad\quad\quad \searrow^{+} \\
\quad\quad \text{H} \quad\quad\quad\quad \text{O}{=}^{+}\text{H}
\end{array}
$$

The oxidation numbers are Cl = −1, H = +1, and O = −2. The oxidation number of the left-hand carbon atom is $-1 - 1 + 1 + 0 = -1$. The oxidation number of the right-hand carbon is +3. Notice that in acetic acid the oxidation number of the left-hand carbon atom was −3. Since Cl is more electronegative than H, it has the effect of making the oxidation number of the left-hand carbon atom in chloroacetic acid more positive (−1 is more positive than −3). The oxidation number of the right-hand carbon hasn't changed when compared to acetic acid, because the atoms bonded to it haven't changed. ∎

As we said earlier, carbon atoms can have oxidation numbers ranging from +4 to −4. If you look at the Lewis formula of a carbon atom, $\overset{..}{\text{C}}\cdot$, you see that carbon can "lose" up to four electrons and "gain" up to four electrons in compounds. This is why the oxidation number of carbon can range from +4 to −4.

PROBLEMS

KEYED PROBLEMS

1. Which element has a greater tendency to attract electrons toward itself in a covalent bond, oxygen (O) or hydrogen (H)?

2. Of all the elements in the periodic table, which element has the second greatest tendency to attract electrons toward itself in a covalent bond?

3. In the molecule HF, which atom has "lost" electrons and which has "gained" electrons? Which atom has the more positive oxidation number?

4. What are the oxidation numbers of each atom in the following: Ne, O_2, Ca, Cl_2, S_8, O_3?

5. What is sum of the oxidation numbers in each of the following: H_2O, CO, HNO_3, $CaCl_2$, H_2SO_4, HSO_4^-, PO_4^{3-}, NO^+, NH_3, NH_4^+, Br^-, PCl_3?

6. What is the oxidation number of each atom in HBr?

7. What is the oxidation number of each atom in lithium hydride, LiH?

8. What is the oxidation number of each atom in aluminum nitride, AlN?

9. What is the oxidation number of each atom in hydrogen sulfide, H_2S?

10. What is the oxidation number of each atom in chlorine trioxide, ClO_3?

11. What is the oxidation number of each atom in chlorine dioxide, ClO_2?

12. What is the oxidation number of each atom in H_2Se?

13. What is the oxidation number of each atom in the perchlorate ion, ClO_4^-?

14. What is the oxidation number of each atom in the tetrathionate ion, $S_4O_6^{2-}$?

15. What is the oxidation number of each atom in ClF_3?

16. What is the oxidation number of each atom in $NaClO_3$?

17. What is the oxidation number of each atom in sodium peroxide, Na_2O_2?

18. What is the oxidation number of each atom in fluoroacetic acid, whose structural formula is

$$
\begin{array}{ccc}
& H & \\
& | & O \\
F - & C - C & \diagup \\
& | & \diagdown \\
& H & O-H \\
\end{array}
$$

SUPPLEMENTAL PROBLEMS

19. What is the oxidation number of each atom in the following?

a. BrCl   c. BrF_3 e. ClF
b. BrF   d. BrF_5 f. ClF_5

20. What is the oxidation number of each atom in the following?

a. $HClO_4$ c. $HClO_2$ e. Cl_2
b. $HClO_3$ d. HClO   f. HCl

21. What is the oxidation number of each atom in the following?

a. N_2O c. N_2O_3 e. N_2O_5 g. NH_2OH
b. NO   d. NO_2 f. NH_3 h. N_2H_4

NOTE: If you do all of Problem 21, you will see that the N atom can have oxidation numbers ranging from +5 to −3. If you look at the Lewis formula of a nitrogen atom, $\cdot \ddot{N} \cdot$, you see that nitrogen can "lose" up to five electrons or "gain" up to three electrons in compounds. This is why the oxidation number of nitrogen can go from +5 to -3.

22. What is the oxidation number of each atom in the following?

a. H_3PO_4 c. HPO_4^{2-} e. H_3PO_2

b. $H_2PO_4^-$ d. PO_4^{3-} f. $H_2PO_2^-$

23. What is the oxidation number of each atom in the following?

a. MnO c. Mn_2O_7 e. $MnCl_3$

b. Mn_2O_3 d. $MnCl_2$ f. $KMnO_4$

24. What is the oxidation number of each atom in the following?

a. $CuCl$ c. Hg_2Cl_2 e. $AuCl_4^-$

b. $CuCl_2$ d. $HgCl_2$ f. $AuCl$

25. What is the oxidation number of each atom in the following?

a.   b.   c.

NOTE: As the number of oxygen bonds attached to a carbon atom increases, the oxidation number of this carbon atom increases. The more oxygen atoms attached to a carbon atom, the more the carbon atom is oxidized.

26. What is the oxidation number of each atom in the following?

a.   b.   c.

d.

27. The superoxide ion, O_2^-, is important in some biological systems. What is the oxidation number of the oxygen atoms in the superoxide ion? (Hint: Contrary to what we said in this chapter, here is a case where a fractional oxidation number exists.)

28. As Examples 10 through 14 indicate, sulfur can have an oxidation number that ranges from -2 to $+6$. Draw the Lewis formula of a sulfur atom and convince yourself that this range is reasonable.

29. Calculate the oxidation number of carbon in formaldehyde, whose structural formula is

$$\begin{array}{c} H \\ \diagdown \\ \diagup C=O \\ H \end{array}$$

(Here is a compound in which carbon actually has a zero oxidation number.)

18

BALANCING OXIDATION–REDUCTION EQUATIONS

In our discussion of oxidation numbers, we mentioned a chemical reaction that involves the transfer of electrons from one substance to another. The example chosen was the chemistry of a mercury battery.

The reaction in which electrons are lost is

$$Zn + 2OH^- \rightarrow ZnO + H_2O + 2e^- \tag{1}$$

The reaction in which electrons are gained is

$$HgO + H_2O + 2e^- \rightarrow Hg + 2OH^- \tag{2}$$

Together these reactions make up the overall reaction involved in a mercury battery. The overall reaction

$$Zn + HgO \rightarrow ZnO + Hg \tag{3}$$

is obtained by adding reactions 1 and 2 and subtracting H_2O, OH^-, and electrons from both sides. This chapter will teach you how to write and balance reactions that involve the transfer of electrons from one substance to another.

(Opposite) The Shasta Dam and Powerplant. Much of the electricity generated by hydroelectric power is used in the production of magnesium and aluminum metals in a redox reaction.

18-1
OXIDATION AND REDUCTION IN CHEMICAL REACTIONS INVOLVE THE LOSS AND GAIN OF ELECTRONS

Before we start to balance equations, we must define the terms oxidation and reduction. **Oxidation** is the loss of electrons, whereas **reduction** is the gain of electrons. In the mercury battery reactions, reaction 1 represents an oxidation:

$$Zn + 2OH^- \rightarrow ZnO + H_2O + 2e^- \tag{1}$$

Notice that the electrons appear on the right side of the arrow. The electrons are products of the reaction. The Zn metal has lost two electrons to form ZnO. Reaction 2 represents a reduction:

$$HgO + H_2O + 2e^- \rightarrow Hg + 2OH^- \tag{2}$$

The electrons appear on the left side of the arrow. The electrons are reactants. The mercury in the HgO has gained two electrons to become the mercury metal, Hg.

The combination of the oxidation reaction (1) and the reduction reaction (2) gives the overall reaction, reaction 3:

$$Zn + HgO \rightarrow ZnO + Hg \tag{3}$$

Even though the electrons are not explicitly written, reaction 3 involves both an oxidation and a reduction. Therefore, it is called an **oxidation–reduction reaction.** The abbreviated name is **redox,** from **red**uction and **ox**idation.

In every redox reaction, the oxidation numbers of some elements are changed. Let's assign oxidation numbers to each element in reaction 3:

$$\overset{0}{Zn} + \overset{+2\;-2}{Hg\,O} \rightarrow \overset{+2\;-2}{Zn\,O} + \overset{0}{Hg} \tag{3}$$

The zinc goes from zero to +2. The mercury goes from +2 to zero. The zinc is oxidized (its oxidation number increases from zero to +2) and the mercury is reduced (its oxidation number decreases from +2 to zero). The way to tell whether a chemical reaction is a redox reaction is to calculate the oxidation number of each atom in the reaction. If there is a change in oxidation numbers in a reaction, the reaction is a redox reaction.[1]

The design of batteries and fuel cells, electroplating of metals, production of metals such as aluminum and magnesium, production of hydrogen

[1]Sometimes only one substance will be both oxidized and reduced. An example is in the reaction $2Cu^+ \rightarrow Cu^{2+} + Cu^0$. This is called a *disproportionation* or a *dismutation*. We will not discuss such reactions further except to offer Problem 17.

from sunlight for energy storage, some analytical chemistry techniques, and many reactions in biological systems involve redox reactions.

EXAMPLE 1   Is the following a redox reaction?

$$Fe + Cu^{2+} \rightarrow Fe^{2+} + Cu$$

Solution: Assign oxidation numbers to each atom:

$$\overset{0}{Fe} + \overset{+2}{Cu^{2+}} \rightarrow \overset{+2}{Fe^{2+}} + \overset{0}{Cu}$$

Since oxidation numbers change, the reaction *is* a redox reaction. ■

EXAMPLE 2   Is the following a redox reaction?

$$HCl + NaOH \rightarrow NaCl + HOH$$

Solution: Assign oxidation numbers to each atom:

$$\overset{+1-1}{HCl} + \overset{+1-2+1}{Na\ O\ H} \rightarrow \overset{+1-1}{NaCl} + \overset{+1-2+1}{H\ O\ H}$$

Since there is no change in oxidation numbers, the reaction is *not* a redox reaction. ■

18-2
WHEN BALANCING REDOX EQUATIONS USING HALF-REACTIONS, LEAVE OUT THE SPECTATOR IONS

Now we will give the details of how to balance redox reactions. To start, we will use the reaction in Example 1, since it illustrates many of the features of redox reactions. The overall reaction, which takes place in a water solution, is

$$Fe + Cu^{2+} \rightarrow Fe^{2+} + Cu$$

We can write this reaction in two parts—the oxidation reaction and the reduction reaction:

$$Fe \rightarrow Fe^{2+} + 2e^{-} \qquad \text{oxidation (electrons lost)}$$
$$Cu^{2+} + 2e^{-} \rightarrow Cu \qquad \text{reduction (electrons gained)}$$

Each of these reactions is called a **half-reaction,** since each is one-half of the overall reaction.

Let's look at the oxidation half-reaction in some detail. The Fe goes to Fe^{2+} by giving up two electrons. Since each electron has a charge of -1, two of them are lost when Fe goes from an oxidation number of zero to $+2$. The total charge on each side of the arrow is the same. The total charge on the left-hand side is zero because Fe is neutral. The total charge on the right-hand side is zero:

$$+2 + (-2) = +2 - 2 = 0$$

from Fe^{2+} from $2e^-$

Now look at the reduction half-reaction. The two electrons lost by the iron are gained by the copper in going from $+2$ to zero. The total charge on the left-hand side of the arrow is zero:

$$+2 + (-2) = +2 - 2 = 0$$

from Cu^{2+} from $2e^-$

The total charge on the right-hand side is also zero because Cu is neutral.

Let's see how to add the two half-reactions to get the overall reaction. Line up the arrows and then add everything on each side of the arrow. The order of the terms on each side of the arrow doesn't matter.

$Fe \rightarrow Fe^{2+} + 2e^-$	oxidation
$Cu^{2+} + 2e^- \rightarrow Cu$	reduction
$Fe + Cu^{2+} + 2e^- \rightarrow Fe^{2+} + Cu + 2e^-$	sum of half-reactions

Since two electrons are on each side of the arrow and we can always subtract the same thing from both sides of an equation, we can subtract two electrons from each side (we will cross them out),

$$Fe + Cu^{2+} + 2\!\!\!/e^- \rightarrow Fe^{2+} + Cu + 2\!\!\!/e^-$$

and obtain the overall reaction

$$Fe + Cu^{2+} \rightarrow Fe^{2+} + Cu \quad \text{overall balanced equation}$$

EXAMPLE 3   Write the two half-reactions of the following unbalanced overall reaction. This reaction doesn't take place in water; instead, it occurs in the molten (melted) state:

$$Na + Mg^{2+} \rightarrow Na^+ + Mg$$

Solution: The two half-reactions are

$$Na \rightarrow Na^+ + e^- \qquad \text{oxidation}$$

$$Mg^{2+} + 2e^- \rightarrow Mg \qquad \text{reduction}$$

Before we can add these up, we must equalize the number of electrons that are lost and gained. Thus we need to multiply the oxidation half-reaction by 2:

$$2Na \rightarrow 2Na^+ + 2e^- \qquad \text{oxidation}$$

$$\underline{Mg^{2+} + 2e^- \rightarrow Mg \qquad \text{reduction}}$$

$$2Na + Mg^{2+} + 2e^- \rightarrow 2Na^+ + Mg + 2e^- \qquad \text{sum of half-reactions}$$

$$2Na + Mg^{2+} \rightarrow 2Na^+ + Mg \qquad \text{balanced overall equation} \quad \blacksquare$$

You may wonder about the equations in Example 3. Where are the negative ions, which keep the solution electrically neutral? Well, they are there, but we didn't write them since they don't take an active part in the chemical reaction. Ions that don't take an active part in a chemical reaction are called **spectator ions.**

EXAMPLE 4   Write the equation from Example 3 as a molecular equation and as a complete ionic equation. For simplicity, use Cl^- as the spectator ion.

Solution: The molecular equation is

$$2Na + MgCl_2 \rightarrow 2NaCl + Mg$$

The complete ionic equation is

$$Na + Mg^{2+} + 2Cl^- \rightarrow 2Na^+ + 2Cl^- + Mg$$

In the complete ionic equation, we can see that Cl^- doesn't undergo a change in oxidation number and can be subtracted from both sides. When we do this, we get the overall balanced equation without spectator ions:

$$Na + Mg^{2+} \rightarrow 2Na^+ + Mg \quad \blacksquare$$

It is common to leave out spectator ions when writing redox reactions because doing so makes it easier for us. Our attention is focused entirely on the species that undergo change.

In many reactions in water solution, oxygen and hydrogen are present in the reactants and products. It is thus necessary to have a convenient way to balance oxygen and hydrogen. The most realistic way is to use molecules and ions commonly present in water. They are H_2O, H^+, and OH^-. The H^+ and OH^- come from water or from acid (H^+) or base (OH^-) put into the water. If the solution is acidic, there are many more H^+ ions than OH^- ions. If the solution is basic, there are many more OH^- ions than H^+ ions. So in acidic solution we will use H_2O and H^+ to balance oxygen and hydrogen. Similarly, in basic solution we will use H_2O and OH^-. Electrons (e^-) are used to balance charge.

Many reactions that occur in water solution are written as ionic reactions because the substances involved actually exist in water in an ionized form.

At this point, it might be useful to summarize the ideas we have presented for balancing redox reactions.

18-3
THE RULES FOR BALANCING REDOX EQUATIONS ARE BASED ON MASS AND ELECTRICAL BALANCE

The rules for balancing redox equations are as follows.

1. Calculate the oxidation number of each element to decide which ones undergo oxidation and reduction.

2. If it isn't already written in this way, write the reaction such that the spectator ions are eliminated.

3. Write the two half-reactions—one for the oxidation and the other for the reduction.

4. Balance each half-reaction **materially;** that is, the number of atoms of each element must be equal on both sides of the equation. This procedure is also referred to as **mass balance.** (You learned how to balance equations materially in Chapter 3.)

5. Balance each half-reaction **electrically;** that is, the charge must be the same on each side of the equation. To balance the charge, add electrons (e^-) to the more positive side.

6. Equalize the number of electrons in each half-reaction by taking multiples of one or both half-reactions.

7. Add both balanced half-reactions. Subtract the electrons (which must be the same on each side of the equation) and chemical species that are common to both sides.

8. Check the final balanced overall equation to see whether it is materially and electrically balanced.

18-4
BALANCE REDOX EQUATIONS IN ACIDIC SOLUTION BY USING H$^+$ AND H$_2$O TO BALANCE HYDROGEN AND OXYGEN ATOMS

Let's look at the following reaction that takes place in acid solution:

unbalanced molecular equation:

$$KMnO_4 + FeCl_2 \rightarrow MnCl_2 + FeCl_3 + KCl$$

unbalanced ionic equation:

$$K^+ + MnO_4^- + Fe^{2+} + 2Cl^- \rightarrow Mn^{2+} + Fe^{3+} + 6Cl^- + K^+$$

The spectator ions are K$^+$ and Cl$^-$. In a redox reaction, the spectator ions don't change oxidation number. Notice that the oxidation numbers of both K$^+$ and Cl$^-$ don't change. Since the spectator ions are K$^+$ and Cl$^-$, we remove them to get the redox reaction without spectator ions (Rule 2):

$$MnO_4^- + Fe^{2+} \rightarrow Mn^{2+} + Fe^{3+}$$

The oxidation numbers of each atom are Mn (in MnO$_4^-$) = +7, Fe^{2+} = +2, Mn^{2+} = +2, and Fe^{3+} = +3 (Rule 1). The unbalanced half-reactions are (Rule 3)

$$MnO_4^+ \rightarrow Mn^{2+} \qquad \text{reduction (Mn goes from +7 to +2)}$$

$$Fe^{2+} \rightarrow Fe^{3+} \qquad \text{oxidation (Fe goes from +2 to +3)}$$

Now we must balance each half-reaction materially (Rule 4). In the reduction half-reaction, there are four oxygen atoms in the MnO$_4^-$ ion on the left-hand side of the equation. This oxygen must be accounted for on the right-hand side of the equation. Since the reaction takes place in a water solution, it seems reasonable that H$_2$O can be used as a source of oxygen. So we will use four H$_2$O molecules to put four oxygen atoms on the right-hand side of the equation:

$$MnO_4^- \rightarrow Mn^{2+} + 4H_2O$$

But now we have added eight H atoms to the right side. Since there are no H atoms on the left-hand side, we must add eight. The best way to do this, since the reaction takes place in acidic solution, is to use eight H$^+$ ions. We now have

$$MnO_4^- + 8H^+ \rightarrow Mn^{2+} + 4H_2O$$

The next thing to do is to balance the half-reaction electrically (Rule 5). A good way to do this is to

1. Add up the charges on the left-hand side of the arrow.
2. Add up the charges on the right-hand side of the arrow.
3. Add enough electrons (e⁻) to the more *positive* side so that both sides have the same charge.

On the left-hand side of the arrow, the MnO_4^- has one negative charge. The eight H^+ contribute eight positive charges. We have $-1 + 8 = +7$ charges on the left-hand side. On the right-hand side the Mn^{2+} has two positive charges. The four H_2O molecules are neutral and contribute no charge. Thus we have a +2 charge on the right-hand side:

	Left-hand side	**Right-hand side**
total charge:	+7	+2

The left-hand side is more positive. If we add five electrons to the left-hand side, then the charge on the left side will equal +2, the same as the charge on the right-hand side:

	Left-hand side	**Right-hand side**
total charge after adding five electrons	$+7 + (-5) =$ $+7 - 5 = +2$	+2

The balanced reduction half-reaction now reads

$$MnO_4^- + 8H^+ + 5e^- \rightarrow Mn^{2+} + 4H_2O$$

The oxidation half-reaction is easier to balance. It is already balanced materially. All we have to do is add one electron to the right-hand side to balance the charge:

$$Fe^{2+} \rightarrow Fe^{3+} + e^-$$

The total charge on each side of the arrow is now +2.
Both balanced half-reactions now read

$$MnO_4^- + 8H^+ + 5e^- \rightarrow Mn^{2+} + 4H_2O \qquad \text{reduction}$$
$$Fe^{2+} \rightarrow Fe^{3+} + e^- \qquad \text{oxidation}$$

Before we can add the two half-reactions, we must equalize the number of electrons (Rule 6). To do this, we notice that the reduction gains *five* electrons and that the oxidation loses only *one* electron. So we must multiply the oxidation half-reaction by *five*. Writing both half-reactions and adding them (Rule 7), we get

$$MnO_4^- + 8H^+ + 5e^- \rightarrow Mn^{2+} + 4H_2O \qquad \text{reduction}$$

$$\underline{5Fe^{2+} \rightarrow 5Fe^{3+} + 5e^-} \qquad \text{oxidation}$$

$$MnO_4^- + 8H^+ + 5e^- + 5Fe^{2+} \rightarrow Mn^{2+} + 4H_2O + 5Fe^{3+} + 5e^- \qquad \text{sum of half-reactions}$$

We can now subtract $5e^-$ from each side:

$$MnO_4^- + 8H^+ + 5Fe^{2+} \rightarrow Mn^{2+} + 4H_2O + 5Fe^{3+} \qquad \text{overall balanced equation}$$

Since there are no chemical species common to both sides that can be subtracted, this is the final overall balanced equation.

You might be interested in seeing how the equation looks as a molecular equation after putting back all the spectator ions (K^+ and Cl^-) and writing molecular formulas. K^+ ions are used to combine with the negative ions, and Cl^- ions are used to combine with the positive ions. You are not expected to be able to do the following from what you have learned so far:

$$KMnO_4 + 8HCl + 5FeCl_2 \rightarrow MnCl_2 + 4H_2O + 5FeCl_3 + KCl$$

18-5
BALANCE REDOX EQUATIONS IN BASIC SOLUTION BY USING OH⁻ AND H₂O TO BALANCE HYDROGEN AND OXYGEN ATOMS

Now we will illustrate how to balance a redox reaction in basic solution. The unbalanced overall equation without spectator ions (the reactants and products are in ionic form) is

$$CrO_4^{2-} + SO_3^{2-} \rightarrow CrO_2^- + SO_4^{2-}$$

The oxidation number of each element is

$$\overset{+6-2}{CrO_4^{2-}} + \overset{+4-2}{SO_3^{2-}} \rightarrow \overset{+3-2}{CrO_2^-} + \overset{+6-2}{SO_4^{2-}}$$

The Cr goes from $+6$ to $+3$; thus it is reduced. The S goes from $+4$ to $+6$; thus it is oxidized. The two unbalanced half-reactions are

$$CrO_4^{2-} \rightarrow CrO_2^- \qquad \text{reduction}$$

$$SO_3^{2-} \rightarrow SO_4^{2-} \qquad \text{oxidation}$$

Let's balance the reduction half-reaction first. There are four oxygen atoms on the left-hand side and two on the right-hand side. Two oxygen atoms must be added to the right-hand side. This is done by adding $2H_2O$ to the right-hand side:

$$CrO_4^{2-} \rightarrow CrO_2^- + 2H_2O$$

Now we must add four H atoms to the left-hand side to compensate for the four H atoms in the $2H_2O$. Since the reaction takes place in basic solution, we can add four H atoms to the left-hand side by adding $4H_2O$ to the left-hand side and $4OH^-$ to the right-hand side. Notice what this does by looking at the individual atoms in H_2O and OH^- and subtracting common atoms (we have crossed them out):

$$H_2\cancel{O} \rightleftharpoons \cancel{O}H^- \quad \text{or} \quad H\!-\!\cancel{O}\!-\!\cancel{H} \rightleftharpoons \cancel{O}\!-\!\cancel{H}^-$$

Adding an H_2O to the left-hand side and an OH^- to the right-hand side of the equation is like adding one H atom to the left-hand side. Thus the reduction half-reaction becomes

$$CrO_4^{2-} + 4H_2O \rightarrow CrO_2^- + 2H_2O + 4OH^-$$

We can subtract $2H_2O$ from each side:

$$CrO_4^{2-} + 2H_2O \rightarrow CrO_2^- + 4OH^-$$

To balance the charges, note that the right-hand side has a -5 charge and that the left-hand side has a -2 charge. The left-hand side is more positive, so we add $3e^-$ to it. This completes the balancing of the reduction half-reaction:

$$CrO_4^{2-} + 2H_2O + 3e^- \rightarrow CrO_2^- + 4OH^-$$

The oxidation half-reaction is balanced as follows:

$$SO_3^{2-} \rightarrow SO_4^{2-} \qquad \text{unbalanced}$$

$$\left.\begin{array}{c} SO_3^{2-} + H_2O \rightarrow SO_4^{2-} \\ SO_3^{2-} + H_2O + 2OH^- \rightarrow SO_4^{2-} + 2H_2O \\ SO_3^{2-} + 2OH^- \rightarrow SO_4^{2-} + H_2O \end{array}\right\} \qquad \text{mass balance}$$

$$SO_3^{2-} + 2OH^- \rightarrow SO_4^{2-} + H_2O + 2e^- \qquad \text{electrical balance}$$

Writing both half-reactions together, we see that the oxidation gives up $2e^-$ and that the reduction takes $3e^-$:

$$CrO_4^{2-} + 2H_2O + 3e^- \rightarrow CrO_2^- + 4OH^- \qquad \text{reduction}$$

$$SO_3^{2-} + 2OH^- \rightarrow SO_4^{2-} + H_2O + 2e^- \qquad \text{oxidation}$$

Before we can add both half-reactions, we must ensure that the number of electrons lost equals the number of electrons gained. We can do this by multiplying the reduction half-reaction by 2 and by multiplying the oxidation half-reaction by 3:

$$2 \times [CrO_4^{2-} + 2H_2O + 3e^- \rightarrow CrO_2^- + 4OH^-]$$
$$3 \times [SO_3^{2-} + 2OH^- \rightarrow SO_4^{2-} + H_2O + 2e^-]$$

equalizing the number of electrons in each half-reaction

By multiplying out we get

$$2CrO_4^{2-} + 4H_2O + 6e^- \rightarrow 2CrO_2^- + 8OH^-$$

$$3SO_3^{2-} + 6OH^- \rightarrow 3SO_4^{2-} + 3H_2O + 6e^-$$

Adding both half-reactions gives

$$2CrO_4^{2-} + 4H_2O + 6e^- + 3SO_3^{2-} + 6OH^- \rightarrow$$ sum of half-reactions
$$2CrO_2^- + 8OH^- + 3SO_4^{2-} + 3H_2O + 6e^-$$

Subtracting $6e^-$, $3H_2O$, and $6OH^-$ from each side, we arrive at the overall balanced equation:

$$2CrO_4^{2-} + H_2O + 3SO_3^{2-} \rightarrow 2CrO_2^- + 3SO_4^{2-} + 2OH^-$$

overall balanced equation

18-6
CHECK THE BALANCED REDOX EQUATION FOR MASS AND ELECTRICAL BALANCE IN THE FOLLOWING FOUR EXAMPLES

In the four examples that follow, what we have done at each step will be noted after that step. While you are following these examples and working the problems at the end of the chapter, it would be an excellent idea if you *checked* the final balanced equation to be sure it is correct. You can do this in two steps. (1) Count the number of each kind of atom on each side of the arrow and make sure the respective numbers are equal; (2) count the total charge on each side of the arrow and make sure the charges are the same.

EXAMPLE 5   Balance the following redox reaction in acidic solution. No spectator ions are written.

$$Cu + NO_3^- \rightarrow Cu^{2+} + NO$$

Solution:

$$\overset{0}{Cu} + \overset{+5\,-2}{NO_3^-} \rightarrow \overset{+2}{Cu^{2+}} + \overset{+2\,-2}{NO} \qquad \text{oxidation number determination}$$

The two unbalanced half-reactions are

$$Cu \rightarrow Cu^{2+} \qquad \text{oxidation}$$
$$NO_3^- \rightarrow NO \qquad \text{reduction}$$

Balancing the oxidation half-reaction gives

$$Cu \rightarrow Cu^{2+} + 2e^- \qquad \begin{array}{l}\text{final mass and}\\ \text{electrical balance}\end{array}$$

Balancing the reduction half-reaction gives

$$NO_3^- \rightarrow NO + 2H_2O \qquad \qquad \text{O balance}$$
$$NO_3^- + 4H^+ \rightarrow NO + 2H_2O \qquad \begin{array}{l}\text{H balance, final}\\ \text{mass balance}\end{array}$$
$$NO_3^- + 4H^+ + 3e^- \rightarrow NO + 2H_2O \qquad \begin{array}{l}\text{final mass and}\\ \text{electrical balance}\end{array}$$

$$\left.\begin{array}{l} 3 \times [Cu \rightarrow Cu^{2+} + 2e^-] \\ 2 \times [NO_3^- + 4H^+ + 3e^- \rightarrow NO + 2H_2O] \end{array}\right\} \qquad \begin{array}{l}\text{equalizing the}\\ \text{number}\\ \text{of electrons}\\ \text{in each}\\ \text{half-reaction}\end{array}$$

$$3Cu + 2NO_3^- + 8H^+ + 6e^- \rightarrow 3Cu^{2+} + 6e^- + 2NO + 4H_2O \qquad \begin{array}{l}\text{sum of half-}\\ \text{reactions}\end{array}$$

$$3Cu + 2NO_3^- + 8H^+ \rightarrow 3Cu^{2+} + 2NO + 4H_2O \qquad \begin{array}{l}\text{overall balanced}\\ \text{equation} \quad \blacksquare\end{array}$$

EXAMPLE 6   Balance the following in acidic solution:

$$HCl + KMnO_4 \rightarrow MnCl_2 + Cl_2 + KCl$$

Solution: The ionic equation, without the spectator ions (K^+ and Cl^-—notice that Cl^- is both a reactant and a spectator ion), is

$$Cl^- + MnO_4^- \rightarrow Mn^{2+} + Cl_2$$

Assigning oxidation numbers, we have

$$\overset{-1}{Cl^-} + \overset{+7\ -2}{MnO_4^-} \rightarrow \overset{+2}{Mn^{2+}} + \overset{0}{Cl_2}$$

The two half-reactions are

$$Cl^- \rightarrow Cl_2 \qquad \text{oxidation}$$
$$MnO_4^- \rightarrow Mn^{2+} \qquad \text{reduction}$$

Balancing the oxidation half-reaction, we get

$$2Cl^- \rightarrow Cl_2 \qquad \text{Cl balance, final mass balance}$$
$$2Cl^- \rightarrow Cl_2 + 2e^- \qquad \text{final mass and electrical balance}$$

The reduction reaction was balanced in Section 18-4, where we obtained

$$MnO_4^- + 8H^+ + 5e^- \rightarrow Mn^{2+} + 4H_2O \qquad \text{final mass and electrical} \\ \text{balance}$$

Equalizing electrons and adding both half-reactions gives

$$\left.\begin{array}{l} 5 \times [2Cl^- \rightarrow Cl_2 + 2e^-] \\ 2 \times [MnO_4^- + 8H^+ + 5e^- \rightarrow Mn^{2+} + 4H_2O] \end{array}\right\}$$

equalizing the number of electrons in each half-reaction

$$10Cl^- + 2MnO_4^- + 16H^+ + 10e^- \rightarrow$$
$$5Cl_2 + 10e^- + 2Mn^{2+} + 8H_2O$$

sum of half-reactions

$$10Cl^- + 2MnO_4^- + 16H^+ \rightarrow 5Cl_2 + 2Mn^{2+} + 8H_2O$$

overall balanced equation

To get the balanced molecular equation, use the spectator ions to neutralize charge. K^+ is combined with the negative ions, and Cl^- is combined with the positive ions. (Again notice that in this reaction, Cl^- is both a reactant and a spectator ion.) In this reaction, since both Cl^- and H^+ are on the left-hand side of the arrow, they also combine to form HCl. You are not expected to know how to do this from what you have learned so far.

$$16HCl + 2KMnO_4 \rightarrow 5Cl_2 + 2MnCl_2 + 8H_2O + 2KCl \quad \blacksquare$$

EXAMPLE 7   Balance the following in basic solution. No spectator ions are written.

$$CN^- + MnO_4^- \rightarrow OCN^- + MnO_2$$

Solution: The oxidation numbers of N and C in OCN^- are not obvious. If you could draw the Lewis formula for OCN^-, you could figure out the oxidation numbers of C and N using the methods of Chapter 17. The Lewis formula is $—\overset{..}{\underset{..}{O}}—C{\equiv}N{:}$, which is slightly different than the ones you have seen because one of the lone pairs of electrons on the oxygen has been written as a line rather than as two dots. This was done because it really isn't a lone pair—it is a bond to a metal atom such as Na in a compound like NaOCN. Putting in the "+" and "−" signs, we get

$$\overset{+\,-}{{}}\overset{..}{\underset{..}{O}}\overset{-\,+}{{-}}C\overset{\overset{+\,-}{\equiv}}{\equiv}N:$$

Counting the "+" and "−" signs gives N = −3 and C = +4. The oxygen has two "−" signs; the one on the left comes from its bond with a metal atom such as Na in a compound like NaOCN. We leave assigning oxidation numbers in CN^- for Problem 33.

You could not be expected to be able to figure out the oxidation numbers in CN^- and OCN^- unless you were given the Lewis formulas and had studied Section 17-4. However, you could still figure out what the half-reactions are and balance the equation without knowing the oxidation numbers, since CN^- must go to OCN^- and MnO_4^- must go to MnO_2:

$$\overset{+2\,-3}{C\,N^-} + \overset{+7\,-2}{MnO_4^-} \rightarrow \overset{-2+4\,-3}{O\,C\,N^-} + \overset{+4\,-2}{MnO_2} \qquad \text{oxidation number determination}$$

The half-reactions are

$$CN^- \rightarrow OCN^- \qquad \text{oxidation}$$

$$MnO_4^- \rightarrow MnO_2 \qquad \text{reduction}$$

Balancing the oxidation half-reaction gives

$CN^- + H_2O \rightarrow OCN^-$	O balance
$CN^- + H_2O + 2OH^- \rightarrow OCN^- + 2H_2O$	H balance
$CN^- + 2OH^- \rightarrow OCN^- + H_2O$	final mass balance
$CN^- + 2OH^- \rightarrow OCN^- + H_2O + 2e^-$	final mass and electrical balance

Balancing the reduction half-reaction, we have

$$MnO_4^- \rightarrow MnO_2 + 2H_2O \qquad \text{O balance}$$

$$MnO_4^- + 4H_2O \rightarrow MnO_2 + 2H_2O + 4OH^- \qquad \text{H balance}$$

$$MnO_4^- + 2H_2O \rightarrow MnO_2 + 4OH^- \qquad \text{final mass balance}$$

$$MnO_4^- + 2H_2O + 3e^- \rightarrow MnO_2 + 4OH^- \qquad \begin{array}{l} \text{final mass and} \\ \text{electrical balance} \end{array}$$

Equalizing electrons and adding the balanced half-reactions gives

$$\left. \begin{array}{l} 3 \times [CN^- + 2OH^- \rightarrow OCN^- + H_2O + 2e^-] \\ 2 \times [MnO_4^- + 2H_2O + 3e^- \rightarrow MnO_2 + 4OH^-] \end{array} \right\} \qquad \begin{array}{l} \text{equalizing the} \\ \text{number of electrons} \\ \text{in each half-reaction} \end{array}$$

$$3CN^- + 6OH^- + 2MnO_4^- + 4H_2O + 6e^- \rightarrow \qquad \begin{array}{l} \text{sum of half-} \\ \text{reactions} \end{array}$$

$$3OCN^- + 3H_2O + 6e^- + 2MnO_2 + 8OH^-$$

$$3CN^- + 2MnO_4^- + H_2O \rightarrow 3OCN^- + 2MnO_2 + 2OH^- \qquad \begin{array}{l} \text{overall balanced} \\ \text{equation} \quad \blacksquare \end{array}$$

EXAMPLE 8   Balance the following in acidic solution. No spectator ions are written.

$$S_2O_3^{2-} + I_2 \rightarrow I^- + S_4O_6^{2-}$$

Solution: The oxidation numbers are

$$\overset{+2\ -2}{S_2O_3^{2-}} + \overset{0}{I_2} \rightarrow \overset{-1}{I^-} + \overset{+2.5-2}{S_4O_6^{2-}} \qquad \text{oxidation number determination}$$

Don't worry about the fractional oxidation number of S in the tetrathionate ion, $S_4O_6^{2-}$. The S atoms are bonded differently, just as were some of the carbon atoms you may have studied in Section 17-4. Here we are only interested in the change in oxidation numbers, so the details of the bonding don't concern us. The half-reactions are

$$S_2O_3^{2-} \rightarrow S_4O_6^{2-} \qquad \text{oxidation}$$

$$I_2 \rightarrow I^- \qquad \text{reduction}$$

Balancing the oxidation half-reaction, we have

$$2S_2O_3^{2-} \rightarrow S_4O_6^{2-}$$ S and O balance, final mass balance

$$2S_2O_3^{2-} \rightarrow S_4O_6^{2-} + 2e^-$$ final mass and electrical balance

Balancing the reduction half-reaction gives

$$I_2 \rightarrow 2I^-$$ I balance, final mass balance

$$I_2 + 2e^- \rightarrow 2I^-$$ final mass and electrical balance

Equalizing electrons and adding the balanced half-reactions, we have

$$2S_2O_3^{2-} \rightarrow S_4O_6^{2-} + 2e^-$$
$$I_2 + 2e^- \rightarrow 2I^-$$

equalizing the number of electrons (the electrons are already equal)

$$2S_2O_3^{2-} + I_2 + 2e^- \rightarrow S_4O_6^{2-} + 2e^- + 2I^-$$ sum of half-reactions

$$2S_2O_3^{2-} + I_2 \rightarrow S_4O_6^{2-} + 2I^-$$ overall balanced equation   ∎

18-7
OXIDIZING AND REDUCING AGENTS CAUSE OXIDATION AND REDUCTION

At this point we will introduce and define two important terms: **oxidizing agent** and **reducing agent.** Earlier in the chapter we defined **oxidation** as the loss of electrons, and we defined **reduction** as the gain of electrons. The name oxidation comes from the effect that oxygen has when it combines with most substances. The rusting of iron to give iron(III) oxide is an example. Oxygen increases the metal's oxidation number. Reduction is an old name for removing oxygen from metallic ores (many are oxides) to give the pure metals. The ores are "reduced" to the metal. Removing oxygen reduces the metal's oxidation number.

Using these definitions we will now define an **oxidizing agent** as something that causes oxidation. A **reducing agent** is something that causes reduction.

For example, in the reaction from Example 1, $Fe + Cu^{2+} \rightarrow Fe^{2+} + Cu$, the half-reactions are

$$Fe \rightarrow Fe^{2+} + 2e^-$$ oxidation

$$Cu^{2+} + 2e^- \rightarrow Cu$$ reduction

The Cu^{2+} causes the Fe to be oxidized. So in this reaction, the Cu^{2+} is the oxidizing agent. At the same time, the Fe causes the Cu^{2+} to be reduced, so in this reaction the Fe is a reducing agent. **Notice that the oxidizing agent is reduced and the reducing agent is oxidized.** The following table will illustrate these definitions.

(**NOTE:** An "agent" is always a reactant.)

SUBSTANCE	TYPE OF REACTION	SUBSTANCE IS	TYPE OF AGENT
Fe	oxidation	oxidized	reducing agent
Cu^{2+}	reduction	reduced	oxidizing agent

EXAMPLE 9   For the reaction from Example 3, $Na + Mg^{2+} \rightarrow Na^+ + Mg$, fill in the following table.

SUBSTANCE	TYPE OF REACTION	SUBSTANCE IS	TYPE OF AGENT
Na			
Mg^{2+}			

Solution: First write down the half-reactions to see what the oxidation and reduction steps are. Then fill in the table.

$$Na \rightarrow Na^+ + e^- \qquad \text{oxidation}$$

$$Mg^{2+} + 2e^- \rightarrow Mg \qquad \text{reduction}$$

SUBSTANCE	TYPE OF REACTION	SUBSTANCE IS	TYPE OF AGENT
Na	oxidation	oxidized	reducing agent
Mg^{2+}	reduction	reduced	oxidizing agent

EXAMPLE 10   For the reaction from Example 6, $Cl^- + MnO_4^- \rightarrow Cl_2 + Mn^{2+}$, fill in a table as was done in Example 9.

Solution: The half-reactions are

$$2Cl^- \rightarrow Cl_2 + 2e^- \qquad \text{oxidation}$$

$$2MnO_4^- + 8H^+ + 5e^- \rightarrow Mn^{2+} + 4H_2O \qquad \text{reduction}$$

SUBSTANCE	TYPE OF REACTION	SUBSTANCE IS	TYPE OF AGENT
Cl^-	oxidation	oxidized	reducing agent
MnO_4^-	reduction	reduced	oxidizing agent

Don't get the idea that a certain substance is always an oxidizing or reducing agent in every reaction. For instance, oxygen, O_2, is usually an oxidizing agent. It oxidizes things very well, as shown in the following reactions:

$$2H_2 + O_2 \rightarrow 2H_2O$$

$$4Fe + 3O_2 \rightarrow 2Fe_2O_3$$

$$C + O_2 \rightarrow CO_2$$

But if oxygen reacts with fluorine, F_2, we get

$$2F_2 + O_2 \rightarrow 2OF_2$$

In this reaction, oxygen is the *reducing agent* and fluorine is the *oxidizing agent*. Look at the oxidation numbers. The F goes from zero to -1. The O goes from zero to $+2$. The oxygen is thus oxidized. Since fluorine is the most electronegative element, it *always* acts as an oxidizing agent. Oxygen, the second most electronegative element, always acts as an oxidizing agent *except* when reacting with fluorine.

PROBLEMS

KEYED PROBLEMS

1. Is the following a redox reaction: $3K + Al^{3+} \rightarrow 3K^+ + Al$?

2. Is the following a redox reaction: $H_2SO_4 + Ca(OH)_2 \rightarrow CaSO_4 + 2H_2O$?

3. Write the two half-reactions of the following unbalanced overall reaction that takes place in a molten salt solution: $Li + Ca^{2+} \rightarrow Li^+ + Ca$.

4. Write the equation from Problem 3 as a molecular equation and as a complete ionic equation. For simplicity, use Cl^- as the spectator ion.

5. Balance the following redox reaction in acidic solution: $Cu + NO_3^- \rightarrow Cu^{2+} + NO_2$.

6. Balance the following redox reaction in acidic solution: $C_2O_4^{2-} + MnO_4^- \rightarrow CO_2 + Mn^{2+}$.

7. Balance the following redox reaction in basic solution: $MnO_4^- + NH_3 \rightarrow MnO_2 + NO_3^-$.

8. Balance the following redox reaction in acidic solution: $I^- + NO_2^- \rightarrow I_2 + NO$.

9. For the reaction in Problem 3, fill in the following table.

SUBSTANCE	TYPE OF REACTION	SUBSTANCE IS	TYPE OF AGENT
Li			
Ca^{2+}			

10. For the reaction in Problem 6, fill in the following table.

SUBSTANCE	TYPE OF REACTION	SUBSTANCE IS	TYPE OF AGENT
$C_2O_4^{2-}$			
MnO_4^-			

SUPPLEMENTAL PROBLEMS

11. Balance the following redox reactions.

   a. $Ni + F_2 \rightarrow Ni^{2+} + F^-$ c. $Co^{2+} + Cl_2 \rightarrow Co^{3+} + Cl^-$
   b. $Fe + O_2 \rightarrow Fe^{2+} + O^{2-}$

12. Balance the following redox reactions in acidic solution.

   a. $ClO_3^- + SO_3^{2-} \rightarrow Cl^- + SO_4^{2-}$ c. $Cr_2O_7^{2-} + I^- \rightarrow Cr^{3+} + I_2$
   b. $MnO_2 + I^- \rightarrow Mn^{2+} + I_2$

13. Balance the following redox reaction in acidic solution: $MnO_4^- + H_2S \rightarrow Mn^{2+} + S$. Which element is being oxidized? Which element is being reduced? What substance is the oxidizing agent? What substance is the reducing agent?

14. Balance the following redox reactions in basic solution.

   a. $MnO_4^- + H_2O \rightarrow MnO_4^{2-} + O_2$ c. $Cl_2 + IO_3^- \rightarrow IO_4^- + Cl^-$
   b. $ClO_2 + H_2O_2 \rightarrow ClO_2^- + O_2$

15. Balance the following redox reactions in basic solution.

   a. $Mn^{2+} + H_2O_2 \rightarrow MnO_2 + H_2O$ c. $MnO_2 + SO_3^{2-} \rightarrow Mn(OH)_2 + SO_4^{2-}$
   b. $Bi(OH)_3 + SnO_2^- \rightarrow SnO_3^{2-} + Bi$

Problems 16 and 17 are harder to balance. Balance them under the conditions stated.

16. $CH_3Cl + MnO_4^- \rightarrow Cl_2 + CO_2 + Mn^{2+}$ (acidic solution). (Hint: The balanced equation will contain large coefficients.)

17. $P_4 \rightarrow PH_3 + H_2PO_2^-$ (basic solution). (Hint: P is both oxidized and reduced.)

18. Balance the following redox reactions in basic solution.

   a. $MnO_4^- + V^{2+} \rightarrow VO_2^+ + Mn^{2+}$ c. $NO_3^- + Zn \rightarrow NH_4^+ + Zn^{2+}$
   b. $Fe^{2+} + ClO_2^- \rightarrow Fe^{3+} + Cl^-$

19. Balance the following redox reactions in acidic solution.

   a. $Cr_2O_7^{2-} + C_2O_4^{2-} \rightarrow Cr^{3+} + CO_2$ c. $Sb + NO_3^- \rightarrow Sb_2O_3 + NO$
   b. $H_2O_2 + Fe^{2+} \rightarrow Fe^{3+} + H_2O$

   Problems 20 through 26 describe the chemistry of various batteries. The equations vary in difficulty.

20. The usual dry cell (flashlight battery): $Zn + NH_4^+ + MnO_2 \rightarrow Zn^{2+} + NH_3 + Mn_2O_3$. Balance in basic solution. (Hint: The NH_4^+ and the NH_3 don't take part in the redox.)

21. Mercury battery: $Zn + HgO \rightarrow ZnO + Hg$. Balance in basic solution.

22. Lead storage battery (automobile battery): $Pb + PbO_2 + SO_4^{2-} \rightarrow PbSO_4$. Notice that both Pb and PbO_2 react to form $PbSO_4$.

23. Hydrogen–oxygen fuel cell: $H_2 + O_2 \rightarrow H_2O$. Balance in basic solution. This is the reaction that supplies electricity and drinking water to most space vehicles, including the Apollo moon ship and the space shuttle. The overall equation itself is easy to balance. But try doing it using the following half-reactions:

$$H_2 + OH^- \rightarrow H_2O \quad \text{and} \quad O_2 + H_2O \rightarrow OH^-$$

24. Propane–oxygen fuel cell: $C_3H_8 + O_2 \rightarrow CO_2 + H_2O$. Balance in basic solution. The overall equation is easy to balance. But try doing it using the following half-reactions:

$$C_3H_8 + H_2O \rightarrow CO_2 + OH^- \quad \text{and} \quad O_2 + OH^- \rightarrow H_2O$$

25. Nickel–cadmium (Nicad) battery: $Cd + NiO_2 \rightarrow Cd(OH)_2 + Ni(OH)_2$. Balance in basic solution.

26. Lithium–iodide ("Lithium") battery: $Li + I_2 \rightarrow Li^+ + I^-$.

27. Calculate the oxidation number of each atom in CN^- given that the Lewis formula is $—C{\equiv}N:$ and a compound containing the CN^- ion is NaCN.

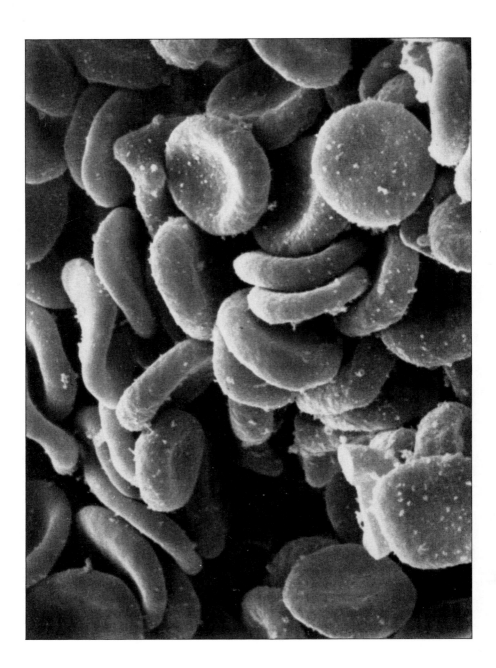

19

LOGARITHMS AND pH

As you know from our discussion of Avogadro's number, chemists work with very large and very small numbers. One area in which the numbers can get very small is in the calculation of hydrogen ions (H^+) in aqueous (water) solutions. The higher the concentration of H^+ ions, the more acidic the solution.

For example, the H^+ concentration in moles per liter in pure water is 1×10^{-7} mol/L. Instead of always writing "H^+ concentration in mol/L," chemists use the notation $[H^+]$. Square brackets around any substance denote concentration in mol/L. So we can write that, in pure water, $[H^+] = 1 \times 10^{-7}$ mol/L or $[H^+] = 1 \times 10^{-7}$ M, where M stands for molarity, another name for mol/L.

EXAMPLE 1   The H^+ concentration in normal human blood is 3.98×10^{-8} M. Write this as an equation with square brackets.

Solution: $[H^+] = 3.98 \times 10^{-8}$ M.   ■

Scientists found that using such small numbers was awkward, so they decided to make things easier. Thus they turned to logarithms. You will understand why after reading this chapter.

(Opposite) The efficient function of red blood cells depends on the pH of the blood.

19-1
THE LOGARITHM OF A NUMBER IS THE POWER TO
WHICH 10 MUST BE RAISED TO GIVE THAT NUMBER

Logarithms were invented by John Napier in 1614 and were improved by Henry Briggs a few years later. The word comes from the Greek "logos," meaning calculation, and "arithmos," meaning number. In this chapter, we will mainly discuss a special case of logarithms. These are called logarithms (abbreviated log) to the base 10. With this in mind, we can define the **logarithm** of a number as *the power to which 10 must be raised to give that number*. For example, what is the log of 100? This is the same as asking, "To what power must 10 be raised to give 100?" The answer is 2. Writing this statement as an equation, we have

$$10^2 = 100$$

From the definition of logarithms, it follows that

$$\log 100 = 2$$

Read this equation as "The logarithm of one hundred equals two."

EXAMPLE 2   What are the logs of the following numbers: 10; 100; 1000; 10,000; 100,000; 1,000,000?

Solution:

$10 = 10^1$	$\log 10 = 1$
$100 = 10^2$	$\log 100 = 2$
$1000 = 10^3$	$\log 1000 = 3$
$10,000 = 10^4$	$\log 10,000 = 4$
$100,000 = 10^5$	$\log 100,000 = 5$
$1,000,000 = 10^6$	$\log 1,000,000 = 6$ ■

EXAMPLE 3   What are the logs of the following numbers: 1, 0.1, 0.01, 0.001, 0.0001?

Solution: Convert each number to scientific notation:

$1 = 10^0$	$\log 1 = 0$
$0.1 = 10^{-1}$	$\log 0.1 = -1$
$0.01 = 10^{-2}$	$\log 0.01 = -2$
$0.001 = 10^{-3}$	$\log 0.001 = -3$
$0.0001 = 10^{-4}$	$\log 0.0001 = -4$ ■

Let's plot a graph of numbers versus their logarithms. The equation we'll plot is $y = \log x$. The horizontal axis will be x. The vertical axis will be y or $\log x$. See Figure 19-1.

Notice some interesting characteristics of logs in Figure 19-1. They seem to compress numbers. For example, the range of numbers from 0.001 to 100 spans a factor of 100,000. This means that 100 is 100,000 times larger than 0.001. The logs of 0.001 and 100 are "compressed" into a span between -3 and $+2$. There are also no points plotted to the left of the y-axis. This means that there are no logs of negative numbers.

You might wonder why scientists went through all the trouble of devising logs if the only advantage they had was to make expressing very large and very small numbers easier. Well, in the days of Napier and Briggs, and even as recently as 1970, there were no hand-held calculators. The computers that were available after 1945 were large and expensive. Doing multiplication and division could be a tedious job, even with the help of mechanical calculators, which had been around for a century or so. Logarithms were

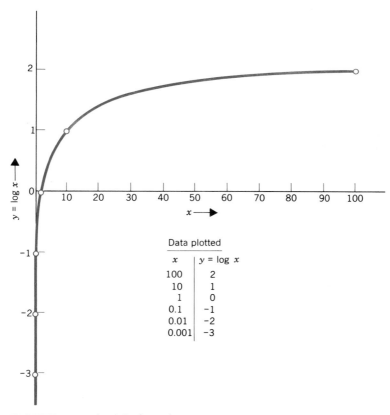

Data plotted

x	$y = \log x$
100	2
10	1
1	0
0.1	−1
0.01	−2
0.001	−3

FIGURE 19-1 A plot of $y = \log x$.

very useful in helping scientists do these calculations because when they added the logs of numbers, the numbers were multiplied. And when they subtracted the logs of numbers, the numbers were divided. It is much easier and faster to add and subtract numbers than to multiply and divide. Obtaining roots and powers is also easier with logarithms.

But now, scientific calculators and computers are available to all. And multiplications, divisions, powers, and roots are easy to do. Thus we will use logs only to make things easier and because they come up naturally in certain equations. We don't need them as calculation aids.

To obtain a log on your calculator, just enter the number and push the **log** button.

EXAMPLE 4   Obtain the logs of the following numbers on your calculator: 100, 2.4, 58.5, 3.7×10^6, 1.33, 8.86×10^4.

Solution:

NUMBER	BUTTON SEQUENCE ON CALCULATOR	LOG OF NUMBER (ROUNDED TO TWO DECIMAL PLACES)
100	**100, log**	2.00
2.4	**2.4, log**	0.38
58.5	**58.5, log**	1.77
3.7×10^6	**3.7, EE, 6, log**	6.57
1.33	**1.33, log**	0.12
8.86×10^4	**8.86, EE, 4, log**	4.95

NOTE: Some calculators use **exp** instead of **EE.** ∎

EXAMPLE 5   Obtain the logs of the following numbers on your calculator: 1, 0.1, 0.46, 0.0037, 4.2×10^{-5}, 8.36×10^{-8}.

Solution:

NUMBER	BUTTON SEQUENCE ON CALCULATOR	LOG OF NUMBER (ROUNDED TO TWO DECIMAL PLACES)
1	**1, log**	0.00
0.1	**0.1, log**	−1.00
0.46	**0.46, log**	−0.34
0.0037	**0.0037, log**	−2.43
4.2×10^{-5}	**4.2, EE, +/−, 5, log**	−4.38
8.36×10^{-8}	**8.36, EE, +/−, 8, log**	−7.08

EXAMPLE 6   Obtain the log of -2.4 on your calculator.

Solution: The calculator button sequence is **2.4, +/−, log.** The answer is E or ERROR. As we said before, you cannot obtain the log of a negative number—it does not exist.   ■

19-2
THE ANTILOGARITHM IS THE NUMBER CORRESPONDING TO A GIVEN LOGARITHM

Suppose you have the log of a number and want to find the number. On a calculator all you have to do is enter the log of the number, push the following button(s), and read the number. The buttons are either **INV, log,** or **10^x,** depending upon which brand of calculator you have. These operations convert the log of a number into the number. The mathematical operation is called **taking the antilogarithm.** An **antilogarithm** is the number corresponding to a given logarithm.

Before we work out a few examples, it would be instructive to see why taking $10^{(\log \text{ of a number})}$ gives the number. The reason is that

$$10^{\log x} = x$$

for any positive number x. Let's see how this works. Take the number $x = 100$. Then $\log x = \log 100 = 2$. Now

$$10^{\log x} = 10^{\log 100} = 10^2 = 100$$

You see, the log operation and raising to a power of 10 are "opposites" of each other and "cancel" each other out. They are called **inverses** of each other. This is the reason for the **INV** button on some calculators.

EXAMPLE 7   What is the antilogarithm (abbreviated antilog) of 3?

Solution: To find the antilog, raise 10 to the power 3:

$$10^3 = 1000$$

The antilog of 3 is 1000. And conversely, the log of 1000 is 3.   ■

In the following examples, we take antilogs using the **INV** and **log** buttons on a calculator. If your calculator has a **10^x** button, use it instead.

EXAMPLE 8   Find the antilogs of the following logarithms: 0, 0.45, 1, 1.5, 2, 2.7, 3.

Solution:

LOG	BUTTON SEQUENCE ON CALCULATOR	ANTILOG OF NUMBER
0	**0, INV, log**	1
0.45	**0.45, INV, log**	2.82
1	**1, INV, log**	10
1.5	**1.5, INV, log**	31.62
2	**2, INV, log**	100
2.7	**2.7, INV, log**	501.19
3	**3, INV, log**	1000

EXAMPLE 9   Find the antilogs of the following logarithms: 0, -0.5, -1, -1.9, -2, -2.2, -3.

Solution:

LOG	BUTTON SEQUENCE ON CALCULATOR	ANTILOG OF NUMBER
0	**0, INV, log**	1
-0.5	**0.5, +/−, INV, log**	0.316
-1	**1, +/−, INV, log**	0.1
-1.9	**1.9, +/−, INV, log**	0.0126
-2	**2, +/−, INV, log**	0.01
-2.2	**2.2, +/−, INV, log**	0.00631
-3	**3, +/−, INV, log**	0.001

EXAMPLE 10   Find the antilogs of the following logarithms: 5.8, 8, 10, 12.7, -7, -9.43, -13.11.

Solution:

LOG	BUTTON SEQUENCE ON CALCULATOR	ANTILOG OF NUMBER
5.8	**5.8, INV, log**	630957.34
8	**8, INV, log**	1×10^8
10	**10, INV, log**	1×10^{10}
12.7	**12.7, INV, log**	5.01×10^{12}
-7	**7, +/−, INV, log**	$0.0000001 = 10^{-7}$
-9.43	**9.43, +/−, INV, log**	3.72×10^{-10}
-13.11	**13.11, +/−, INV, log**	7.76×10^{-14}

19-3
THE NATURAL LOGARITHM IS USEFUL IN ADVANCED SCIENTIFIC WORK

In looking at your calculator, you may have noticed another button labeled **ln x.** This is the natural logarithm button and takes natural logs of numbers. The base of the natural log system is the number 2.7182818. . . . (Remember the base of the log system we have been using is 10.) This number is so important in science and mathematics that it has been given a special symbol, e. Thus, e = 2.7182818. . . . The inverse of ln x is e^x. If you want to display the value of e on your calculator, push **1, e^x** or push **1, INV, ln x,** depending on the model you have. The value for e that you will get will be 2.7182818. When we wrote this number earlier, we put three dots after it. That's because it doesn't end with eight digits but goes on forever. We won't discuss the natural log system further in this book, but if you take more chemistry and mathematics (especially calculus), you will surely learn its importance.

19-4
pH IS AN APPLICATION OF LOGARITHMS TO ACIDIC AND BASIC SOLUTIONS

In Example 1, we said that in normal human blood, $[H^+] = 3.98 \times 10^{-8}$ M. We promised to show you how scientists decided to make it easier to discuss such numbers. As you can see from Example 5, the logs of very small numbers become rather simple negative numbers ($\log 8.36 \times 10^{-8} = -7.08$). So scientists decided to take logs of $[H^+]$. But they didn't like the negative numbers they got. So they put a "$-$" sign in front of the log:

$$-\log [H^+]$$

The two minuses became a plus and all was well. Let's try it for blood. For normal human blood, $[H^+] = 3.98 \times 10^{-8}$ M. Thus

$$-\log [H^+] = -\log (3.98 \times 10^{-8}) = -(-7.4) = 7.4$$

The pH is defined as

$$\textbf{pH} = -\textbf{log} [\textbf{H}^+]$$

It is read: "pH equals the negative logarithm of the hydrogen ion concentration in mol/L (or molarity)." Don't worry about what to do with the unit M (molarity, or mol/L). You cannot take the log of a unit. (How would you enter it on a calculator?) Note that pH and all other logs don't have units; they are dimensionless. Anyway, the calculation we just performed shows that the pH of normal human blood is 7.4.

EXAMPLE 11   What is the calculator button sequence for calculating the pH of normal human blood, whose $[H^+]$ is 3.98×10^{-8} M?

Solution: pH $= -\log [H^+] = -\log (3.98 \times 10^{-8}) = -(-7.4) = 7.4$. The calculator button sequence is **3.98, EE, +/−, 8, log, +/−**. ∎

The symbol pH was first used by the Danish biochemist Soren Sorensen (1868–1939). The "p" stands for "power." Sorensen meant the strength, or "power," of the acid solution. The "p" can also mean "power" in the following sense (using the pH of human blood as the example): $[H^+] = 10^{-pH} = 10^{-7.4} = 3.98 \times 10^{-8}$ M. Thus pH is the negative power to which 10 must be raised to get $[H^+]$.

EXAMPLE 12   Calculate the pH of a milk of magnesia solution, $Mg(OH)_2$, where $[H^+] = 2.51 \times 10^{-11}$ M.

Solution: pH $= -\log [H^+] = -\log (2.51 \times 10^{-11}) = 10.6$. The calculator sequence is **2.51, EE, +/−, 11, log, +/−**. ∎

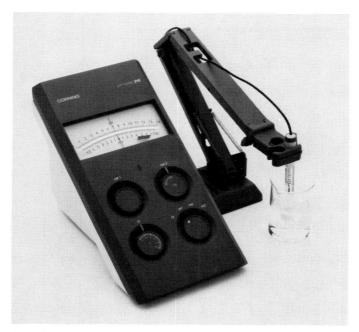

A modern pH meter. Notice that the pH scale goes from 0 to 14.

EXAMPLE 13   Calculate the pH of orange juice. The $[H^+] = 3.16 \times 10^{-4}$ M.

Solution: $pH = -\log [H^+] = -\log (3.16 \times 10^{-4}) = 3.5$. The calculator sequence is **3.16, EE, +/−, 4, log, +/−.** ∎

Now that you know how to calculate pH, you might like to look at Table 19-1, which lists the pH values for some common substances. The pH of a solution can be measured directly with an instrument called a pH meter. An example is shown in the photo above.

If you know the pH of a solution, it is sometimes desirable to calculate the $[H^+]$. This can be done as follows. Since $pH = -\log [H^+]$, we first put the minus sign on the other side by multiplying both sides of the equation by -1 and then raise both sides to a power of 10:

$$pH = -\log [H^+]$$

$$-pH = \log [H^+]$$

$$10^{-pH} = 10^{\log[H^+]}$$

$$10^{-pH} = [H^+]$$

TABLE 19-1

[H⁺] AND THE pH OF SOME COMMON SUBSTANCES[a]		
$[H^+]$	pH	SUBSTANCE
10	-1	
1	0	
10^{-1}	1	stomach acid, pH = 1
10^{-2}	2	lemon juice, pH = 2.2
		vinegar, pH = 2.4
10^{-3}	3	carbonated soft drinks, pH = 3
		grapefruit juice, pH = 3.2
		orange juice, pH = 3.5
10^{-4}	4	club soda, pH = 4
		tomato juice, pH = 4.2
10^{-5}	5	urine, pH = 4.8–7.5,
10^{-6}	6	depending on what you eat
		and your metabolism
10^{-7}	7	pure water, pH = 7
10^{-8}	8	human blood, pH = 7.4
10^{-9}	9	
10^{-10}	10	milk of magnesia, pH = 10.6
10^{-11}	11	household ammonia, pH = 11.8
10^{-12}	12	

[a]A solution with a pH of 7 is called a **neutral solution.** As the pH gets smaller than 7, the acidity increases. As the pH gets larger than 7, the solution becomes more basic.

EXAMPLE 14 Calculate the [H⁺] in human erythrocytes (red blood cells), whose pH is 7.25.

Solution: $[H^+] = 10^{-pH} = 10^{-7.25} = 5.62 \times 10^{-8}$ M. The calculator sequence is **7.25, +/−, INV, log** or **7.25, +/−, 10ˣ.** ∎

EXAMPLE 15 Calculate the [H⁺] of the urine of a person who takes large amounts of ascorbic acid (vitamin C). The pH of the urine is 4.5.

Solution: $[H^+] = 10^{-pH} = 10^{-4.5} = 3.16 \times 10^{-5}$ M. The calculator sequence is **4.5, +/−, INV, log** or **4.5, +/−, 10ˣ.** ∎

PROBLEMS

KEYED PROBLEMS

1. The H^+ concentration in human saliva is about 2.51×10^{-7} M. Write this as an equation with square brackets around the H^+.

2. What are the logs of 10,000,000 and 100,000,000?

3. What are the logs of 0.00001 and 0.000001?

4. Take the logs of 962 and 7.76×10^5 on your calculator.

5. Take the logs of 0.087 and 3.24×10^{-6} on your calculator.

6. Take the log of any negative number.

7. What is the antilog of 4?

8. Find the antilogs of 0.86 and 4.66.

9. Find the antilogs of -0.67 and -2.33.

10. Find the antilogs of 7.3 and -7.3.

11. What is your calculator button sequence for calculating the pH of blood in a person who has serious acidosis? The $[H^+]$ is 8.91×10^{-8} M.

    (**NOTE:** Your button sequence will depend on which calculator you have.)

12. Calculate the pH of household ammonia. The $[H^+]$ is 1.58×10^{-12} M.

13. Calculate the pH of lemon juice. The $[H^+]$ is 3.98×10^{-3} M.

14. Calculate the $[H^+]$ of milk whose pH is 6.5.

15. Calculate the $[H^+]$ of human tears whose pH is 7.4.

SUPPLEMENTAL PROBLEMS

16. What is the logarithm of the following numbers?

    a. 1×10^8 b. 1   c. 1×10^{-8}

17. Find the logarithms of each of the following.

    a. 327   c. 4.76×10^3 e. 6.5×10^{13}
    b. 8262   d. 9.8×10^{23} f. 4.2×10^9

18. Find the logarithm of each of the following.

    a. 0.134   c. 6.22×10^{-5} e. 9.21×10^{-18}
    b. 0.00467   d. 5.43×10^{-10} f. 1.01×10^{-23}

19. Find the antilogarithm of each of the following logarithms.

    a. 5   b. 0   c. -5

20. Find the antilog of each of the following.

    a. 3.22   c. 10.99   e. 0.432
    b. 7.58   d. 13.76   f. 0.30103

21. Find the antilog of each of the following.

    a. -0.3010 c. -0.0058 e. -22.37
    b. -4.53 d. -11.45 f. -1.11

22. What is the pH of tomato juice whose $[H^+]$ is 7.94×10^{-5} M?

23. What is the pH of wine whose $[H^+]$ is 3.16×10^{-4} M?

24. What is the $[H^+]$ of black coffee whose pH is 5.1?

25. What is the $[H^+]$ of milk whose pH is 6.9?

GLOSSARY

Absolute temperature scale A temperature scale that uses absolute zero (zero kelvins) as the lowest temperature.

Absolute zero The lowest temperature possible. It cannot be reached by any apparatus.

Abundance The amount of a substance present in a mixture. See also *percent abundance* and *fractional abundance*.

Accuracy The closeness of a measurement to the true value.

Acid A substance that yields one or more hydrogen ions (H^+) when dissolved in water.

Acidic solution A water solution in which the hydrogen ion concentration is greater than the hydroxide ion concentration. Its pH is less than 7.

Actinoid series Elements 90 (Th) through 103 (Lr). Also called the actinide series.

Alkali metals Metals in group IA of the periodic table. These are Li, Na, K, Rb, Cs, and Fr. All have one s electron in their valence shell.

Alkaline Basic. Pertaining to a water solution having more hydroxide ions than hydrogen ions. Its pH is greater than 7.

Alkaline earth metals Metals in group IIA of the periodic table. These are Be, Mg, Ca, Sr, Ba, and Ra. All have two s electrons in their valence shell.

Alpha particle A helium nucleus, $^4_2He^{2+}$.

Anhydride (anhydrous) Without water. For example, $CuSO_4$ is sometimes called anhydrous copper(II) sulfate, whereas $CuSO_4 \cdot 5H_2O$ is called copper(II) sulfate pentahydrate.

Antilogarithm The number obtained by raising 10 to a power. If $y = 10^x$, then y is the antilogarithm of x. Another way of saying this is the following: If $\log y = x$, then y is the antilogarithm of x.

Aqueous Relating to water; or "dissolved in water."

Atmosphere A unit of pressure based on standard atmospheric pressure at sea level; 1 atm = 760 torr. Also, the atmosphere is the air surrounding the earth.

Atmospheric pressure The pressure on objects resulting from the layer of air surrounding our earth. At the earth's surface, this pressure is about 760 torr.

Atom The basic unit of an element that can enter into chemical combination (unless it is one of the three noble gases that do not form compounds: He, Ne, and Ar).

Atomic mass unit (u) One-twelfth the mass of an atom of the isotope of carbon, carbon-12.

Atomic number The number of protons in an atom.

Atomic theory The theory stating that substances are composed of atoms.

Atomic weight The weighted average mass of all the naturally occurring isotopes of an element.

Avogadro's number The number of particles in one mole. Its numerical value is 6.02×10^{23}.

Balanced chemical equation A chemical equation having the same number of each kind of atom and the same total electrical charge on each side of the equation.

Balancing numbers (or coefficients) Numbers placed in front of atoms or molecules in a chemical equation to balance the equation. The term "coefficients" is preferred.

Barometer An instrument for measuring atmospheric pressure.

Base A substance that produces one or more hydroxide ions (OH^-) when dissolved in water. Another definition, which is generally more useful, is this: A base is a substance that can take hydrogen ions (H^+) from an acid.

Basic solution A water solution containing more hydroxide ions than hydrogen ions. Its pH is greater than 7.

Battery An electrochemical cell that can be used as a source of current and voltage (in other words, a battery is a source of electricity).

Beta particle An electron, $_{-1}^{0}e$.

Binding energy Energy equivalent to the mass difference between the sum of the masses of the individual free protons and neutrons in a nucleus and the actual mass of the nucleus.

Bond See *chemical bond*.

Bonding electrons Electrons transferred or shared in forming a chemical bond. The valence electrons.

Boyle's law $P_1V_1 = P_2V_2$ at constant temperature and amount of gas.

Buret A device designed to allow the delivery of an accurate quantity of liquid. Any amount up to the buret's capacity can be accurately delivered.

Calibrate To compare your instrument with a standard instrument so that you know it is working correctly. You can also run a known sample through your instrument to see whether you get the correct value.

Celsius scale (°C) The temperature scale in which water freezes at 0 °C and boils at 100 °C at 1 atm pressure.

Centrifugal force A force directed outward from an object moving in a curved path. This is the force that throws you to the side when your car goes around a turn.

Charge An electrical property. There are positive charges and negative charges.

Charles's law $V_1/T_1 = V_2/T_2$ at constant pressure and amount of gas.

Chemical bond The attractive force that holds atoms together.

Chemical change A change that produces substances different from those originally present.

Chemical equation An expression showing the reactants and products in a chemical reaction.

Chemical family Elements appearing in the same column in the periodic table. They have similar valence electron configurations and therefore have similar chemical properties.

Chemical formula An expression showing the number of atoms in a molecule. It can also show the simplest ratio of atoms in a compound where no individual molecules are found (such as a "giant array"). Chemical formulas use the symbols of the elements.

Chemical property Any property of a substance that can be studied only by changing the substance to another substance by a chemical reaction.

Chemical reaction The process by which one or more substances are converted to different substances by making and breaking chemical bonds.

Chemistry The study of the composition, structure, properties, and reactions of atoms and molecules.

Coefficients Whole numbers or fractions placed before the chemical formulas in a chemical equation. When this is done correctly, the equation is said to be balanced. An older name is "balancing numbers."

Combined gas law $P_1V_1/T_1 = P_2V_2/T_2$ at constant amount of gas.

Combustion Usually the process of burning a substance with oxygen. The term can also apply to burning substances with other oxidizing agents such as F_2, Cl_2, N_2O_4, and H_2O_2.

Common name (of a compound) A very old name that is commonly used.

It usually doesn't follow any of the nomenclature rules.

Compound Two or more atoms of different elements held together with one or more chemical bonds.

Concentrated solution A solution containing a relatively large amount of solute.

Concentration of a solution An expression describing the amount of dissolved solute in a certain quantity of solvent or solution.

Constant A number in a mathematical expression that doesn't change its value under the specified conditions.

Constant of proportionality A number that relates two different quantities. It generally has a number part and a unit part.

Cosmic rays Nuclei that come from outer space at extremely high speed. They collide with the air atoms in the atmosphere and produce, among other things, tritium and carbon-14.

Covalent bond A chemical bond in which two electrons are shared by two atoms.

Crystalline solid A solid in which the component atoms or molecules are arranged in an orderly, three-dimensional, repetitive structural pattern.

Dalton's law of partial pressures The total pressure exerted by a mixture of gases is equal to the sum of the partial pressure of each gas in the mixture.

Decay (radioactive) The spontaneous disintegration of an atomic nucleus by the emission of particles and/or radiation.

Decompose Break up into one or more parts.

Density The mass of a substance divided by its volume. $D = m/V$, where D = density, m = mass, and V = volume.

Deuterium An isotope of hydrogen, written as ^2_1H or ^2_1D.

Diatomic molecule A molecule consisting of two atoms.

Diffusion The process by which atoms or molecules of one substance gradually mix with another substance. The kinetic energy of the atoms or molecules gives rise to the motion that causes the diffusion.

Dilute solution A solution containing a relatively small amount of solute.

Directly proportional Pertaining to two quantities interacting so that one gets larger as the other gets larger. These quantities are said to be directly proportional to each. It is also implied that as one quantity gets smaller, the other gets smaller.

Dismutation See *disproportionation*.

Disproportionation The process in which a substance spontaneously reacts to produce one atom in a higher oxidation state and another atom of the same element in a lower oxidation state. Also called dismutation.

Dissociation The process by which a compound separates into individual ions when dissolved in water. Example: NaCl dissolved in water dissociates to give Na^+ and Cl^- ions. Dissociation also refers to molecules breaking apart such as $H_2 \rightarrow H + H$.

Dissolve To go into solution.

Distributive law of algebra $a(b + c) = ab + ac$ for any numbers a, b, and c.

Double bond Two covalent bonds joining two nuclei. Each bond consists of two electrons, giving a total of four electrons joining the two nuclei together.

Dry cell A battery in which the electrolyte solution is a paste rather than a liquid. A regular flashlight battery (a "D" cell) is a dry cell. An automobile battery is called a wet cell because it contains liquid sulfuric acid as the electrolyte.

Einstein's mass–energy equation An equation, published by Einstein in 1906, showing the relationship between mass and energy: $E = mc^2$, where E = energy, m = mass, c = speed of light. Since the speed of light squared is so large, a small amount of mass can be converted into a large amount of energy.

Electrolyte A substance whose water solution will conduct electricity.

Electron A subatomic particle with low mass and unit negative charge.

Electron cloud A region of space around or between atoms that is occupied by electrons.

Electron configuration A listing of the occupied orbitals of an atom, going from the first energy level to the highest occupied energy level.

Electronegativity The ability of an atom to attract electrons toward itself in a covalent bond.

Electron orbit The circular or elliptical path followed by an electron around a nucleus. This term is not used anymore and has been replaced by electron orbital.

Electron orbital A description of the shape of the space in which an electron exists.

Electron shell See *energy levels of electrons*.

Electron pair repulsion The principle stating that molecular geometry is determined by the repulsion between electron pairs around a bonded atom.

Electrostatic force The force of attraction or repulsion between two electrically charged objects. Like charges repel and unlike charges attract each other.

Element A collection of atoms all having the same atomic number. Also, a pure substance that cannot be separated into simpler substances by chemical means.

Elemental molecules Molecules consisting of atoms that all have the same atomic number.

Empirical formula A chemical formula that gives the smallest whole number ratio of atoms in a substance.

Empirical weight The molecular (or formula) weight of an empirical formula.

Energy The ability to do work.

Energy levels of electrons Areas in which electrons are located at various distances from the nucleus.

Energy sublevels The orbitals that are in an energy level. Examples are the *s*, *p*, *d*, and *f* orbitals.

Exact number A number that has an infinite number of significant figures.

Excited state An energy state of a system that is higher than the normal or ground state. An example of this is when an electron in an atom or molecule absorbs energy and is raised to a higher-energy orbital.

Exponent A superscript written after a quantity to indicate the power to which it is to be raised. Example: In 10^n, n is an exponent.

Exponential notation A notation in which exponents are used. See *exponent*.

Family See *chemical family*.

Fahrenheit scale (°F) The temperature scale in which water freezes at 32 °F and boils at 212 °F at 1 atm pressure.

Fission, nuclear The splitting of a heavy atomic nucleus into two or more pieces.

Formula, chemical See *chemical formula*.

Formula unit The smallest number of ions that represent the composition of an ionic compound. Example: NaCl. For other "giant arrays," the formula unit is the simplest formula that represents the composition.

Formula weight The mass of one mole of a substance. For a substance that consists of individual molecules, the formula weight and the molecular weight are the same. For a substance that is an ionic crystal or other "giant array," the term "formula weight" is used for the mass of one mole of the simplest formula.

Fractional abundance The decimal equivalent of the percent abundance. See *percent abundance*.

Fusion, nuclear The combination of small nuclei at a high temperature to form a larger, more stable nucleus.

Gas A state of matter that has neither definite shape nor definite volume.

Giant array A crystalline substance in which individual atoms or molecules don't exist in a separate state. The whole crystal can be considered a molecule or an array of molecules. Metals can also be considered as giant arrays of atoms.

GIGO rule Garbage In, Garbage Out. If you enter incorrect data into a calculator or computer, the results you get from calculation will be wrong.

Graduated cylinder A piece of apparatus that can contain a given amount of liquid. It is not used for the most accurate measurements.

Gram (g) A unit of mass in the metric system that is 0.001 of a kilogram.

Graph A pictorial representation of two or more variables.

Ground state The lowest energy state of an electron, an atom, or a molecule.

Group (periodic table) The elements in a vertical column in the periodic table. See *chemical family*.

Half-reaction Either the oxidation or the reduction step in a redox reaction.

Halide ion A negative ion of the halogen family, which consists of the elements F, Cl, Br, I, and At. The corresponding negative ions are F^-, Cl^-, Br^-, I^-, and At^-.

Halogen An element in group VIIA of the periodic table. These elements are F, Cl, Br, I, and At.

Homogeneous Having uniform properties (such as the same composition) throughout a sample.

Horizontal orbital diagram An orbital diagram drawn horizontally rather than vertically.

Hormone A chemical substance that acts as a control or regulatory agent in the body.

Hydrate A substance that contains water as part of the formula. Example: $CuSO_4 \cdot 5H_2O$ is called copper(II) sulfate pentahydrate.

Hydration A process in which water molecules are added to a substance.

Hydrocarbon Molecules consisting only of carbon and hydrogen atoms.

Hypo As a prefix it means "less than." It is also the common name for the "fixer" in photography.

Ideal gas A gas that can be described by the ideal gas equation. In an ideal gas, the molecules are assumed to be points with no force of attraction or repulsion between them, except when they collide. No real gas is ideal.

Ideal gas equation $PV = nRT$.

Ideal gas law A law described by the ideal gas equation.

Imaginary number The square root of a negative number.

Indicator A dye whose color changes over a specific pH range. Used as an aid in doing titrations and for measuring the pH of solutions. There are also indicators for other reactions, such as redox reactions.

Inorganic chemistry The branch of chemistry that, in general, studies all the elements and their compounds except for most carbon compounds.

Integer A whole number such as 1, 2, 3, 4, 5, and so on.

Inversely proportional Pertaining to two quantities interacting so that one becomes larger as the other becomes smaller.

Ion A charged particle formed by the loss or gain of electrons by a neutral atom or group of atoms.

Ionic bond A bond resulting from the electrostatic attraction between the positive and negative ions in an ionic compound.

Ionic compound A substance that contains positive and negative ions.

Ionic equation A chemical equation in which, as far as possible, ions, rather than molecules, are written.

Ionization The formation of an ion or ions from an atom or molecule.

Isotonic solution A solution that can be injected into the blood without damaging blood cells or blood vessels.

Isotopes Atoms of the same element having a different number of neutrons.

Isotopic symbol A notation for writing isotopes. The atomic number is put at the lower left of the element's symbol. The mass number is put at the upper left of the symbol.

Kelvin temperature scale A temperature scale starting at absolute zero, the lowest possible temperature. Zero kelvins equals $-273.15\ °C$.

Kinetic energy Energy of motion equal to $\frac{1}{2}mv^2$, where m = mass and v = velocity.

Lanthanoid series Elements 58 (Ce) through 71 (Lu). Also called the rare earth elements or the lanthanide series.

Law of conservation of energy Energy can neither be created nor destroyed, but it can be converted from one form to another. See *Einstein's mass–energy equation* for an extension of this law.

Lewis formula Representation of atoms, molecules, and ions in which the valence electrons are represented as dots and/or lines.

Limiting reagent The reactant that is used up completely in a chemical reaction.

Liquid One of the three states of matter. A liquid can move about freely while it still retains a definite volume. Thus liquids flow and take the shape of their containers.

Liter (L) A unit of volume. The volume of one kilogram of water at 4° C.

Log See *logarithm.*

Logarithm A power to which 10 must be raised to give the number of interest. Example: log 100 = 2. Thus 2 is the logarithm of 100. Note that 10^2 = 100.

Mantissa The number written before the "×" sign in scientific notation. In 3.5×10^6, 3.5 is the mantissa. In logarithms, the mantissa is the number written after the decimal point. The number before the decimal point is called the "characteristic."

Mass The quantity of matter in an object which causes the object to have weight in a gravitational field.

Mass defect The difference between the actual mass of an atom and the calculated mass of the protons, neutrons, and electrons that make up that atom.

Mass number The sum of the number of protons and neutrons in the nucleus of an atom.

Matter Anything that has mass and occupies space.

Measurement The amount of something; this is indicated by a number followed by a unit.

Mechanism of a reaction The route or steps by which a reaction takes place. The mechanism describes the manner in which atoms or molecules are transformed from reactants into products.

Metal An element that conducts electricity and heat. Metals are usually shiny and tend to lose valence electrons and become positive ions. Combinations of metallic elements are also called metals or alloys.

Metalloid The older name for semimetal.

Meter (m) The standard of length in the metric system of units.

Metric system A system of measurement based on factors of 10 between different units.

Mixture Two or more substances combined without a chemical bond. They can be mixed in any proportion.

Molarity The number of moles of solute per liter of solution.

Molar solution A solution containing one mole of solute per liter of solution.

Mole An amount of substance containing 6.02×10^{23} particles.

Molecular compound A compound made up of individual discrete molecules.

Molecular formula An expression showing the number of atoms of each element in a molecule.

Molecular geometry The bond lengths and bond angles of the atoms in a molecule. The three-dimensional structure of a molecule with definite bond lengths and bond angles.

Molecular orbital A description of a chemical bond in which the atomic orbitals combine in a special way.

Molecular weight The sum of the atomic weights of all the atoms in a molecule.

Molecule Two or more atoms held together with one or more chemical bond(s).

Molten state The liquid state of a substance. Water is the molten state of ice. Liquid sodium chloride is the molten state of solid table salt.

Monatomic One atom.

Net ionic equation A chemical equation in which only the species (many of which are ions) that actually take part in the chemical reaction are shown.

Neutralization The reaction of an acid and base to form a salt and water.

Neutron A subatomic particle with a mass of about 1 u and no charge.

Noble gas An element in group VIIIA of the periodic table. These elements are He, Ne, Ar, Kr, Xe, and Ra. He, Ne, and Ar are unreactive chemically.

Nomenclature A systematic method of naming things.

Nonbonded electron pair An electron pair that doesn't take part in a chemical bond. These pairs of electrons play a role in determining molecular geometry.

Nonmetal Elements that are generally poor conductors of heat and electricity. They are not shiny and tend to form negative ions. Many nonmetals are gases.

Nuclear fission See *fission, nuclear.*

Nuclear fusion See *fusion, nuclear.*

Nuclear power Controllable energy produced by the use of nuclear fission or, probably in the future, nuclear fusion.

Nuclear reactor A device that uses fissionable elements (such as uranium-235) to produce large amounts of neutrons and heat.

Nucleon A nuclear particle such as a proton or a neutron.

Nucleus The central part of an atom that contains the protons and neutrons. The nucleus has most of the mass and all the positive charge of the atom.

Octahedron (regular) An eight-sided solid geometric figure with all edges and faces equal.

Octet rule Many atoms other than hydrogen tend to form bonds until they are surrounded by eight electrons in their valence shell.

Orbit See *electron orbit.*

Orbital See *electron orbital.*

Orbital diagram A diagram showing the kinds of orbitals and their relative energies for the electrons of an atom.

Organic chemistry The chemistry of carbon compounds.

Orthomolecular nutrition A nutritional theory stating that, in general, each person's enzyme systems have different properties and that supplying the optimum levels of nutrients for a person can help lead to optimal health for that person. Literally, orthomolecular means "the right molecule."

Oxidation The process in which a substance loses electrons.

Oxidation number A number assigned to each atom in a substance (or ion), using somewhat arbitrary rules; this number helps to keep track of what happens in redox reactions.

Oxidation–reduction reaction A reaction in which electrons are both lost and gained. Also called "redox" reaction.

Oxidizing agent The species that takes electrons in a redox reaction. The oxidizing agent is reduced.

Oxyacid An acid that contains oxygen.

Partial pressure The pressure of one component in a mixture of gases.

Pascal (Pa) A unit of pressure; $101,325$ Pa = 1 atm.

Pauli exclusion principle A principle of quantum mechanics which states that no more than two electrons (each having opposite spin) can occupy the same atomic or molecular orbital.

Per As part of a chemical name, "the most" or the "greatest."

Percent Parts per hundred.

Percent abundance The percent by number of atoms of an isotope that makes up an element.

Percent composition of a compound The percent by weight of each element in a compound.

Percent uncertainty The error of a measurement divided by the value of the measurement and then multiplied by 100.

Period (periodic table) A horizontal row of the periodic table.

Periodic law The properties of the elements are periodic functions of their atomic number. This means that the properties repeat themselves regularly.

Periodic table A table arranging the elements in order of increasing atomic number. Elements with similar valence electron configurations and similar chemical and physical properties are arranged in columns.

pH A way of expressing acidity or alkalinity; pH $= -\log [H^+]$.

Photon The particle associated with a light wave.

Photosynthesis The process that green plants use to make sugars and elemental oxygen. In nature, the energy for this process comes from sunlight.

Physical change An alteration in form, such as size, shape, or physical state, without a change in composition. A change without chemical reaction.

Physical property A property of a substance that can be studied without causing a chemical reaction. Examples of physical properties are color, density, melting point, and boiling point.

Pipet A device used to deliver a known volume of a liquid.

Planar molecule A molecule whose atoms are all in the same plane.

Polyatomic ion An ion consisting of more than one atom.

Potential energy Stored energy or the energy an object has because of its position, chemical state, or electrical condition.

Precipitate An insoluble solid that can be separated from a solution.

Precision The closeness with which individual measurements agree with one another. This quality is related to how good an experimenter you are. In general, the greater the number of significant figures in a measurement, the greater the precision.

Pressure Force per unit area. Pressure can be expressed in the following units: pounds per square inch, atmospheres, torr, millimeters of mercury, and pascal.

Principal energy levels The main energy levels of electrons in an atom. These energy levels are quantized and are referred to by a set of integers, the principal quantum numbers n, which can take the value 1, 2, 3, 4, 5, 6, 7, and so on.

Product (of a chemical reaction) A substance produced from the reactants in a chemical reaction.

Properties The characteristics, or traits, of substances which can be observed or measured. Properties are classified as physical or chemical.

Proportion A mathematical relationship between two quantities. See *directly proportional* and *inversely proportional*.

Proportionality constant A constant that turns a proportional relationship into an equality.

Protium A somewhat uncommon name for an isotope of hydrogen, written as ${}_{1}^{1}H$.

Proton A subatomic particle having a mass of about 1 u and a charge of $+1$.

Pure substance A substance that consists of only one kind of compound, molecule, or atom.

Pyramid (trigonal) A geometric figure that has four triangular sides. In general, the triangles are not equilateral triangles.

Pyramidal molecule A molecule that has the structure of a pyramid.

Quantization of energy The existence of certain discrete (separate) energy levels in an atom or molecule.

Quantum mechanics The current theory of atomic structure which has the uncertainty principle and quantization of energy as basic principles.

Quotient The number produced by dividing a given number by another number.

Radical A reactive species that usually exists for only a short time during a chemical reaction. Also called a "free radical."

Radioactivity The spontaneous disintegration or decay of a nucleus by the emission of a particle and/or a wave.

Radioactive decay See *radioactivity*.

Radioisotope An isotope that undergoes radioactive decay.

Ratio identity A ratio made from an identity.

Reactant A starting substance in a chemical reaction.

Reaction equivalent (REQ) The number of moles of a reactant divided by the coefficient of that substance in a balanced chemical equation. The REQ is used to determine the limiting reagent.

Redox reaction A reaction in which there is a transfer of one or more electrons from one substance to another.

Reducing agent A substance that can donate electrons to another substance.

Reduction The process in which a substance gains electrons.

Regular notation The usual way of writing numbers, as contrasted to the scientific notation way of writing numbers.

Relative atomic weight The weighted average mass of all the naturally occurring isotopes of an element relative to an atom of ${}_{6}^{12}C$, which is given a mass of exactly 12 u.

Rounding error A small error introduced into a calculation caused by rounding some of the numbers before the calculation is finished.

Salt A compound formed by the reaction between an acid and a base. In the reaction KOH + HCl → KCl + H_2O, the KCl is the salt.

Saturated solution A solution that has as much solute in it as it can hold at a given temperature.

Scientific notation A means of expressing very small and very large numbers in two parts: a number between 1 and 10 (or 1) multiplied by 10 raised to some exponent or power.

Semimetal An element having properties between that of a metal and a nonmetal. Also called a "metalloid."

Shell Another name for an energy level of an atom.

Shielding A decrease in the nuclear charge felt by outer electrons owing to the presence of inner electrons.

SI system of units A system of units based on the metric system, which uses seven basic units. The seven basic units are the *meter, kilogram, second, kelvin, mole, ampere,* and *candela.*

Significant figures The number of digits in a number that are known for sure, plus one more that is uncertain.

Single bond A covalent bond consisting of two electrons.

Solid A state of matter that has definite shape and volume.

Solubility The maximum amount of solute that will dissolve in a given quantity of solvent.

Soluble Capable of being dissolved.

Solute The substance that is dissolved in a solvent to form a solution.

Solution A homogeneous mixture of two substances.

Solvent The component of a solution that is present in the largest amount. In a solution consisting of a solid dissolved in a liquid, the liquid is the solvent.

Spectator ion An ion that does not undergo chemical change during a chemical reaction.

Spin (of an electron) A quantity that is one of the properties of an electron. Two electrons with different spins can be in the same orbital.

Spin state A description of the spin of an electron. The two spin states are referred to as "up" and "down."

Square pyramid A geometric figure with a square base and four triangular sides. The Egyptian pyramids are square pyramids.

Stoichiometry The mass relationships of reactants and products in a chemical reaction.

Structural formula An expression showing the number of atoms in a molecule and giving information about how the atoms are bonded to one another.

Substance A form of matter that has a definite composition and distinct properties.

Superconductor A material that conducts electricity with no apparent resistance.

Supersaturated solution A solution that contains more solute than a saturated solution at the same temperature. These solutions tend to be unstable, and the excess solute can easily crystallize out.

Suspension A mixture that gradually separates on standing.

Symbol (for an element) An abbreviation for the name of an element.

Temperature A measure of how hot or cold a system is.

Tetrahedral angle The angle that the corners of a regular tetrahedron make with one another. The value of the angle is 109.5°.

Tetrahedron (regular) A regular four-sided solid. Each side consists of congruent (identical) equilateral triangles.

Theory A unifying principle that explains a large number of facts or observations.

Titration The process of measuring the volume of one reagent required to react with a measured weight or volume of another reagent.

Torr A unit of pressure equal to 1 mmHg or 1/760 atm.

Total ionic equation An equation that shows substances in the form in which they actually exist.

Transition elements The elements in groups IB through VIIB and in group VIII in the periodic table.

Transition metal See *transition elements*.

Transuranium elements Elements with atomic numbers greater than that of uranium.

Trigonal bipyramid A geometric figure consisting of two trigonal pyramids connected by a common face.

Trigonal pyramid A four-sided geometric figure whose sides are triangles. In general, the sides are not equilateral triangles.

Triple bond A covalent bond that consists of three pairs of electrons that are shared between two atoms.

Tritium An isotope of hydrogen, written as ^3_1H or ^3_1T.

Uncertainty principle A basic principle of quantum mechanics which states that it is impossible to determine the position and the velocity of an electron accurately at the same time.

Unionized equation (molecular equation) A chemical equation that shows the substances as molecules, even if this isn't the form in which they actually react.

Unit The scale of a measurement. For example, a unit of length is the meter.

Unsaturated solution A solution that contains less solute than its corresponding saturated solution at the same temperature.

Valence The combining power of an atom or the ability of an atom to form chemical bonds.

Valence electrons Electrons in the outermost occupied energy level that are available for forming chemical bonds.

Valence shell The energy level containing the valence electrons.

Variable A mathematical quantity that can have different values in a formula or equation.

Vitamins Organic compounds that are needed by the body in small amounts and that perform vital biological functions. Many of the vitamins act as coenzymes; that is, they help enzymes to work.

Volume The amount of space an object occupies.

Volumetric flask A flask that is designed to contain accurately a specific quantity of liquid.

Volumetric pipet A pipet that is designed to deliver accurately a specific quantity of liquid.

Water of crystallization Water molecules that are included as part of the structural parts of crystals formed from water solutions.

Water of hydration See *water of crystallization*.

Wavelength The distance between identical points on successive waves.

Weight The force that gravity exerts on an object.

Weight percent solution The grams of solute in 100 g of solution.

Weight–volume percent solution The grams of solute in 100 mL of solution.

Weighted average (to calculate atomic weight) The sum of all the following terms: the fractional abundance multiplied by the mass of an isotope for each naturally occurring isotope of an element. (Note: Weighted average can be used for calculating things other than atomic weights.)

SOLUTIONS TO PROBLEMS

CHAPTER 1

1. $^{85}_{37}Rb$

2. Atomic number = 79, mass number = 197.

3. Neutrons = 41 − 19 = 22, electrons = 19.

4. $\left(\begin{array}{c}9p^+\\10n\end{array}\right)9e^-$ $^{19}_9F$

5. $\left(\begin{array}{c}7p^+\\7n\end{array}\right)7e^-$ $^{14}_7N$, $\left(\begin{array}{c}7p^+\\8n\end{array}\right)7e^-$ $^{15}_7N$

6. $\left(\begin{array}{c}86p^+\\136n\end{array}\right)86e^-$ $^{222}_{86}Rn$, $\left(\begin{array}{c}86p^+\\134n\end{array}\right)86e^-$ $^{220}_{86}Rn$,

$\left(\begin{array}{c}86p^+\\133n\end{array}\right)86e^-$ $^{219}_{86}Rn$

7. See glossary.

8. a. $\left(\begin{array}{c}10p^+\\10n\end{array}\right)10e^-$ $^{20}_{10}Ne$, $\left(\begin{array}{c}10p^+\\11n\end{array}\right)10e^-$ $^{21}_{10}Ne$,

$\left(\begin{array}{c}10p^+\\12n\end{array}\right)10e^-$ $^{22}_{10}Ne$

b. $\left(\begin{array}{c}16p^+\\16n\end{array}\right)16e^-$ $^{32}_{16}S$, $\left(\begin{array}{c}16p^+\\17n\end{array}\right)16e^-$ $^{33}_{16}S$,

$\left(\begin{array}{c}16p^+\\18n\end{array}\right)16e^-$ $^{34}_{16}S$, $\left(\begin{array}{c}16p^+\\20n\end{array}\right)16e^-$ $^{36}_{16}S$

c. $\left(\begin{matrix}22p^+\\26n\end{matrix}\right)$ 22e$^-$ $^{48}_{22}$Ti, $\left(\begin{matrix}22p^+\\24n\end{matrix}\right)$ 22e$^-$ $^{46}_{22}$Ti,

$\left(\begin{matrix}22p^+\\25n\end{matrix}\right)$ 22e$^-$ $^{47}_{22}$Ti, $\left(\begin{matrix}22p^+\\27n\end{matrix}\right)$ 22e$^-$ $^{49}_{22}$Ti,

$\left(\begin{matrix}22p^+\\28n\end{matrix}\right)$ 22e$^-$ $^{50}_{22}$Ti

d. $\left(\begin{matrix}29p^+\\34n\end{matrix}\right)$ 29e$^-$ $^{63}_{29}$Cu, $\left(\begin{matrix}29p^+\\36n\end{matrix}\right)$ 29e$^-$ $^{65}_{29}$Cu

e. $\left(\begin{matrix}34p^+\\46n\end{matrix}\right)$ 34e$^-$ $^{80}_{34}$Se, $\left(\begin{matrix}34p^+\\44n\end{matrix}\right)$ 34e$^-$ $^{78}_{34}$Se,

$\left(\begin{matrix}34p^+\\42n\end{matrix}\right)$ 34e$^-$ $^{76}_{34}$Se, $\left(\begin{matrix}34p^+\\48n\end{matrix}\right)$ 34e$^-$ $^{82}_{34}$Se,

$\left(\begin{matrix}34p^+\\43n\end{matrix}\right)$ 34e$^-$ $^{77}_{34}$Se, $\left(\begin{matrix}34p^+\\40n\end{matrix}\right)$ 34e$^-$ $^{74}_{34}$Se

f. $\left(\begin{matrix}39p^+\\50n\end{matrix}\right)$ 39e$^-$ $^{89}_{39}$Y   g. $\left(\begin{matrix}53p^+\\74n\end{matrix}\right)$ 53e$^-$ $^{127}_{53}$I

h. $\left(\begin{matrix}63p^+\\90n\end{matrix}\right)$ 63e$^-$ $^{153}_{63}$Eu, $\left(\begin{matrix}63p^+\\88n\end{matrix}\right)$ 63e$^-$ $^{151}_{63}$Eu

i. $\left(\begin{matrix}77p^+\\116n\end{matrix}\right)$ 77e$^-$ $^{193}_{77}$Ir, $\left(\begin{matrix}77p^+\\114n\end{matrix}\right)$ 77e$^-$ $^{191}_{77}$Ir

j. $\left(\begin{matrix}83p^+\\126n\end{matrix}\right)$ 83e$^-$ $^{209}_{83}$Bi, $\left(\begin{matrix}83p^+\\127n\end{matrix}\right)$ 83e$^-$ $^{210}_{83}$Bi,

$\left(\begin{matrix}83p^+\\128n\end{matrix}\right)$ 83e$^-$ $^{211}_{83}$Bi, $\left(\begin{matrix}83p^+\\129n\end{matrix}\right)$ 83e$^-$ $^{212}_{83}$Bi,

$\left(\begin{matrix}83p^+\\131n\end{matrix}\right)$ 83e$^-$ $^{214}_{83}$Bi

9. $\left(\begin{matrix}105p^+\\157n\end{matrix}\right)$ 105e$^-$ $^{262}_{105}$Ha, $\left(\begin{matrix}97p^+\\152n\end{matrix}\right)$ 97e$^-$ $^{249}_{97}$Bk,

$\left(\begin{matrix}8p^+\\10n\end{matrix}\right)$ 8e$^-$ $^{18}_{8}$O

10. $\left(\begin{array}{c}103p^+\\158n\end{array}\right)$ 103e$^-$ $^{261}_{103}$Lr, $\left(\begin{array}{c}103p^+\\159n\end{array}\right)$ 103e$^-$ $^{262}_{103}$Lr,

$\left(\begin{array}{c}99p^+\\155n\end{array}\right)$ 99e$^-$ $^{254}_{99}$Es, $\left(\begin{array}{c}10p^+\\12n\end{array}\right)$ 10e$^-$ $^{22}_{10}$Ne

11.

ISOTOPIC SYMBOL	NUMBER OF PROTONS	NUMBER OF NEUTRONS	NUMBER OF ELECTRONS	ATOMIC NUMBER	MASS NUMBER
$^{24}_{12}$Mg	12	12	12	12	24
$^{31}_{15}$P	15	16	15	15	31
$^{40}_{20}$Ca	20	20	20	20	40
$^{37}_{17}$Cl	17	20	17	17	37
$^{40}_{18}$Ar	18	22	18	18	40
$^{72}_{32}$Ge	32	40	32	32	72

CHAPTER 2

1. $(41,900)(0.26) + (49,900)(0.74) =$ $47,820.

2. $(34.96885)(0.7577) + (36.96590)(0.2423) = 35.453$ u. Atomic weight from Table 2.1 is 35.453 u.

3. $(19.99244)(0.9051) + (20.99385)(0.0027) + (21.99138)(0.0922) = 20.179$ u. Atomic weight from Table 2.1 is 20.179 u.

4. $(35)(0.7577) + (37)(0.2423) = 35.48$ u. Atomic weight from Table 2.1 is 35.453 u.

5. Br = 79.904 u, He = 4.00260 u, Co = 58.9332 u, Ag = 107.868 u, Sb = 121.75 u.

6. See glossary.

7. a. $(1.007825)(0.99985) + (2.01410)(0.00015) = 1.00798$ u. Atomic weight from Table 2.1 is 1.0079 u.
   b. $(10.01294)(0.20) + (11.00931)(0.80) = 10.81$ u. Atomic weight from Table 2.1 is 10.81 u.
   c. 12.011 u. Atomic weight from Table 2.1 is 12.011 u.
   d. 55.846 u. Atomic weight from Table 2.1 is 55.847 u.
   e. 118.69 u. Atomic weight from Table 2.1 is 118.69 u.
   f. 238.029 u. Atomic weight from Table 2.1 is 138.029 u.

8. a. $(1)(0.99985) + (2)(0.00015) = 1.00015$ u. Atomic weight from Table 2.1 is 1.0079 u.
   b. $(10)(0.20) + (11)(0.80) = 10.80$ u. Atomic weight from Table 2.1 is 10.81 u.
   c. 12.011 u. Atomic weight from Table 2.1 is 12.011 u.
   d. 55.911 u. Atomic weight from Table 2.1 is 55.847 u.
   e. 118.783 u. Atomic weight from Table 2.1 is 118.69 u.
   f. 237.978 u. Atomic weight from Table 2.1 is 238.029 u.

9. The mass of one proton is 1.0072765 u, the mass of one neutron is 1.0086650 u, and the mass of one electron is 0.00054858 u. Thus the mass of eight protons is 8.058212 u, the mass of eight neutrons is 8.06932 u, and the mass of eight electrons is 0.00438864 u. The total mass of eight protons, eight neutrons, and eight electrons is 16.13192 u. The

isotopic mass of oxygen-18 = 15.99491 u. The difference is 16.13192 u − 15.99491 u = 0.137 u.

10. Move the decimal point two places to the right. In other words, multiply by 100:

a. 0.75 = 75%   d. 0.031 = 3.1%
b. 0.42 = 42%   e. 0.101 + 10.1%
c. 1.61 = 161%   f. 0.002 = 0.2%

11. Move the decimal point two places to the left. In other words, divide by 100:

a. 38% = 0.38   d. 0.24% = 0.0024
b.   7% = 0.07   e. 1.46% = 0.0146
c. 135% = 1.35   f. 11.77% = 0.1177

12. (70)(0.25) = 17.5.

13. (250)(0.072) = 18.

14. (150)(0.0002) = 0.03.

15. (30)(1.50) = 45.

16. $7/12 \times 100 = 58.3\%$.

17. $30/180 \times 100 = 16.7\%$.

18. $0.03/7 \times 100 = 0.43\%$.

CHAPTER 3

1. N_2 is the reactant and 2N are the products.

2. $2Cl \rightarrow Cl_2$

3. In one NH_3 molecule, there is one N atom and three H atoms. In two NH_3 molecules, there are two N atoms and six H atoms.

4. $NO + \frac{1}{2}O_2 \rightarrow NO_2$ or $2NO + O_2 \rightarrow 2NO_2$

5. $C_2H_6 + \frac{7}{2}O_2 \rightarrow 2CO_2 + 3H_2O$ or $2C_2H_6 + 7O_2 \rightarrow 4CO_2 + 6H_2O$

6. $2Fe + \frac{3}{2}O_2 \rightarrow Fe_2O_3$ or $4Fe + 3O_2 \rightarrow 2Fe_2O_3$

7. $C_{12}H_{26} + \frac{37}{2}O_2 \rightarrow 12CO_2 + 13H_2O$ or $2C_{12}H_{26} + 37O_2 \rightarrow 24CO_2 + 26H_2O$

8. In the formula $Fe_2(SO_3)_3$, there are 2 Fe atoms, 3 S atoms, and 9 O atoms.

9. $2Al(OH)_3 + 3H_2SO_4 \rightarrow Al_2(SO_4)_3 + 6HOH$

10. See glossary.

11. a. 1 K atom, 1 Cl atom
b. 1 K atom, 1 Cl atom, 4 O atoms
c. 1 Ag atom, 1 N atom, 3 O atoms
d. 6 C atoms, 14 H atoms
e. 4 C atoms, 8 H atoms, 2 O atoms
f. 2 H atoms, 1 S atom
g. 2 N atoms, 8 H atoms, 1 C atom, 3 O atoms
h. 1 Zn atom, 2 N atoms, 6 O atoms
i. 1 Mg atom, 2 C atoms, 2 N atoms
j. 1 Ca atom, 2 H atoms, 2 C atoms, 6 O atoms

12. a. $SO_2 + \frac{1}{2}O_2 \rightarrow SO_3$
b. $PCl_5 \rightarrow PCl_3 + Cl_2$
c. $CaH_2 + 2H_2O \rightarrow Ca(OH)_2 + 2H_2$
d. $(NH_4)_2Cr_2O_7 \rightarrow Cr_2O_3 + N_2 + 4H_2O$
e. $4Na + O_2 \rightarrow 2Na_2O$
f. $H_2 + Cl_2 \rightarrow 2HCl$
g. $4P + 3O_2 \rightarrow 2P_2O_3$
h. $2NH_3 + H_2SO_4 \rightarrow (NH_4)_2SO_4$
i. $Zn + Pb(NO_3)_2 \rightarrow Zn(NO_3)_2 + Pb$
j. $2Cu + S \rightarrow Cu_2S$
k. $2Al + 2H_3PO_4 \rightarrow 3H_2 + 2AlPO_4$
l. $NaNO_3 \rightarrow NaNO_2 + \frac{1}{2}O_2$
m. $H_2O_2 \rightarrow H_2O + \frac{1}{2}O_2$
n. $BaO_2 \rightarrow BaO + \frac{1}{2}O_2$
o. $2Al + 3Cl_2 \rightarrow 2AlCl_3$
p. $P_4 + 5O_2 \rightarrow P_4O_{10}$
q. $3H_2 + N_2 \rightarrow 2NH_3$
r. $BaCl_2 + (NH_4)_2CO_3 \rightarrow BaCO_3 + 2NH_4Cl$
s. $PbO_2 \rightarrow PbO + \frac{1}{2}O_2$
t. $2Al + 6HCl \rightarrow 2AlCl_3 + 3H_2$
u. $Fe_2(SO_4)_3 + 3Ba(OH)_2 \rightarrow 3BaSO_4 + 2Fe(OH)_3$
v. $2KClO_3 \rightarrow 2KCl + 3O_2$
w. $3Mg + N_2 \rightarrow Mg_3N_2$
x. $C_3H_7CHO + \frac{11}{2}O_2 \rightarrow 4CO_2 + 4H_2O$
y. $NaHCO_3 + HCl \rightarrow NaCl + H_2O + CO_2$
z. $Zn(OH)_2 + H_2SO_4 \rightarrow ZnSO_4 + 2HOH$
aa. $C_4H_9OH + 6O_2 \rightarrow 4CO_2 + 5H_2O$
bb. $CaC_2 + 2H_2O \rightarrow C_2H_2 + Ca(OH)_2$
cc. $3CaCO_3 + 2H_3PO_4 \rightarrow Ca_3(PO_4)_2 + 3CO_2 + 3H_2O$

dd. $C_3H_7COOH + 5O_2 \rightarrow 4CO_2 + 4H_2O$

13. $N_2H_4 + 2H_2O_2 \rightarrow N_2 + 4H_2O$

14. $C_6H_{12}O_6 \rightarrow 2C_2H_5OH + 2CO_2$

15. $CO_2 + 2LiOH \rightarrow Li_2CO_3 + H_2O$

16. $CaO + H_2O \rightarrow Ca(OH)_2$

17. $Fe_2O_3 + 3CO \rightarrow 2Fe + 3CO_2$

CHAPTER 4

1. Elements: F_2, P_4, Na, Cr, Br_2. Compounds: N_2O_5, C_3H_8, PCl_5, SO_2, CO_2.

2. Mixtures: salt and pepper; iron powder and charcoal; polluted air; cherry soda. Compounds: sugar; SF_6; steam (gaseous H_2O). Elements: liquid nitrogen (liquid N_2); rubidium metal; P_4.

3. See glossary.

4. a. mercury and bromine
   b. He and Ne
   c. N_2 and O_2
   d. H_2O and ethanol
   e. SO_2 and NO_2
   f. NaCl and KCl
   g. carbon (diamond); metals such as Fe or Na

CHAPTER 5

1. 5×10^7

2. $2 \times 5 \times 10 = 2 \times 10 \times 5 = 5 \times 2 \times 10 = 5 \times 10 \times 2 = 10 \times 2 \times 5 = 10 \times 5 \times 2 = 100$

3. $6 \times 7 \times 10 = (6 \times 7) \times 10 = 42 \times 10 = 420$; $6 \times 7 \times 10 = 6 \times (7 \times 10) = 6 \times 70 = 420$

4. $\frac{1}{4} \times 16 \times 5 = (\frac{1}{4} \times 16) \times 5 = 4 \times 5 = 20$; $\frac{1}{4} \times 16 \times 5 = \frac{1}{4} \times (16 \times 5) = \frac{1}{4} \times 80 = 20$

5. $20 \times \frac{1}{4} = \frac{20}{1} \times \frac{1}{4} = \frac{20 \times 1}{1 \times 4} = \frac{20 \times 1}{4 \times 1} = \frac{20}{4} \times \frac{1}{1} = 5$

6. 3.8423×10^4

7. 2.56×10^1

8. 7.360000×10^6

9. 820,000

10. 572,000,000

11. 588,500,000

12. $10^6 \times 10^{13} = 10^{6+13} = 10^{19}$

13. $8 \times 10^5 \times 3 \times 10^{15} = 8 \times 3 \times 10^5 \times 10^{15} = 24 \times 10^{20} = 2.4 \times 10^{21}$

14. $5 \times 7 \times 10^{12} \times 10^6 = 35 \times 10^{18} = 3.5 \times 10^{19}$

15. $10^7/10^5 = 10^{7-5} = 10^2$

16. $\frac{4 \times 10^8}{3 \times 10^5} = \frac{4}{3} \times \frac{10^8}{10^5} = 1.33 \times 10^3$

17. $\frac{5 \times 8 \times 10^5 \times 10^8}{9 \times 3 \times 10^2 \times 10^4} = \frac{40 \times 10^{13}}{27 \times 10^6} = 1.48 \, 10^7$

18. $10^7/10^8 = 10^{7-8} = 10^{-1}$

19. $10^9/10^9 = 10^{9-9} = 10^0 = 1$

20. $\frac{8 \times 10^{-6}}{6 \times 10^5} = \frac{8}{6} \times \frac{10^{-6}}{10^5} = 1.33 \times 10^{-11}$

21. $\frac{4}{3} \times \frac{10^8}{10^{-7}} = 1.33 \times 10^{15}$

22. $\frac{7}{3} \times \frac{10^{-4}}{10^{-9}} = 2.33 \times 10^5$

23. $\frac{9.3}{3} \times 10^{-15} = 3.1 \times 10^{-15}$

24. $8.34 \times \frac{10^{-12}}{10^{-9}} = 8.34 \times 10^{-3}$

25. $\frac{9}{4} \times \frac{1}{10^{15}} = 2.25 \times 10^{-15}$

26. $0.72 = 7.2 \times 10^{-1}$

27. $0.00048 = 4.8 \times 10^{-4}$

28. $4.25 \times 10^{-3} = 0.00425$

29. $(3 \times 27)^{1/2} = 81^{1/2} = 9$

30. $(8 \times 2)^2 = 16^2 = 256$

31. **5, √** gives 2.24

32. **27, y^x, 0.666 . . ., =** gives 9. Note: $27^{1/3} = 3$ and $3^2 = 9$.

33. **7, y^x, 0.666 . . ., =** gives 3.66

34. **5, y^x, 6.71, =** gives 48987.9

35. **17, y^x, 0.055, +/−, =** gives 0.856

36. $(10^4)^5 = 10^{20}$

37. $(10^{16})^{1/4} = 10^{16 \times 1/4} = 10^4$

38. $(25 \times 10^{12})^{1/2} = 25^{1/2} \times (10^{12})^{1/2}$
$= 5 \times 10^6$

39. $(64 \times 10^{21})^{1/3} = 64^{1/3} \times 10^{21 \times 1/3}$
$= 4 \times 10^7$

40. **7.2, EE, 9,** $\sqrt{\ }$ gives 8.48×10^4

41. **3.6, EE, +/−,** $\sqrt{\ }$ gives 6×10^{-4}. By
hand: $(3.6 \times 10^{-7})^{1/2} = (36 \times 10^{-8})^{1/2} =$
$36^{1/2} \times (10^{-8})^{1/2} = 6 \times 10^{-4}$

42. **1, EE, 15, y^x, 0.25, =** gives 5.62
$\times 10^3$

43. **5.73, EE, 15, y^x, 0.2, =** gives 1.42
$\times 10^3$

44. **7.47, EE, +/−, 16, y^x, 5.41, =** gives
1.46×10^{-82}

45. a. 3.264×10^3 e. 4.67×10^6
    b. 5.82×10^2 f. 9×10^{-6}
    c. 4.3×10^{-2} g. 6×10^9
    d. 5.72×10^{-4} h. 7.001×10^3

46. a. 370   e. 511,700
    b. 48,900,000   f. 32,400
    c. 0.051   g. 0.0001
    d. 0.00000000892 h. 3200

47. a. $4.2 \times 3.6 \times 10^2 \times 10^8 = 15.12 \times$
       $10^{10} = 1.512 \times 10^{11}$
    b. $8 \times 6 \times 10^{15} \times 10^{23} = 48 \times 10^{38} =$
       4.8×10^{39}
    c. $5.3 \times 6 \times 10^{-2} \times 10^5 = 31.8 \times 10^3$
       $= 3.18 \times 10^4$
    d. $3.1 \times 2 \times 10^{-5} \times 10^{-10} = 6.2$
       $\times 10^{-15}$
    e. $4.9 \times 8 \times 10^6 \times 10^{-12} = 39.2 \times$
       $10^{-6} = 3.92 \times 10^{-5}$
    f. $3 \times 4 \times 10^{-10} = 12 \times 10^{-10} = 1.2$
       $\times 10^{-9}$

48. a. $8/4 \times 10^5/10^2 = 2 \times 10^3$
    b. $6/4 \times 10^{-2}/10^7 = 1.5 \times 10^{-9}$
    c. $4.7/8.2 \times 10^{-12}/10^{-15} = 0.573 \times 10^3$
       $= 5.73 \times 10^2$
    d. $7.43/2 \times 10^{10}/10^{-4} = 3.715 \times 10^{14}$
    e. $3.2/4 \times 1/10^3 = 0.8 \times 10^{-3} = 8$
       $\times 10^{-4}$
    f. $2.7/4 \times 10^{-2} = 0.675 \times 10^{-2} = 6.75$
       $\times 10^{-3}$

49. **4.07, EE, +/−, 8, x, 3.26, EE, +/−,
5, ÷, 8.99, EE, +/−, 7, =** gives 1.48
$\times 10^{-6}$

50. **5.88, EE, 5, ÷, 3.16, EE, 7, ÷, 7.02,
EE, +/−, 6 =** gives 2.65×10^3.

51. **3.27, EE, 4, ×, 8.53, EE, 7, ÷, 5.55,
EE, 8, ÷, 7.76, EE, +/−, 5, =** gives 6.48
$\times 10^7$

52. $7.61 \times 10^4 + 92.3 \times 10^4 = 99.9$
$\times 10^4$

53. $2.21 \times 10^{-5} - 0.890 \times 10^{-5} = 1.32$
$\times 10^{-5} = 99.9 \times 10^5$

54. 5.66×10^8

55. a. **4.26, y^x, 0.8, =** gives 3.19
    b. **8.99, y^x, 1.26, =** gives 15.9
    c. **6.25, y^x, 0.011, +/−, =** gives 0.980
    d. **5.42, EE, +/−, 11, y^x, 4.45, +/−,
       =** gives 4.83×10^{45}
    e. **4.77, EE, 9, y^x, 0.76, =** gives 2.27
       $\times 10^7$
    f. **3.2, EE, +/−, 4, y^x, 0.015, +/−, =**
       gives 1.13

CHAPTER 6

1. 2.3

2. 59.9 ± 0.2

3. 59.3

4. 2.001 has four significant figures;
97.300 has five significant figures;
0.001161 has four significant figures.

5. 1.26 (Yes, to three significant figures,
not two.)

6. 1.0342 g   9. 75.9   12. 34.0

7. 1.134 g   10. 28.9   13. 96.8

8. 437.2   11. 4.89   14. 440

15. a. 4.27   d. 3.99×10^{12}
    b. 25.1   e. 9.23×10^{-4}
    c. 0.0306   f. 3.26×10^4

16. a. 27.1   d. 2.63×10^{-1}
    b. 2.21×10^3 or 0.263
    c. 1.21×10^3 e. 4.02×10^{11}
                         f. 1.91×10^{-11}

17. a. 1.79   d. 1.91
    b. 1.27   e. 1.18×10^{-11}
    c. 173   f. 3.16×10^{-2}

18. a. 15   d. 0.078
    b. 78.3   e. 1.36×10^9
    c. 31.6   f. 3.8×10^{-6}

19. a. 0.45
    b. 0.0633
    c. 1.18
    d. 1.2×10^2
    e. 2.0×10^{-1}, or 0.20
    f. 3.65×10^3

20. a. 7.3
    b. 13.77
    c. 252.7
    d. 12.3
    e. 129.6
    f. 197.00

21. The coefficients are exact numbers because they represent individual molecules.

CHAPTER 7

1. 20 mi/5 mi = 4 times farther

2. 200 g × 1 lb/454 g = 0.441 lb

3. 5.0 lb × 454 g/1 lb = 2.3×10^3 g (to two significant figures)

4. 40 min × 60 s/1 min = 2.4×10^3 s

5. 8.0 in. × 2.54 cm/1 in. = 2.0×10^1 cm (to two significant figures)

6. 50 cm × 1 in./2.54 cm = 20 in.

7. 200 m × 1000 mm/1 m = 2.00×10^5 mm

8. 200 g × 1 kg/1000 g = 0.200 kg

9. 725 mL × 1 L/1000 mL = 0.725 L

10. 24 hr × 60 min/1 hr × 60 s/1 min = 8.6×10^4 s

11. 0.200 gal × 4 qt/1 gal × 1 L/1.057 qt × 1000 mL/1 L = 7.57×10^2 mL

12. 60 mg × 1 g/1000 mg = 0.060 g

13. 3 g × 1000 mg/1 g = 3×10^3 mg

14. 3 μg × 1 g/10^6 μg = 3×10^{-6} g

15. 3.0×10^3 mi × 5280 ft/1 mi × 12 in./1 ft × 1 m/39.37 in. × 1 km/1000 m = 4.8×10^3 km

16. 150 lb × 1 kg/2.2 lb = 68.2 kg

17. 38 in. × 2.54 cm/1 in. = 97 cm; 26 in. = 66 cm; 34 in. = 86 cm

18. 26 mi × 5280 ft/1 mi × 12 in./1 ft = 1.65×10^6 in. 385 yd × 3 ft/1 yd × 12 in./1 ft = 1.39×10^4 in. Total = 1.66×10^6 in.

19. 1000 s × 1 min/60 s × 1 hr/60 min = 0.278 hr

20. 300 mL × 1 L/1000 mL × 1.057 qt/1 L = 0.317 qt

21. 400 IU × 1 mg/1.49 IU = 268 mg

22. 0.1 lb × 454 g/1 lb = 5×10^1 g

23. $\frac{1}{2}$ gal × 4 qt/1 gal × 1 L/1.057 qt = 1.89 L (The number of significant figures in $\frac{1}{2}$ is unknown.)

24. 750 mL × 1 L/1000 mL × 1.057 qt/1 L = 0.793 qt; $\frac{1}{5}$ gal × 4 qt/1 gal = 0.800 qt. Thus you're getting 0.007 qt less now than before.

25. 48 L × 1.057 qt/1 L × 1 gal/4 qt = 13 gal

26. 2500 kcal × 4.184 kJ/1 kcal = 1.046×10^4 kJ; 1.046×10^4 kJ × 1000 J/1 kJ = 1.046×10^7 J

27. 9200 kJ × 1 kcal/4.184 kJ = 2199 kcal; 2199 kcal × 0.18 = 396 kcal/day from refined sugar.

28. 589 nm × 1 m/10^9 nm × 100 cm/1 m = 5.89×10^{-5} cm

29. 240 mg/dL × [(0.02586 mmol/L)/(1 mg/dL)] = 6.21 mmol/L

CHAPTER 8

1. 40/12 = 3.3

2. The atomic weights are Na = 22.98977 u; Xe = 131.30 u; W = 183.85 u; Pb = 207.2 u.

3. 55.847 g Fe = 1 atomic weight of Fe = 6.02×10^{23} Fe atoms

4. 6.02×10^{23} Pd atoms = 106.4 g Pd

5. 1 mol Ni weighs 58.70 g.

6. 1 mol Ni contains 6.02×10^{23} Ni atoms.

7. Since the author doesn't know the date you are reading this, here is the calculation for the number of seconds in one year. All you need to do is multiply by the number of years. If you count leap years, notice that 1900 was *not* a leap year. 1 year × 365 days/1 year × 24 h/1 day × 60 min/1 h × 60 s/1 min = 31,536,000 s

8. 2.0 mol Si × 28.1 g Si/1 mol Si = 56 g Si

9. 14 g Si × 1 mol Si/28.1 g Si = 0.50 mol Si

10. 14 g Si × 6.02 × 10^{23} Si atoms/28.1 g Si = 3.0 × 10^{23} Si atoms

11. 2.0 mol Si × 6.02 × 10^{23} Si atoms/1 mol Si = 1.2 × 10^{24} Si atoms

12. 3.0 × 10^{23} Si atoms × 1 mol Si/6.02 × 10^{23} Si atoms = 0.50 mol Si

13. 1.2 × 10^{24} Si atoms × 28.1 g Si/6.02 × 10^{23} Si atoms = 56 g Si

14. 19.0 g × 2 = 38.0 g

15. 14.0 g × 3(1.01 g) = 17.0 g

16. 2(12.0 g) + 6(1.01 g) + 16.0 g = 46.1 g

17. 2(27.0 g) + 3(32.1 g) + 12(16.0 g) = 342 g

18. 34 g NH_3 × 1 mol NH_3/17.0 g NH_3 = 2.0 mol NH_3; 34 g NH_3 × 6.02 × 10^{23} NH_3 molecules/17.0 g NH_3 = 1.2 × 10^{14} NH_3 molecules

19. 23 g ethanol × 1 mol ethanol/46.1 g ethanol = 0.50 mol ethanol

20. 7 mol CH_4 × 1 mol C/1 mol CH_4 = 7 mol C; 7 mol CH_4 × 4 mol H/1 mol CH_4 = 28 mol H

21. 5 C_2H_6 molecules × 2 C atoms/1 molecule C_2H_6 = 10 C atoms; 5 C_2H_6 molecules × 6 H atoms/1 C_2H_6 molecule = 30 H atoms

22. 8 mol C_6H_{14} × 14 mol H/1 mol C_6H_{14} × 6.02 × 10^{23} H atoms/1 mol H = 6.7 × 10^{25} H atoms

23. 49 g H_3PO_4 × 6.02 × 10^{23} H_3PO_4 molecules/98.0 g H_3PO_4 × 4 O atoms/1 H_3PO_4 molecule = 1.2 × 10^{24} O atoms

24. 2.0 mol e^- × 6.02 × 10^{23} e^-/1 mol e^- = 1.2 × 10^{24} e^-

25. 5.0 mol photons × 6.02 × 10^{23} photons/1 mol photons = 3.0 × 10^{24} photons

26. 2 mol NO + 1 mol O_2 → 2 mol NO_2

27. 1 mol $Mg(OH)_2$ + 2 mol HCl → 1 mol $MgCl_2$ + 2 mol H_2O

28. 19 g F × 6.02 × 10^{23} F atoms/19.0 g F = 6.0 × 10^{23} F atoms

29. 40.1 g + 32.1 g + 4(16.0 g) = 136.2 g

30. 25 g MnO_2 × 1 mol MnO_2/86.9 g MnO_2 = 0.29 mol MnO_2

31. 4.0 mol $Na_2S_2O_3$ × 158 g $Na_2S_2O_3$/1 mol $Na_2S_2O_3$ = 632 g $Na_2S_2O_3$

32. 3.0 mol SO_3 × 6.02 × 10^{23} SO_3 molecules/1 mol SO_3 × 1 S atom/1 SO_3 molecule = 1.8 × 10^{24} S atoms; 3.0 mol SO_3 × 6.02 × 10^{23} SO_3 molecules/1 mol SO_3 × 3 O atoms/1 SO_3 molecule = 5.4 × 10^{24} O atoms

33. 50 g ClF_3 × 6.02 × 10^{23} ClF_3/92.5 g ClF_3 × 1 atom Cl/1 molecule ClF_3 = 3.3 × 10^{23} Cl atoms; 50 g ClF_3 × 6.02 × 10^{23} ClF_3 molecules/92.5 g ClF_3 × 3 F atoms/1 ClF_3 molecule = 9.8 × 10^{23} F atoms

34. 86 g HN_3 × 1 mol HN_3/43.0 g HN_3 = 2.0 mol HN_3

35. 9.50 × 10^{24} CO_2 molecules × 44.0 g CO_2/6.02 × 10^{23} CO_2 molecules = 694 g CO_2

36. 3.2 × 10^{21} $C_4H_{10}O$ molecules × 74.1 g $C_4H_{10}O$/6.02 × 10^{23} $C_4H_{10}O$ molecules = 0.394 g $C_4H_{10}O$

37. 4.0 g AA × 1 mol AA/176 g AA = 0.023 mol AA

38. 1.0 × 10^{-10} mol B_{12} × 63 mol C/1 mol B_{12} × 6.02 × 10^{23} C atoms/1 mol C = 3.8 × 10^{15} C atoms

39. 5 μg T_4 × 1 g T_4/10^6 μg T_4 × 1 mol T_4/777 g T_4 = 6 × 10^{-9} mol T_4

40. Abbreviate progesterone as P: 1.0 mg P × 1 g P/1000 mg P × 1 mol P/314 g P = 3.2 × 10^{-6} mol P

41. 4 × 10^{-3} mol K × 39.1 g K/1 mol K = 0.2 mol K

42. 4 mol NH_3 + 5 mol O_2 → 4 mol NO + 6 mol H_2O

43. 1 mol PBr_3 + 3 mol HOH → 1 mol $P(OH)_3$ + 3 mol HBr

44. 1 Au atom × 1 mol Au/6.02 × 10^{23} Au atoms × 197 g Au/1 mol Au = 3.27 × 10^{-22} g Au

CHAPTER 9

1. 1 mol S = 1 mol O_2

2. 2 mol SO_2 = 1 mol O_2

3. 4 mol Al = 3 mol O_2 = 2 mol Al_2O_3

4. 5.0 mol Al × 3 mol O_2/4 mol Al = 3.8 mol O_2

5. 10 mol Cl_2 × 2 mol HCl/1 mol Cl_2 = 20 mol HCl

6. 128 g SO_2 × 1 mol SO_2/64.0 g SO_2 × 1 mol O_2/2 mol SO_2 × 32.0 g O_2/1 mol O_2 = 32.0 g O_2

7. 0.50 mol octane × 9 mol H_2O/1 mol octane × 18.0 g H_2O/1 mol H_2O = 81 g H_2O

8. 23 g ethanol × 1 mol ethanol/46.0 g ethanol × 2 mol CO_2/1 mol ethanol × 44.0 g CO_2/1 mol CO_2 = 44 g CO_2

9. 40.0 g HCl × 1 mol HCl/36.5 g HCl × 1 mol $Mg(OH)_2$/2 mol HCl × 58.3 g $Mg(OH)_2$/1 mol $Mg(OH)_2$ = 31.9 g $Mg(OH)_2$

10. 4 mol S/1 = 4 and 3 mol O_2/1 = 3. Thus O_2 is the limiting reagent and is completely used up. There is 1 mol of S left over; 3 mol of SO_2 are produced.

11. 10 mol A/3 = 3.3, and 8 mol B/7 = 1.1. B will be completely used up—it is the limiting reagent.

12. 40 mol Al/2 = 20, and 15 mol Fe_2O_3/1 = 15. Fe_2O_3 is completely used up—it is the limiting reagent.

13. 8.0 g CH_4 × 1 mol CH_4/16.0 g CH_4 = 0.50 mol CH_4; 64 g O_2 × 1 mol O_2/32.0 g O_2 = 2.0 mol O_2; 0.50 mol CH_4/1 = 0.50; and 2.0 mol O_2/2 = 1.0. Thus CH_4 is the limiting reagent.

14. 8.0 g CH_4 × 1 mol CH_4/16.0 g CH_4 × 2 mol O_2/1 mol CH_4 × 32.0 g O_2/1 mol O_2 = 32 g O_2 used up. There were 64 g O_2 to start with, so 32 g O_2 are left over after the reaction. 8.0 g CH_4 × 1 mol CH_4/16.0 g CH_4 × 1 mol CO_2/1 mol CH_4 × 44.0 g CO_2/1 mol CO_2 = 22 g CO_2 formed. 8.0 g CH_4 × 1 mol CH_4/16.0 g CH_4 × 2 mol H_2O/1 mol CH_4 × 18.0 g H_2O/1 mol H_2O = 18 g H_2O formed.

15. 20 g acid × 1 mol acid/60.0 g acid × 1 mol bicarb/1 mol acid × 84.0 g bicarb/1 mol bicarb = 28 g bicarb

16. Abbreviate salicylic acid as SA: 100 g SA × 1 mol SA/138 g SA × 1 mol aspirin/1 mol SA × 180 g aspirin/1 mol aspirin = 130 g aspirin

17. 7.0 g N_2 × 1 mol N_2/28.0 g N_2 × 3 mol H_2/1 mol N_2 × 2.02 g H_2/1 mol H_2 = 1.5 g H_2

18. 1.0 g $CaCO_3$ × 1 mol $CaCO_3$/100 g $CaCO_3$ × 2 mol HCl/1 mol $CaCO_3$ × 36.5 g HCl/1 mol HCl = 0.73 g HCl

19. First determine the limiting reagent: 10 g Na × 1 mol Na/23 g Na = 0.44 mol Na; 10 g $AlCl_3$ × 1 mol $AlCl_3$/134 g $AlCl_3$ = 0.075 mol $AlCl_3$; 0.44 mol Na/3 = 0.15; and 0.075 mol $AlCl_3$/1 = 0.075. Thus $AlCl_3$ is the limiting reagent. 10 g $AlCl_3$ × 1 mol $AlCl_3$/134 g $AlCl_3$ × 1 mol Al/1 mol $AlCl_3$ × 27.0 g Al/1 mol Al = 2.0 g Al metal produced.

20. 0.50 g AgBr × 1 mol AgBr/188 g AgBr × 2 mol $Na_2S_2O_3$/1 mol AgBr × 158 g $Na_2S_2O_3$/1 mol $Na_2S_2O_3$ = 0.84 g $Na_2S_2O_3$

21. 5.0 g CaO × 1 mol CaO/56.1 g CaO = 0.089 mol CaO; 10 g H_2O × 1 mol H_2O/18.0 g H_2O = 0.56 mol H_2O. Since the coefficients of both reactants in the balanced chemical equation are 1, CaO is the limiting reagent. No CaO will be left after the reaction. 5.0 g CaO × 1 mol CaO/56.1 g CaO × 1 mol H_2O/1 mol CaO × 18.0 g H_2O/1 mol H_2O = 1.6 g H_2O used up. Therefore, 10 g − 1.6 g = 8.4 g H_2O left after the reaction.

22. First determine the limiting reagent: 25 g C_3H_4 × 1 mol C_3H_4/40.1 g C_3H_4 = 0.62 mol C_3H_4; 20 g H_2 × 1 mol H_2/2.02 g H_2 = 9.9 g H_2; 0.62 mol C_3H_4/1 = 0.62; and 9.9 mol H_2/2 = 5.0. Thus propyne is the limiting reagent. 25 g C_3H_4 × 1 mol C_3H_4/40.0 g C_3H_4 × 1 mol C_3H_8/1 mol C_3H_4 × 44.0 g C_3H_8/1 mol C_3H_8 = 28 g C_3H_8

23. 50 g CaC_2 × 1 mol CaC_2/64.1 g CaC_2 × 1 mol C_2H_2/1 mol CaC_2 × 26.0 g C_2H_2/1 mol C_2H_2 = 20 g C_2H_2

24. 100 g alcohol × 1 mol alcohol/46.1 g alcohol × 1 mol glucose/2 mol alcohol ×

180 g glucose/1 mol glucose = 196 g glucose

25. 250 g Al × 1 mol Al/27.0 g Al × 3 mol Br_2/2 mol Al × 160 g Br_2/1 mol Br_2 = 2.22×10^3 g Br_2

CHAPTER 10

1. The fraction correct is 88/100 = 0.88. The percent correct is 88/100 × 100 = 88%.

2. 72/89 × 100 = 81%

3. You got 40 − 12 = 28 correct. Your grade is 28/40 × 100 = 70%.

4. 32.1/64.1 × 100 = 50.1%

5. 32.0/64. × 100 = 49.9%

6. The percent composition of SO_2 is 50.1% sulfur and 49.9% oxygen. See Problems 4 and 5.

7. %H = 2.02/34.0 × 100 = 5.94%; %O = 32.0/34.0 × 100 = 94.1%

8. %C = 48.0/74.1 × 100 = 64.8%; %H = 10.1/74.1 × 100 = 13.6%; %O = 16.0/74.1 × 100 = 21.6%

9. Using the results of Problem 8, we have 50.0 g × 0.648 = 32.4 g C; 50.0 g × 0.136 = 6.80 g H; 50.0 g × 0.216 = 10.8 g O

10. The empirical weight of CH is 12.0 g + 1.01 g = 13.0 g, and 78.1 g/13.0 g = 6. Thus the molecular formula of benzene is C_6H_6.

11. The empirical weight of NO_2 is 14.0 g + 2(16.0) = 46.0 g, and 46.0 g/92.0 g = 2. Thus the molecular formula of the oxide of nitrogen is N_2O_4, which is called dinitrogen tetraoxide.

12. For carbon: 15.78 g C × 1 mol C/12.01 g C = 1.314 mol C. For sulfur: 84.22 g S × 1 mol S/32.06 g S = 2.627 mol S. Divide each mole value by the smaller of the two: 1.314/1.314 = 1; and 2.627/1.314 = 1.999, which can be rounded to 2. The empirical formula is CS_2.

13. For carbon: 8.56 g C × 1 mol C/12.01 g C = 7.13 mol C. For hydrogen: 14.4 g

H × 1 mol H/1.01 g H = 14.3 mol H. Divide each mole value by the smaller of the two: 7.13/7.13 = 1; and 14.3/7.13 = 2.01, or 2. The empirical formula is CH_2. The empirical weight is 14.0 g. Dividing the molecular weight by the empirical weight, we get 65.1 g/14.0 g = 4.01, or 4. The molecular formula is 4 × CH_2 = C_4H_8.

14. For carbon: 54.5 g C × 1 mol C/12.0 g C = 4.54 mol C. For oxygen: 36.3 g O × 1 mol O/16.0 g O = 2.27 mol O. For hydrogen: 9.5 g H × 1 mol H/1.01 g H = 9.06 mol H. Divide by the smallest mole value: 4.54/2.27 = 2; 2.27/2.27 = 1; and 9.06/2.27 = 3.99, or 4. The empirical formula is C_2H_4O. The empirical weight is 44.1 g. Dividing the molecular weight by the empirical weight, we get 88.1 g/44.1 g = 2. The molecular formula is 2 × C_2H_4O = $C_4H_8O_2$.

15. For sulfur: 25.2 g S × 1 mol S/32.1 g S = 0.785 mol S. For fluorine: 74.8 g F × 1 mol F/19.0 g F = 3.94 mol F. Divide by the smallest mole value: 0.785/0.785 = 1; and 3.94/0.785 = 5.02, or 5. The empirical formula is SF_5. The empirical weight is 127 g. Dividing the molecular weight by the empirical weight, we get 254 g/127 g = 2. The molecular formula is 2 × SF_5 = S_2F_{10}.

16. For ClF_3: %Cl = 35.5 g/92.5 g × 100 = 38.4%; %F = 57.0 g/92.5 g × 100 = 61.6%. For BrF_5: %Br = 79.9 g/174.9 g × 100 = 45.7%; %F = 95.0 g/174.9 g × 100 = 54.3%. For IF_7: %I = 127 g/260 g × 100 = 48.8%; %F = 133 g/260 g × 100 = 51.2%.

17. The formula weight of $(CaSO_4)_2 \cdot H_2O$ is 290 g. %Ca = 80.2 g/290 g × 100 = 27.7%; %S = 64.2 g/290 g × 100 = 22.1%; %O = 144 g/290 g × 100 = 49.7%; %H = 2.02 g/290 g × 100 = 0.70%. The sum of the percents is 100.2%. This is due to rounding error. The percent of water is 18.0 g/290 g × 100 = 6.21%.

18. The formula weight of $CuSO_4 \cdot 5H_2O$ is 250 g. %Cu = 63.5 g/250 g × 100 =

25.4%; %S = 32.1 g/250 g × 100 = 12.8%; %O = 144 g/250 g = 57.6%; %H = 10.1 g/250 g × 100 = 4.04%. The sum of the percents is 99.8%. This is due to rounding error.

19. The formula weight of Ca_3N_2 is 148 g. %Ca = 120 g/148 g × 100 = 81.1%; %N = 28.0 g/148 g × 100 = 18.9%.

20. For carbon: 40.9 g C × 1 mol C/12.0 g C = 3.41 mol C. For oxygen: 54.5 g O × 1 mol O/16.0 g O = 3.41 mol O. For hydrogen: 4.55 g H × 1 mol H/1.01 g H = 4.50 mol H. Divide by the smallest mole value: 3.41/3.41 = 1, 3.41/3.41 = 1, and 4.50/3.41 = 1.32. Multiplying by 3 gives the empirical formula $C_3H_4O_3$. The empirical weight is 88.0 g. Dividing the molecular weight by the empirical weight gives 176 g/88.0 g = 2. Thus the molecular formula is 2 × $C_3H_4O_3$ = $C_6H_8O_6$.

21. For hydrogen: 2.04 g H × 1 mol H/1.01 g H = 2.02 mol H. For sulfur: 32.65 g S × 1 mol S/32.1 g S = 1.02 mol S. For oxygen: 65.31 g O × 1 mol O/16.0 g O = 4.08 mol O. Divide by the smallest mole value: 2.02/1.02 = 1.98, 1.02/1.02 = 1, and 4.08/1.02 = 4. The empirical formula is H_2SO_4. The empirical weight is 98.1 g, which is the same as the molecular weight. Thus the molecular formula is H_2SO_4.

22. For phosphorus: 22.54 g P × 1 mol P/30.97 g P = 0.7278 mol P. For chlorine: 77.46 g Cl × 1 mol Cl/35.45 g Cl = 2.185 mol Cl. Divide by the smallest mole value: 0.7278/0.7278 = 1; and 2.185/0.7278 = 3.002, or 3. The empirical formula is PCl_3; the empirical weight is 137.3 g, which is the same as the molecular weight. Thus the molecular formula is PCl_3.

23. For silver: 87.1 g Ag × 1 mol Ag/108 g Ag = 0.806 mol Ag. For sulfur: 12.9 g S × 1 mol S/32.1 g S = 0.402 mol S. Divide by the smallest mole value: 0.806/0.403 = 2 and 0.403/0.403 = 1. The empirical formula is Ag_2S. The empirical weight is 248 g, which is the same as the

molecular weight. Thus the molecular formula is Ag_2S.

24. %C in CO_2: 12.01 g/44.01 g × 100 = 27.29%. %H in H_2O: 2.016 g/18.02 g × 100 = 11.19%. Grams of C: 13.72 g × 0.2729 = 3.744 g C. Grams of H: 11.23 g × 0.1119 = 1.257 g H. %C in hydrocarbon (HC): 3.744 g/5.000 g × 100 = 74.88%. %H in HC: 1.257 g/5.000 g × 100 = 25.14%. Moles of C in HC: 74.88 g C × 1 mol C/12.01 g C = 6.235 mol C. Moles of H in HC: 25.14 g H × 1 mol H/1.008 g H = 24.94 mol H. Divide by the smallest mole value: 6.235/6.235 = 1 and 24.94/6.235 = 4. The empirical formula of the HC is CH_4.

25. The formula weights are KBr = 119 g, SiC = 40.1 g.

26. All molecular weights are formula weights, but not all formula weights are molecular weights.

CHAPTER 11

1. The salt is the solute, and the water is the solvent.

2. M = mol/L = 1.5 mol sucrose/5.0 L solution = 0.30 M sucrose

3. M = mol/L = 0.250 mol $KMnO_4$/0.350 L solution = 0.714 M $KMnO_4$

4. M = mol/L = 0.400 mol fructose/0.400 L solution = 1.00 M fructose

5. 1 mol $CaSO_4$ = 136 g $CaSO_4$; 8.00 g $CaSO_4$ × 1 mol $CaSO_4$/136 g $CaSO_4$ = 0.0588 mol $CaSO_4$; M = mol/L = 0.0588 mol $CaSO_4$/0.800 L solution = 0.735 M $CaSO_4$

6. M = mol/L and mol = mol/L × L = 0.12 mol KBr/L × 0.075 L = 0.0090 mol KBr. Also, 1 mol KBr = 119 g KBr = 119 g KBr and 0.0090 mol KBr × 119 g KBr/1 mol KBr = 1.1 g KBr.

7. 1 mol $MgCl_2$ = 95.2 g $MgCl_2$; 5.2 g $MgCl_2$ × 1 mol $MgCl_2$/95.2 g $MgCl_2$ = 0.055 mol $MgCl_2$; L = 0.055 mol $MgCl_2$/0.50 M $MgCl_2$ = 1.1 L

8. mol KOH = 0.0950 mol KOH/L × 0.235 L = 0.00223 mol KOH; 1 mol ox-

alic acid = 2 mol KOH; 0.00223 mol KOH × 1 mol oxalic acid/2 mol KOH = 0.00112 mol oxalic acid; M = 0.00112 mol oxalic acid/0.025 L = 0.0446 M oxalic acid

9. The balanced equation is 3NaOH + $H_3PO_4 \rightarrow Na_3PO_4$ + 3HOH. Then we have the following: mol H_3PO_4 = 0.110 mol H_3PO_4/L × 0.175 L = 0.0192 mol H_3PO_4; 3 mol NaOH = 1 mol H_3PO_4; 0.0192 mol H_3PO_4 × 3 mol NaOH/1 mol H_3PO_4 = 0.0578 mol NaOH; and L = 0.0578 mol NaOH/0.250 M NaOH = 0.231 L = 231 mL.

10. The number of moles of $KMnO_4$ used is 0.125 M $KMnO_4$ × 0.0273 L = 0.00341 mol $KMnO_4$. From the balanced chemical equation in Example 10, we have 2 mol $KMnO_4$ = 5 mol $H_2C_2O_4$. The number of moles of $H_2C_2O_4$ is 0.00341 mol $KMnO_4$ × 5 mol $H_2C_2O_4$/2 mol $KMnO_4$ = 0.00853 mol $H_2C_2O_4$. The molarity of the oxalic acid solution is M = 0.00853 mol $H_2C_2O_4$/0.0250 L = 0.341 M $H_2C_2O_4$.

11. $M_iL_i = M_fL_f$ and (0.150 M)(100 mL) = (M_f)(500 mL). M_f = (0.150 M)(100 mL)/500 mL = 0.0300 M.

12. $M_iL_i = M_fL_f$ and (16 M)(L_i) = (0.23 M)(600 mL). L_i = (0.23 M)(600 mL)/16 M = 8.6 mL. You need 8.6 mL of concentrated nitric acid and about 600 mL − 8.6 mL = 591.4 mL of water. You would prepare the solution as described at the end of Example 12.

13. 0.075 M $CuSO_4$·5H_2O × 0.250 L = 0.0188 mol $CuSO_4$·5H_2O; 0.0188 mol $CuSO_4$·5H_2O × 250 g $CuSO_4$·5H_2O/1 mol $CuSO_4$·5H_2O = 4.7 g $CuSO_4$·5H_2O

14. 0.07362 M $FeSO_4$ × 0.05000 L = 0.003681 mol $FeSO_4$; 0.003681 mol $FeSO_4$ × 2 mol $Ce(SO_4)_2$/2 mol $FeSO_4$ = 0.003681 mol $Ce(SO_4)_2$; 0.003681 mol $Ce(SO_4)_2$/0.1422 M $Ce(SO_4)_2$ = 0.02589 L $Ce(SO_4)_2$ = 25.89 mL $Ce(SO_4)_2$

15. This will be solved in one step: 0.1018 M $KMnO_4$ × 0.03500 L × 16 mol HCl/2 mol $KMnO_4$ × 1/0.5125 M HCl × 1000 mL/1 L = 55.62 mL HCl.

16. This will be solved in one step: 0.987 M H_2SO_4 × 0.00934 L × 2 mol NaOH/1 mol H_2SO_4 × 1/0.102 M NaOH × 1000 mL/1 L = 18.1 mL NaOH.

17. 1 mol $ZnCl_2$ = 136 g $ZnCl_2$; 15.2 g $ZnCl_2$ × 1 mol $ZnCl_2$/136 g $ZnCl_2$ = 0.112 mol $ZnCl_2$; 0.112 mol $ZnCl_2$/0.850 M $ZnCl_2$ = 1.31 L = 1.31 × 10³ mL. The final volume of the solution is 1.31 × 10³ mL, so you would probably add slightly less than 1.31 × 10³ mL of water.

18. $M_iL_i = M_fL_f$ and (12 M)(L_i) = (0.50 M)(1.0 L); L_i = (0.50 M)(1.0 L)/12 M = 0.042 L = 42 mL. Add 42 mL of 12 M HCl to a 1-L graduated cylinder. Then add enough water to bring the volume up to 1.0 L. There is no need to use a volumetric flask, since we need the concentration to only two significant figures.

19. $M_iL_i = M_fL_f$ and (1.50 M)(300 mL) = (0.350 M)(L_f); L_f = (1.50 M)(300 mL)/0.350 M = 1286 mL = 1.29 × 10³ mL. The approximate amount of water needed is 1286 mL − 300 mL = 986 mL.

20. 0.85 g NaCl × 1 mol NaCl/58.5 g NaCl = 0.0145 mol NaCl; M = 0.0145 mol NaCl/0.10 L = 0.15 M NaCl

21. 0.5103 M NaOH × 0.01702 L = 0.008685 mol NaOH; 0.008685 mol NaOH × 1 mol acetic acid/1 mol NaOH = 0.008685 mol acetic acid; 0.008685 mol acetic acid/0.01000 L = 0.8685 M acetic acid

22. 0.600 M NaOH × 0.253 L = 0.152 mol NaOH; 0.152 M NaOH × 1 mol Cl_2/2 mol NaOH = 0.0759 mol Cl_2; 0.0759 mol Cl_2 × 70.9 g Cl_2/1 mol Cl_2 = 5.38 g Cl_2

23. 1.32 g Zn × 1 mol Zn/65.4 g Zn × 2 mol HCl/1 mol Zn × 1/0.500 M HCl × 1000 mL/1 L = 80.7 mL HCl

24. 0.4216 M $AgNO_3$ × 0.02632 L × 1 mol Cl^-/1 mol $AgNO_3$ × 1/0.02500 L = 0.4439 M Cl^-

25. 0.296 M H_3PO_4 × 0.0428 L = 0.0127 mol H_3PO_4; 0.0127 mol H_3PO_4 × 3 mol KOH/1 mol H_3PO_4 = 0.0381 mol KOH;

M = 0.0381 mol KOH/0.0249 L = 1.53 M
KOH

CHAPTER 12

1. 1500 torr × 1 atm/760 torr = 1.97 atm

2. 12 lb/in.^{2} × 760 torr/14.7 lb = 620 torr
or 6.2 × 10^2 torr

3. 3 atm × 101 kPa/atm = 303 kPa or 3
× 10^2 kPa

4. °F = 1.8 °C + 32 = (1.8)(25) + 32 =
77 °F

5. °C = (°F − 32)/1.8 = (86 − 32)/1.8 =
30 °C

6. K = °C + 273 = 100 °C + 273 =
373 K

7. K = °C + 273; 77 K = °C + 273; °C =
77 K − 273 = −196 °C

8. $a = qb$

9. $a = b/t$

10. $P \propto n$. If n doubles (10 mol → 20
mol), then P doubles (5 atm → 10 atm).

11. $V \propto n$. If n doubles (3 mol → 6 mol),
then V doubles (15 L → 30 L).

12. $P = \dfrac{nRT}{V} = \dfrac{(2.0 \text{ mol})(0.0821 \text{ atm·L/mol·K})(400 \text{ K})}{7.0 \text{ L}} = 9.4$ atm

13. 80 °C + 273 = 353 K, and 500 torr × 1 atm/760 torr = 0.66 atm; thus

$$V = \frac{nRT}{P} = \frac{(0.25 \text{ mol})(0.0821 \text{ atm·L/mol·K})(353 \text{ K})}{0.66 \text{ atm}} = 11 \text{ L}$$

14. 30 °C + 273 = 303 K, and 80.0 kPa × 1 atm/101 kPa = 0.792 atm; thus

$$n = \frac{PV}{RT} = \frac{(0.792 \text{ atm})(0.200 \text{ L})}{(0.0821 \text{ atm·L/mol·K})(303 \text{ K})} = 6.37 \times 10^{-3} \text{ mol}$$

15. 25 °C + 273 = 298 K; thus $n = \dfrac{PV}{RT} = \dfrac{(1.30 \text{ atm})(1.50 \text{ L})}{(0.0821 \text{ atm·L/mol·K})(298 \text{ K})} = 0.0797$ mol

MW = g/mol = 4.40 g/0.0797 mol = 55.2 g/mol

16. $P_1V_1/T_1 = P_2V_2/T_2$; thus

$$V_2 = \frac{P_1V_1T_2}{T_1P_2} = \frac{(1.5 \text{ atm})(5.0 \text{ L})(200 \text{ K})}{(273 \text{ K})(4.0 \text{ atm})} = 1.4 \text{ L}$$

17. $P_1V_1/T_1 = P_2V_2/T_2$; thus $P_2 = \dfrac{P_1V_1T_2}{T_1V_2} = \dfrac{(0.20 \text{ atm})(5.0 \text{ L})(310 \text{ K})}{(210 \text{ K})(0.70 \text{ L})} = 2.1$ atm

18. 150 torr + 150 torr + 500 torr =
800 torr

19. (7 atm)(0.208) = 1.46 atm; thus 1.46
atm × 560 torr/1 atm = 1107 torr, or 1 ×
10^3 torr

20. 500 kPa × 1 atm/101 kPa = 4.95 atm

21. 0.50 atm × 760 torr/1 atm = 3.8 ×
10^2 torr

22. °C = (°F − 32)/1.8 = (104 °F − 32)/1.8
= 40 °C

23. °F = 1.8 °C + 32 = (1.8)(41) + 32 =
105.8 °F

24. °C = (°F − 32)/1.8 = (68 °F − 32)/1.8
= 20 °C. Then 20 °C + 273 = 293 K.

25. 892 °C + 273 = 1165 K

26. $-195.8\ °C + 273 = 77.2\ K$

27. $2800\ K - 273 = 2527\ °C$; thus $°F = 1.8\ °C + 32 = (1.8)(2527\ °C) + 32 = 4581\ °F$

28. $T = 1.8T + 32$, $T - 1.8T = 32$, $-0.8T = 32$, $T = -32/0.8 = -40$. Therefore, $-40\ °C = -40\ °F$.

29. $°C = (°F - 32)/1.8 = (100\ °F - 32)/1.8 = 37.8\ °C$; $37.8\ °C + 273 = 311\ K$; and 400 torr $\times\ 1$ atm/760 torr $= 0.526$ atm. Also, $8.0\ g\ O_2 \times 1\ mol\ O_2/32.0\ g\ O_2 = 0.25\ mol\ O_2$; thus

$$V = \frac{nRT}{P} = \frac{(0.25\ mol)(0.0821\ atm\cdot L/mol\cdot K)(311\ K)}{0.526\ atm} = 12\ L$$

30. $25\ °C + 273 = 298\ K$, and $15\ g\ F_2 \times 1\ mol\ F_2/38.0\ g\ F_2 = 0.40\ mol\ F_2$; thus

$$P = \frac{nRT}{V} = \frac{(0.40\ mol)(0.0821\ atm\cdot L/mol\cdot K)(298\ K)}{2.0\ L} = 4.9\ atm$$

31. $200\ kPa \times 1$ atm/101 kPa $= 1.98$ atm, and $40.0\ °C + 273 = 313.0\ K$; thus

$$n = \frac{PV}{RT} = \frac{(1.98\ atm)(0.700\ L)}{(0.0821\ atm\cdot L/mol\cdot K)(313.0\ K)} = 0.0539\ mol$$

$0.0539\ mol\ N_2O \times 44.0\ g\ N_2O/1\ mol\ N_2O = 2.37\ g\ N_2O$

32. 300 torr $\times\ 1$ atm/760 torr $= 0.395$ atm, and $22\ °C + 273 = 295\ K$; thus

$$n = \frac{PV}{RT} = \frac{(0.395\ atm)(0.400\ L)}{(0.0821\ atm\cdot L/mol\cdot K)(295\ K)} = 0.00652\ mol$$

$MW = g/mol = 3.00\ g/0.00652\ mol = 460\ g/mol$

33. 1 atm $= 760$ torr, and 50 torr $+ 150$ torr $+ P_{H_2} = 760$ torr; thus $P_{H_2} = 560$ torr

34. $50\ °C + 273 = 323\ K = T_1$; $100\ °C + 273 = 373\ K = T_2$; and $P_1/T_1 = P_2/T_2$; thus $P_2 = P_1T_2/T_1 = (150\ kPa)(373\ K)/323\ K = 173\ kPa$

35. $P_1V_1 = P_2V_2$; thus $V_2 = P_1V_1/P_2 = (800\ torr)(2.0\ L)/760\ torr = 2.1\ L$

36. $23\ °C + 273 = 296\ K = T_1$; $30\ °C + 273 = 303\ K = T_2$; and $P_1V_1/T_1 = P_2V_2/T_2$; thus

$$P_2 = \frac{P_1V_1T_2}{T_1V_2} = \frac{(20.0\ kPa)(800\ mL)(303\ K)}{(296\ K)(400\ mL)} = 40.9\ kPa$$

37. 3.0×10^{10} molecules $\times\ 1$ mol/6.02 $\times 10^{23}$ molecules $= 5.0 \times 10^{-14}$ mol; thus

$$P = \frac{nRT}{V} = \frac{(5.0 \times 10^{-14}\ mol)(0.0821\ atm\cdot L/mol\cdot K)(298\ K)}{0.0030\ L} = 4.1 \times 10^{-10}\ atm$$

38. $PV = (m/MW)RT$, and $95\ kPa \times 1$ atm/101 kPa $= 0.94$ atm; thus

$$MW = \frac{mRT}{PV} = \frac{(7.2 \text{ g})(0.0821 \text{ atm·L/mol·K})(373 \text{ K})}{(0.94 \text{ atm})(2.5 \text{ L})} = 94 \text{ g/mol}$$

39. 1 mol C_2H_4 = 28.0 g C_2H_4, and 50 g $C_2H_4 \times$ 1 mol C_2H_4/28.0 g C_2H_4 = 1.8 mol C_2H_4; thus

$$P = \frac{nRT}{V} = \frac{(1.8 \text{ mol})(0.0821 \text{ atm·L/mol·L})(307 \text{ K})}{0.100 \text{ L}} = 4.5 \times 10^3 \text{ atm}$$

40. 50 torr $\times$ 1 atm/760 torr = 0.066 atm; thus

$$n = \frac{PV}{RT} = \frac{(0.066 \text{ atm})(5.0 \text{ L})}{(0.0821 \text{ atm·L/mol·K})(348 \text{ K})} = 0.012 \text{ mol}$$

Since 1 mol C_6H_6 = 78.1 g C_6H_6, we have 0.012 mol $C_6H_6 \times$ 78.1 g C_6H_6/1 mol C_6H_6 = 0.94 g C_6H_6.

41. $PV = nRT$ and $n = m$/MW. (Here m stands for mass.) Substituting the value of n into $PV = nRT$, we get $PV = (m$/MW)RT. Solving for MW we get MW = mRT/PV.

42. $P_1V_1/T_1 = P_2V_2/T_2$. Let $T_1 = T_2 = T$, since T is constant. Thus $P_1V_1/\cancel{T} = P_2V_2/\cancel{T}$, or $P_1V_1 = P_2V_2$.

43. $P_1V_1/T_1 = P_2V_2/T_2$. Let $P_1 = P_2 = P$, since P is constant. Thus $\cancel{P}V_1/T_1 = \cancel{P}V_2/T_2$, or $V_1/T_1 = V_2/T_2$.

44. The total pressure is about 7 atm. Air is 78.1% N_2. Thus 7 atm $\times$ 0.781 = 5.5 atm of N_2.

CHAPTER 13

1. Uncertainty principle: There is a limit to the accuracy of any measurement, and this limit becomes important when we try to measure (or "see") the electron.

2. Wavelength: The distance between nearest identical parts of a wave.

3. As the photon energy increases, the wavelength of light decreases.

4. Optical effect: To see an electron, the wavelength of the light must be very small. In fact, it must be a few times smaller than the size of the electron. Energy effect: As the wavelength of light gets shorter, the photon energy increases.

Combining these effects, the energetic photon needed to see the electron will clearly change the electron's position.

5. The effects of the uncertainty principle are not noticeable in everyday living because the photon energy needed to see large objects is very low compared to the mass of the object. For example, the uncertainty in measuring a one-kilogram object with photons is a very small fraction of the size of an atom's nucleus. This is not noticeable.

6. An orbital is a description of the shape of the space in which the electrons exist. Each orbital can contain up to two electrons.

7. Rule 1: Electrons surrounding a nucleus are arranged in orbitals. Rule 2: An orbital can contain a maximum of two electrons. Rule 3: Electrons like to be as far apart from one another as possible because they have negative charges and like charges repel. Rule 4: Electrons like to be as close to the nucleus as possible because being close to the nucleus means that they need less energy. Rule 5: The farther away from the nucleus, the more room there is for electrons.

8. Energy level, $n = 8$; number of orbitals, $n^2 = 64$; number of electrons, $2n^2 = 128$.

9. The quantum numbers of the two spin states for an electron are $\frac{1}{2}$ and $-\frac{1}{2}$.

10. Two electrons can be in the same orbital *only* if they have opposite spins.

11.

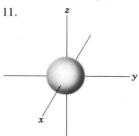

12.

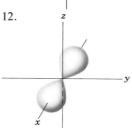 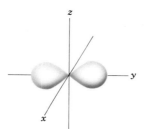 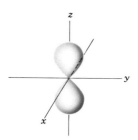

13. a. There are nine *g* orbitals in the fifth and higher energy levels.

b. The nine *g* orbitals can contain up to 18 electrons.

CHAPTER 14

1. A 5*p* orbital.

2. A 6*d* orbital.

3. 4*s*: K, Ca. 3*d*: Sc, Ti, V, Cr, Mn, Fe, Co, Ni, Cu, Zn. 5*f*: Th, Pa, U, Np, Pu, Am, Cm, Bk, Cf, Es, Fm, Md, No, Lr.

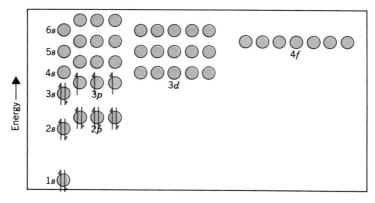

4. P: $1s^2 2s^2 2p_x^2 2p_y^2 2p_z^2 3s^2 3p_x^1 3p_y^1 3p_z^1$

5. $\underset{1s^2}{\uparrow\downarrow}\ \underset{2s^2}{\uparrow\downarrow}\ \underset{2p_x^2}{\uparrow\downarrow}\ \underset{2p_y^2}{\uparrow\downarrow}\ \underset{2p_z^2}{\uparrow\downarrow}\ \underset{3s^2}{\uparrow\downarrow}\ \underset{3p_x^1}{\uparrow}\ \underset{3p_y^1}{\uparrow}$

6. a. This diagram is incorrect because the *p* electrons have different spins.

b. This diagram is correct.

c. This diagram is incorrect because one of the $2p_y$ electrons should be in the $2p_z$ orbital.

d. This diagram is correct. However, we usually draw this with the electrons in the $2p_x$ and $2p_y$ orbitals.

7. Group IIIA: B, Al, Ga, In. Tl. Group VB: V, Nb, Ta. Group VIA: O, S, Se, Te, Po.

8. Period 4: K, Ca, Sc, Ti, V, Cr, Mn, Fe, Co, Ni, Cu, Zn, Ga, Ge, As, Se, Br, Kr. Period 5: Rb, Sr, Y, Zr, Nb, Mo, Tc, Ru, Rh, Pd, Ag, Cd, In, Sn, Sb, Te, I, Xe.

9. $5d$: La, Hf, Ta, W, Re, Os, Ir, Pt, Au, Hg.

10. $5f$: Th, Pa, U, Np, Pu, Am, Cm, Bk, Cf, Es, Fm, Md, No, Lr.

11. Group IIA: 2 electrons. Group VA: 5 electrons. Group VIA: 6 electrons.

12. a. $3p_x^2$: Two electrons in a third energy level p orbital.
b. $4f^7$: Seven electrons in fourth energy level f orbitals.
c. $2s^1$: One electron in a second energy level s orbital.
d. $4d^{10}$: Ten electrons in fourth energy level d orbitals.

13. a. Li; b. C; c. Na; d. S; e. K; f. Co

14. a. C: $\frac{\uparrow\downarrow}{1s^2}\ \frac{\uparrow\downarrow}{2s^2}\ \frac{\uparrow}{2p_x^1}\ \frac{\uparrow}{2p_y^1}$

b. O: $\frac{\uparrow\downarrow}{1s^2}\ \frac{\uparrow\downarrow}{2s^2}\ \frac{\uparrow\downarrow}{2p_x^2}\ \frac{\uparrow}{2p_y^1}\ \frac{\uparrow}{2p_z^1}$

c. Mg: $\frac{\uparrow\downarrow}{1s^2}\ \frac{\uparrow\downarrow}{2s^2}\ \frac{\uparrow\downarrow}{2p_x^2}\ \frac{\uparrow\downarrow}{2p_y^2}\ \frac{\uparrow\downarrow}{2p_z^2}\ \frac{\uparrow\downarrow}{3s^2}$

d. P: $\frac{\uparrow\downarrow}{1s^2}\ \frac{\uparrow\downarrow}{2s^2}\ \frac{\uparrow\downarrow}{2p_x^2}\ \frac{\uparrow\downarrow}{2p_y^2}\ \frac{\uparrow\downarrow}{2p_z^2}\ \frac{\uparrow\downarrow}{3s^2}\ \frac{\uparrow}{3p_x^1}\ \frac{\uparrow}{3p_y^1}\ \frac{\uparrow}{3p_z^1}$

e. Ar: $\frac{\uparrow\downarrow}{1s^2}\ \frac{\uparrow\downarrow}{2s^2}\ \frac{\uparrow\downarrow}{2p_x^2}\ \frac{\uparrow\downarrow}{2p_y^2}\ \frac{\uparrow\downarrow}{2p_z^2}\ \frac{\uparrow\downarrow}{3s^2}\ \frac{\uparrow\downarrow}{3p_x^2}\ \frac{\uparrow\downarrow}{3p_y^2}\ \frac{\uparrow\downarrow}{3p_z^2}$

f. V: $\frac{\uparrow\downarrow}{1s^2}\ \frac{\uparrow\downarrow}{2s^2}\ \frac{\uparrow\downarrow}{2p_x^2}\ \frac{\uparrow\downarrow}{2p_y^2}\ \frac{\uparrow\downarrow}{2p_z^2}\ \frac{\uparrow\downarrow}{3s^2}\ \frac{\uparrow\downarrow}{3p_x^2}\ \frac{\uparrow\downarrow}{3p_y^2}\ \frac{\uparrow\downarrow}{3p_z^2}$
$\frac{\uparrow\downarrow}{4s^2}\ \frac{\uparrow}{3d^1}\ \frac{\uparrow}{3d^1}$

15. a. Li, Na, K, Rb, Cs, Fr.
b. Be, Mg, Ca, Sr, Ba, Ra.
c. F, Cl, Br, I, At.
d. He, Ne, Ar, Kr, Xe, Rn.
e. Ce, Pr, Nd, Pm, Sm, Eu, Gd, Tb, Dy, Ho, Er, Tm, Yb, Lu.
f. Th, Pa, U, Np, Pu, Am, Cm, Bk, Cf, Es, Fm, Md, No, Lr.

16. B, Si, Ge, As, Sb, Te, Po, At.

17. scandium, Sc; titanium, Ti; vanadium, V; chromium, Cr; manganese, Mn; iron, Fe; cobalt, Co; nickel, Ni; copper, Cu; zinc, Zn.

18. carbon, C; silicon, Si; germanium, Ge; tin, Sn; lead, Pb.

19. a. This diagram is incorrect. The $2s$ orbital should have two electrons.
b. This diagram is incorrect. The $2p_x$ orbital should have only one electron.
c. This diagram is incorrect. The $2p_x$ orbital should have only two electrons.
d. This diagram is correct. The usual way of writing it would be to have the $2p$ electrons with spin up.
e. This diagram is correct. The usual way of writing it would be to have the $2p$ electron placed in the $2p_x$ orbital.

CHAPTER 15

1. Li·

2. Mg:

3. K·   :Äs·

4. :Ċl:N̈:Ċl: :Ċl:B̈:Ċl: :Ċl:Ö:Ċl:
   :Ċl: :Ċl:

5. Cl—N—Cl   Cl—B—Cl   Cl—O—Cl
         | |
         Cl   Cl

    :C̈l—N̈—C̈l: :C̈l—B—C̈l: :C̈l—Ö—C̈l:
         | |
        :C̈l: :C̈l:

6. Na → Na$^+$ + e$^-$

7. Ca → Ca^{2+} + 2e$^-$

8. Br + e$^-$ → Br$^-$

9. S + 2e$^-$ → S^{2-}

10. K + Br → KBr. Since bromine occurs in nature as Br$_2$, we should write 2K + Br$_2$ → 2KBr.

11. Ca + Cl$_2$ → CaCl$_2$

12. We mean that electrons tend to exist in the lowest energy level available to them.

13. A molecular orbital is a chemical bond in which the atomic orbitals combine in a special way.

14. Only electrons with opposite spins can be in the same molecular orbital. Since there are only two spin states, up and down, there can be only two electrons in a molecular orbital.

15. A covalent bond is a chemical bond in which two electrons are shared by two atoms.

16. Valence electrons are electrons in the outermost energy level of an atom that are available for forming chemical bonds.

17. Valence electrons are represented as dots. The other electrons, those below the outermost energy level, are not shown.

18. Na·   Mg: :Al·   :Si·
    :P·   :S̈:   :C̈l·   :Ar:

19. a.   Cl   :C̈l:
              | |
         Cl—C—Cl   :C̈l—C—C̈l:
              | |
              Cl   :C̈l:

    b. Br—N—Br   :B̈r—N̈—B̈r:
             | |
             Br   :B̈r:

c. H—S—H   H—S̈—H

d. H—Br   H—B̈r:

e.   O   ·Ö·
        ‖   ‖
    H—C—H   H—C—H
                          :Ï—Ï:

f. I—I   :Ï—Ï:

20. a.   H H
           | |
       H—C—C—H
           | |
           H H

    b. H   H
          \   /
            C═C
          /   \
        H   H

    c. H—C≡C—H

21. :C̈l:P̈:C̈l:
        :C̈l:

22.   :F̈:
    :F̈ :S̈: F̈:
         :F̈:

23. An ionic bond is a bond resulting from the electrostatic attraction between positive and negative ions.

24. Al → Al^{3+} + 3e$^-$

25. Se + 2e$^-$ → Se^{2-}

26. a. 2Al + 3Cl$_2$ → 2AlCl$_3$
    b. 4K + O$_2$ → 2K$_2$O
    c. Be + Br$_2$ → BeBr$_2$
    d. Ba + I$_2$ → BaI$_2$
    e. 16Li + S$_8$ → 8Li$_2$S (Notice that elemental sulfur forms molecules with eight atoms.)
    f. 2Al + 3H$_2$ → 2AlH$_3$

27. a. BeF$_2$ or BeH$_2$; b. OF$_2$ or H$_2$O; c. BF$_3$; d. NH$_3$; e. CH$_4$.

28. a. PCl$_5$; b. SF$_6$; c. ClF$_5$.

CHAPTER 16

1. CsI, cesium iodide

2. $MgBr_2$, magnesium bromide

3. $AlCl_3$, aluminum chloride

4. Ca_3N_2, calcium nitride

5. Al_2S_3, aluminum sulfide

6. Na_2S, sodium sulfide; MgI_2, magnesium iodide; NaH, sodium hydride; RaO, radium oxide; Al_2S_3, aluminum sulfide; NH_4F, ammonium fluoride

7. rubidium nitrate, $RbNO_3$; magnesium oxide, MgO; calcium chloride, $CaCl_2$; sodium oxide, Na_2O; beryllium hydride, BeH_2; barium bromide, $BaBr_2$; aluminum oxide, Al_2O_3

8. potassium nitrate, KNO_3; magnesium hypochlorite, $Mg(ClO)_2$; lithium permanganate, $LiMnO_4$; barium acetate, $Ba(C_2H_3O_2)_2$; calcium hydroxide, $Ca(OH)_2$; cesium peroxide, Cs_2O_2; beryllium perchlorate, $Be(ClO_4)_2$; sodium dichromate, $Na_2Cr_2O_7$; ammonium carbonate, $(NH_4)_2CO_3$; potassium hydrogen sulfite, $KHSO_3$

9. KNO_2; potassium nitrite; Na_2O_2, sodium peroxide; $Mg(OH)_2$, magnesium hydroxide; $Ca(ClO_2)_2$, calcium chlorite; K_3PO_4, potassium phosphate; NaCN, sodium cyanide, $Rb_2S_2O_3$, rubidium thiosulfate; BaC_2O_4, barium oxalate; Li_2CrO_4, lithium chromate

10. gold(III) chloride, $AuCl_3$; nickel(II) hydroxide, $Ni(OH)_2$; stannic fluoride, SnF_4; cadmium thiocyanate, $Cd(SCN)_2$; lead(IV) sulfate, $Pb(SO_4)_2$; chromous carbonate, $CrCO_3$; copper(II) phosphate, $Cu_3(PO_4)_2$; mercurous nitrate, $Hg_2(NO_3)_2$; silver hydrogen carbonate, $AgHCO_3$

11. CoS, cobalt(II) sulfide; $Fe_2(SO_4)_3$, iron(III) sulfate; PbO, lead(II) oxide; $Ni(ClO_4)_3$, nickel(III) perchlorate; $CdBr_2$, cadmium bromide; Cu_2O, copper(I) oxide; $Mn(NO_3)_2$, manganese(II) nitrate; $HgSO_4$, mercury(II) sulfate; $Cr_3(AsO_4)_2$, chromium(II) arsenate; $Au(H_2PO_4)_3$, gold(III) dihydrogen phosphate

12. sulfur dioxide, SO_2; sulfur trioxide, SO_3; hydrogen selenide, H_2Se; dichlorine monoxide, Cl_2O; phosphorus trichloride, PCl_3; carbon tetrafluoride, CF_4; tetraboron decahydride, B_4H_{10}

13. PCl_5, phosphorus pentachloride; CO, carbon monoxide; CO_2, carbon dioxide; N_2O_4, dinitrogen tetraoxide; P_4O_6, tetraphosphorus hexaoxide; I_2O_5, diiodine pentaoxide; BrF_3, bromine trifluoride; Sb_2O_5, diantimony pentaoxide

14. a. CaF_2, calcium fluoride
    b. $AlCl_3$, aluminum chloride
    c. MgO, magnesium oxide
    d. $SrCl_2$, strontium chloride
    e. $CoCl_2$, cobalt(II) chloride
    f. $CoCl_3$, cobalt(III) chloride

15. a. NO, nitric oxide (a common name)
    b. N_2O_3, dinitrogen trioxide
    c. N_2O_5, dinitrogen pentaoxide
    d. SF_4, sulfur tetrafluoride
    e. SF_2, sulfur difluoride
    f. SF_6, sulfur hexafluoride

16. a. SO_3, sulfur trioxide
    b. CO, carbon monoxide
    c. SiF_4, silicon tetrafluoride
    d. N_2O_5, dinitrogen pentaoxide
    e. P_4S_{10}, tetraphosphorus decasulfide
    f. XeF_4, xenon tetrafluoride

17. a. PCl_5, phosphorus pentachloride
    b. IF_7, iodine heptafluoride
    c. P_4O_6, tetraphosphorus hexaoxide
    d. As_4O_{10}, tetraarsenic decaoxide
    e. Cl_2O_3, dichlorine trioxide
    f. Cl_2O_7, dichlorine heptaoxide

18. a. H_3PO_4, phosphoric acid
    b. HNO_2, nitrous acid
    c. H_2CO_3, carbonic acid
    d. $HClO_3$, chloric acid
    e. HIO_3, iodic acid
    f. HNO_3, nitric acid
    g. H_3PO_4, phosphoric acid

19. a. HBr, hydrogen bromide
    b. HBrO, hypobromous acid
    c. $HBrO_2$, bromous acid
    d. $HBrO_3$, bromic acid
    e. $HBrO_4$, perbromic acid

20. a. $PbSO_3$, lead (II) sulfite
    b. Mg_2SiO_4, magnesium silicate
    c. Na_3PO_4, sodium phosphate
    d. $Ca(NO_3)_2$, calcium nitrate
    e. $LiBrO$, lithium hypobromite
    f. NH_4ClO_4, ammonium perchlorate
    g. $Fe_2(SO_4)_3$, iron(III) sulfate
    h. $FeSO_4$, iron(II) sulfate
    i. $CaCO_3$, calcium carbonate
    j. $Ba(NO_2)_2$, barium nitrite

21. a. iron(III) sulfate, $Fe_2(SO_4)_3$
    b. chromium(II) sulfite, $CrSO_3$
    c. cupric sulfide, CuS
    d. silver dihydrogen phosphate, AgH_2PO_4
    e. tellurium hexafluoride, TeF_6
    f. mercury(I) acetate, $Hg_2(C_2H_3O_2)_2$
    g. disulfur dichloride, S_2Cl_2
    h. ammonium nitrate, NH_4NO_3

22. a. cobalt(III) sulfide, Co_2S_3
    b. bromine pentafluoride, BrF_5
    c. ferrous carbonate, $FeCO_3$
    d. gold(I) sulfate, Au_2SO_4
    e. calcium bicarbonate, $Ca(HCO_3)_2$
    f. tetraphosphorus trisulfide, P_4S_3
    g. cuprous cyanide, $CuCN$
    h. lanthanum(III) phosphate, $LaPO_4$

23. a. iron(II) oxide, FeO
    b. potassium oxide, K_2O
    c. magnesium nitride, Mg_3N_2
    d. magnesium nitrite, $Mg(NO_2)_2$
    e. pentasulfur dinitride, S_5N_2
    f. calcium permanganate, $Ca(MnO_4)_2$
    g. ferric sulfite, $Fe_2(SO_3)_3$
    h. lead(II) chromate, $PbCrO_4$
    i. silver dichromate, $Ag_2Cr_2O_7$

24. Na_3N

25. a. XeF_6, xenon hexafluoride; XeF_4, xenon tetrafluoride; XeF_2, xenon difluoride
    b. krypton difluoride, KrF_2; xenon trioxide, XeO_3; xenon tetraoxide, XeO_4

26. Both N and P have three unpaired electrons in their valence shell. These two elements can combine to form a compound with bonding similar to that of N_2; this compound is PN, which is called phosphorus nitride. See Section 16-2 for the list giving the order of the elements when naming this type of compound.

CHAPTER 17

1. oxygen

2. oxygen

3. H has given an electron. F has taken an electron. H has the more positive oxidation number.

4. The oxidation numbers are all zero, since these are all uncombined elements.

5. For H_2O, CO, HNO_3, $CaCl_2$, H_2SO_4, NH_3, and PCl_3: The sum of the oxidation numbers is zero. For HSO_2^- and Br^-: The sum of the oxidation numbers is -1. For PO_4^{3-}: The sum of the oxidation numbers is -3. For NO^+ and NH_4^+: The sum of the oxidation numbers is $+1$.

6. $\overset{+1\ -1}{\text{H Br}}$

7. $\overset{+1\ -1}{\text{Li H}}$

8. $\overset{+3\ -3}{\text{Al N}}$

9. $\overset{\overset{+2\ -2}{+1\ -2}}{\text{H}_2\ \text{S}}$

10. $\overset{\overset{+6\ -6}{+6\ -2}}{\text{Cl O}_3}$

11. $\overset{\overset{+4\ -4}{+4\ -2}}{\text{ClO}_2}$

12. $\overset{\overset{+2\ -2}{+1\ -2}}{\text{H}_2\text{Se}}$

13. $\overset{\overset{+7\ -8\ =\ -1}{+7\ -2}}{\text{ClO}_4^-}$

14. $\overset{\overset{+10\ -12\ =\ -2}{+2.5\ -2}}{\text{S}_4\text{O}_6^{2-}}$

15. $\overset{\overset{+3\ -3}{+3\ -1}}{\text{Cl F}_3}$

16. $\overset{+1\ +5\ -6}{\underset{+1\ +5\ -2}{\text{Na Cl O}_3}}$

17. $\overset{+2\ -2}{\underset{+1\ -1}{\text{Na}_2\text{O}_2}}$

18.

Left-hand C $= -1$; right-hand C $= +3$; F $= -1$; each O $= -2$; each H $= +1$. The sum of the oxidation numbers of all the atoms is zero.

19. a. $\overset{+1 \; -1}{Br \; Cl}$ c. $\overset{+3 \; -3}{\underset{+3 \; -1}{Br \; F_3}}$ e. $\overset{+1 \; -1}{Cl \; F}$ e. $\overset{+3 \; -4 \; = \; -1}{\underset{+3 \; -1}{Au \; Cl_4^-}}$ f. $\overset{+1 \; -1}{Au \; Cl}$

b. $\overset{+1 \; -1}{Br \; F}$ d. $\overset{+5 \; -5}{\underset{+5 \; -1}{Br \; F_5}}$ f. $\overset{+5 \; -1}{\underset{+5 \; -1}{Cl \; F_5}}$

25. a.

H   H
H—C—C—O—H
H   H

Left-hand C = −3; right C = −1;
all H = +1; O = −2.

20. a. $\overset{+1 \; +7 \; -8}{\underset{+1 \; +7 \; -2}{H \; Cl \; O_4}}$ d. $\overset{+1 \; +1 \; -2}{H \; Cl \; O}$

b. $\overset{+1 \; +5 \; -6}{\underset{+1 \; +5 \; -2}{H \; Cl \; O_3}}$ e. $\overset{0}{Cl_2}$

b.

H   O
H—C—C
H   H

Left-hand C = −3; right-hand C =
+1; all H = +1; O = −2.

c. $\overset{+1 \; +3 \; -4}{\underset{+1 \; +3 \; -2}{H \; Cl \; O_2}}$ f. $\overset{+1 \; -1}{H \; Cl}$

21. a. $\overset{+2 \; -2}{\underset{+1 \; -2}{N_2 \; O}}$ e. $\overset{+10 \; -10}{\underset{+5 \; -2}{N_2 \; O_5}}$

c.

H   O
H—C—C
H   O—H

b. $\overset{+2 \; -2}{N \; O}$ f. $\overset{-3 \; +3}{\underset{-3 \; +1}{N \; H_3}}$

Left-hand C = −3; right-hand C =
+3; all H = +1; each O = −2.

c. $\overset{+6 \; -6}{\underset{+3 \; -2}{N_2 O_3}}$ g. $\overset{-1 \; +2 \; -2 \; +1}{\underset{-1 \; +1 \; -2 \; +1}{N \; H_2 O \; H}}$

d. $\overset{+4 \; -4}{\underset{+4 \; -2}{N \; O_2}}$ h. $\overset{-4 \; +1}{\underset{-2 \; +1}{N_2 \; H_4}}$

26. a.

H   H
H—C—O—C—H
H   H

All C = −2; oxygen = −2; all H
= +1.

22. a. $\overset{+3 \; +5 \; -8}{\underset{+1 \; +5 \; -2}{H_3 \; P \; O_4}}$ d. $\overset{+5 \; -8 \; = \; -3}{\underset{+5 \; -2}{P \; O_4{}^{3-}}}$

b. $\overset{+2 \; +5 \; -8 \; = \; -1}{\underset{+1 \; +5 \; -2}{H_2 \; P \; O_4^-}}$ e. $\overset{+3 \; +1 \; -4}{\underset{+1 \; +1 \; -2}{H_3 \; P \; O_2}}$

c. $\overset{+1 \; +5 \; -8 \; = \; -2}{\underset{+1 \; +5 \; -2}{H \; P \; O_4{}^{2-}}}$ f. $\overset{+2 \; +1 \; -4 \; = \; -1}{\underset{+1 \; +1 \; -2}{H_2 \; P O_2^-}}$

b.

F   H
F—C—C—H
F   H

All F = −1; all H = +1; left-hand
C = +3; right-hand C = −3.

23. a. $\overset{+2 \; -2}{Mn \; O}$ d. $\overset{+2 \; -2}{\underset{+2 \; -1}{Mn \; Cl_2}}$

b. $\overset{+6 \; -6}{\underset{+3 \; -2}{Mn_2 \; O_3}}$ e. $\overset{+3 \; -3}{\underset{+3 \; -1}{Mn \; Cl_3}}$

c. $\overset{+14 \; -14}{\underset{+7 \; -2}{Mn_2 \; O_7}}$ f. $\overset{+1 \; +7 \; -8}{\underset{+1 \; +7 \; -2}{K \; Mn \; O_4}}$

24. a. $\overset{+1 \; -1}{Cu \; Cl}$ c. $\overset{+2 \; -2}{\underset{+1 \; -1}{Hg_2 \; Cl_2}}$

b. $\overset{+2 \; -2}{\underset{+2 \; -1}{Cu \; Cl_2}}$ d. $\overset{+2 \; -2}{\underset{+2 \; -1}{Hg \; Cl_2}}$

c.

$$H \overset{+-}{\text{---}} \overset{H}{\underset{H}{\overset{|}{\text{C}}}} \overset{00}{\text{---}} C \overset{+-}{\equiv} N$$

All H = +1; N = −3; left-hand C
= −3; right-hand C = +3.

d.

$$H \overset{+-}{\text{---}} \overset{H}{\underset{H}{\overset{|}{\text{C}}}} \overset{00}{\text{---}} \overset{H}{\underset{H}{\overset{|}{\text{C}}}} \overset{+-}{\text{---}} N \overset{H}{\underset{H}{\diagdown}}$$

Left-hand C = −3; right-hand C =
−1; all H = +1; N = −3.

27. −1

−1/2

O_2^-

28. $:\overset{\cdot\cdot}{\underset{\cdot}{S}}\cdot$ Sulfur can "attract" as many as
two electrons or "give" as many
as six electrons.

29. H

$$\overset{+}{\underset{+}{\diagup}}\overset{\diagdown}{}C \overset{+-}{\equiv} O$$

C = 0; all H = +1;
O = −2.

CHAPTER 18

1. Yes, the reaction *is* a redox reaction,
because there is a change in oxidation
number.

2. No, the reaction is *not* a redox reaction, because there is no change in oxidation number.

3. $Li \rightarrow Li^+ + e^-$ and $Ca^{2+} + 2e^- \rightarrow Ca$

4. The balanced molecular equation is
$2Li + CaCl_2 \rightarrow 2LiCl + Ca$. The balanced ionic equation is $2Li + Ca^{2+} +$
$2Cl^- \rightarrow 2Li^+ + 2Cl^- + Ca$.

5.

$$1 \times [Cu \rightarrow Cu^+ + 2e^-]$$
$$2 \times [e^- + NO_3^- + 2H^+ \rightarrow NO_2 + H_2O]$$
$$\overline{Cu + 2NO_3^- + 4H^+ \rightarrow Cu^{2+} + 2NO_2 + 2H_2O}$$

6.

$$5 \times [C_2O_4^{2-} \rightarrow 2CO_2 + 2e^-]$$
$$2 \times [5e^- + MnO_4^- + 8H^+ \rightarrow Mn^{2+} + 4H_2O]$$
$$\overline{5C_2O_4^{2-} + 2MnO_4^- + 16H^+ \rightarrow 10CO_2 + 2Mn^{2+} + 8H_2O}$$

7.

$$8 \times [3e^- + MnO_4^- + 2H_2O \rightarrow MnO_2 + 4OH^-]$$
$$3 \times [NH_3 + 9OH^- \rightarrow NO_3^- + 6H_2O + 8e^-]$$
$$\overline{8MnO_4^- + 16H_2O + 3NH_3 + 27OH^- \rightarrow 8MnO_2 + 32OH^- + 3NO_3^- + 18H_2O}$$
$$8MnO_4^- + 3NH_3 \rightarrow 8MnO_2 + 5OH^- + 3NO_3^- + 2H_2O$$

8. $2 \times [e^- + NO_2^- + H_2O \rightarrow NO + 2OH^-]$
$$1 \times [2I^- \rightarrow I_2 + 2e^-]$$
$$\overline{2NO_2^- + 2H_2O + 2I^- \rightarrow 2NO + 4OH^- + I_2}$$

9.

SUBSTANCE	TYPE OF REACTION	SUBSTANCE IS	TYPE OF AGENT
Li	oxidation	oxidized	reducing agent
Ca^{2+}	reduction	reduced	oxidizing agent

10.

SUBSTANCE	TYPE OF REACTION	SUBSTANCE IS	TYPE OF AGENT
$C_2O_4^{2-}$	oxidation	oxidized	reducing agent
MnO_4^-	reduction	reduced	oxidizing agent

11. a.

$$Ni \rightarrow Ni^{2+} + 2e^-$$
$$\underline{F_2 + 2e^- \rightarrow 2F^-}$$
$$Ni + F_2 \rightarrow Ni^{2+} + 2F^-$$

   b.

$$2 \times [Fe \rightarrow Fe^{2+} + 2e^-]$$
$$\underline{1 \times [O_2 + 4e^- \rightarrow 2O^{-2}]}$$
$$2Fe + O_2 \rightarrow 2Fe^{2+} + 2O^{2-}$$

   c.

$$2 \times [Co^{2+} \rightarrow Co^{3+} + e^-]$$
$$\underline{1 \times [Cl_2 + 2e^- \rightarrow 2Cl^-]}$$
$$2Co^{2+} + Cl_2 \rightarrow 2Co^{3+} + 2Cl^-$$

12. a.

$$1 \times [6e^- + ClO_3^- + 6H^+ \rightarrow Cl^- + 3H_2O]$$
$$\underline{3 \times [SO_3^{2-} + H_2O \rightarrow SO_4^{2-} + 2H^+ + 2e^-]}$$
$$ClO_3^- + 6H^+ + 3SO_3^{2-} + 3H_2O \rightarrow Cl^- + 3H_2O + 3SO_4^{2-} + 6H^+$$
$$ClO_3^- + 3SO_3^{2-} \rightarrow Cl^- + 3SO_4^{2-}$$

   b.

$$2e^- + MnO_2 + 4H^+ \rightarrow Mn^{2+} + 2H_2O$$
$$\underline{2I^- \rightarrow I_2 + 2e^-}$$
$$MnO_2 + 4H^+ + 2I^- \rightarrow Mn^{2+} + 2H_2O + I_2$$

   c.

$$1 \times [6e^- + Cr_2O_7^{2-} + 14H^+ \rightarrow 2Cr^{3+} + 7H_2O]$$
$$\underline{3 \times [2I^- \rightarrow I_2 + 2e^-]}$$
$$Cr_2O_7^{2-} + 14H^+ + 6I^- \rightarrow 2Cr^{3+} + 7H_2O + 3I_2$$

13. $2 \times [5e^- + MnO_4^- + 8H^+ \rightarrow Mn^{2+} + 4H_2O]$

$\underline{\qquad\qquad 5 \times [H_2S \rightarrow S + 2H^+ + 2e^-] \qquad\qquad}$

$2MnO_4^- + 16H^+ + 5H_2S \rightarrow 2Mn^{2+} + 8H_2O + 5S + 10H^+$

S is oxidized. Mn is reduced. MnO_4^- is the oxidizing agent. H_2S is the reducing agent.

14. a. $4 \times [MnO_4^- + e^- \rightarrow MnO_4^{2-}]$

$\underline{\qquad\quad 1 \times [4OH^- \rightarrow O_2 + 2H_2O + 4e^-] \qquad\quad}$

$4MnO_4^- + 4OH^- \rightarrow 4MnO_4^{2-} + O_2 + 2H_2O$

b. $\qquad\quad 2 \times [ClO_2 + e^- \rightarrow ClO_2^-]$

$\underline{\quad 1 \times [H_2O_2 + 2OH^- \rightarrow O_2 + 2H_2O + 2e^-] \quad}$

$2ClO_2 + H_2O_2 + 2OH^- \rightarrow 2ClO_2^- + O_2 + 2H_2O$

c. $\qquad\quad 2e^- + Cl_2 \rightarrow 2Cl^-$

$\underline{\quad IO_3^- + 2OH^- \rightarrow IO_4^- + H_2O + 2e^- \quad}$

$Cl_2 + IO_3^- + 2OH^- \rightarrow 2Cl^- + IO_4^- + H_2O$

15. a. $\qquad Mn^{2+} + 4OH^- \rightarrow MnO_2 + 2H_2O + 2e^-$

$\underline{\qquad\qquad H_2O_2 + 2e^- \rightarrow 2OH^- \qquad\qquad}$

$Mn^{2+} + 2OH^- + H_2O_2 \rightarrow MnO_2 + 2H_2O$

b. $\qquad 1 \times [Bi(OH)_3 + 3e^- \rightarrow Bi + 3OH^-]$

$\underline{\quad 3 \times [SnO_2^- + 2OH^- \rightarrow SnO_3^{2-} + H_2O + e^-] \quad}$

$Bi(OH)_3 + 3SnO_2^- + 3OH^- \rightarrow Bi + 3SnO_3^{2-} + 3H_2O$

c. $2e^- + MnO_2 + 2H_2O \rightarrow Mn(OH)_2 + 2OH^-$

$\underline{\qquad SO_3^{2-} + 2OH^- \rightarrow SO_4^{2-} + H_2O + 2e^- \qquad}$

$MnO_2 + H_2O + SO_3^{2-} \rightarrow Mn(OH)_2 + SO_4^{2-}$

16. $\quad 14 \times [5e^- + MnO_4^- + 8H^+ \rightarrow Mn^{2+} + 4H_2O]$

$\underline{\quad 5 \times [2CH_3Cl + 4H_2O \rightarrow Cl_2 + 2CO_2 + 14H^+ + 14e^-] \quad}$

$14MnO_4^- + 42H^+ + 10CH_3Cl \rightarrow 14Mn^{2+} + 36H_2O + 5Cl_2 + 10CO_2$

17. $1 \times [12e^- + P_4 + 12H_2O \rightarrow 4PH_3 + 12OH^-]$

$\underline{\qquad 3 \times [P_4 + 8OH^- \rightarrow 4H_2PO_2^- + 4e^-] \qquad}$

$4P_4 + 12H_2O + 12OH^- \rightarrow 4PH_3 + 12H_2PO_2^-$

18. a. $3 \times [5e^- + MnO_4^- + 4H_2O \rightarrow Mn^{2+} + 8OH^-]$

$\underline{\quad 5 \times [V^{2+} + 4OH^- \rightarrow VO_2^+ + 2H_2O + 3e^-] \quad}$

$3MnO_4^- + 2H_2O + 5V^{2+} \rightarrow 3Mn^{2+} + 4OH^- + 5VO_2^+]$

b.
$$4 \times [Fe^{2+} \rightarrow Fe^{3+} + e^-]$$
$$\underline{1 \times [4e^- + ClO_2^- + 2H_2O \rightarrow Cl^- + 4OH^-]}$$
$$4Fe^{2+} + ClO_2^- + 2H_2O \rightarrow 4Fe^{3+} + Cl^- + 4OH^-$$

c. $1 \times [8e^- + NO_3^- + 7H_2O \rightarrow NH_4^+ + 10OH^-]$
$$\underline{4 \times [Zn \rightarrow Zn^{2+} + 2e^-]}$$
$$NO_3^- + 7H_2O + 4Zn \rightarrow NH^+ + 10OH^- + 4Zn^{2+}$$

19. a. $1 \times [6e^- + Cr_2O_7^{2-} + 14H^+ \rightarrow 2Cr^{3+} + 7H_2O]$
$$\underline{3 \times [C_2O_4^{2-} \rightarrow 2CO_2 + 2e^-]}$$
$$Cr_2O_7^{2-} + 14H^+ + 3C_2O_4^{2-} \rightarrow 2Cr^{3+} + 7H_2O + 6CO_2$$

b. $1 \times [2e^- + H_2O_2 + 2H^+ \rightarrow H_2O + H_2O]$
$$\underline{2 \times [Fe^{2+} \rightarrow Fe^{3+} + e^-]}$$
$$H_2O_2 + 2H^+ + 2Fe^{2+} \rightarrow 2H_2O + 2Fe^{3+}$$

c.
$$3 \times [2Sb + 5H_2O \rightarrow Sb_2O_5 + 10H^+ + 10e^-]$$
$$\underline{10 \times [3e^- + NO_3^- + 4H^+ \rightarrow NO + 2H_2O]}$$
$$6Sb + 10NO_3^- + 10H^+ \rightarrow 3Sb_2O_5 + 10NO + 5H_2O$$

20.
$$Zn \rightarrow Zn^{2+} + 2e^-$$
$$\underline{2MnO_2 + 2H_2O + 2e^- \rightarrow Mn_2O_3 + H_2O + 2OH^-}$$
$$Zn + 2MnO_2 + H_2O \rightarrow Zn^{2+} + Mn_2O_3 + 2OH^-$$

Now insert the NH_4^+ and the NH_3. Use H^+ to balance H, since this is easy to do and the H^+ will disappear in the end anyway:

$$Zn + 2MnO_2 + H_2O + NH_4^+ \rightarrow Zn^{2+} + Mn_2O_3 + 2OH^- + NH_3 + H^+$$

Combine H^+ and OH^- to give H_2O:

$$Zn + 2MnO_2 + H_2O + NH_4^+ \rightarrow Zn^{2+} + Mn_2O_3 + OH^- + NH_3 + H_2O$$

Subtract H_2O from each side:

$$Zn + 2MnO_2 + NH_4^+ \rightarrow Zn^{2+} + Mn_2O_3 + OH^- + NH_3$$

21. $Zn + H_2O + 2OH^- \rightarrow ZnO + 2H_2O + 2e^-$
$$\underline{HgO + 2H_2O + 2e^- \rightarrow Hg + H_2O + 2OH^-}$$
$$Zn + HgO \rightarrow ZnO + Hg$$

Notice that the equation was balanced as given.

22.
$$Pb + SO_4^{2-} \rightarrow PbSO_4 + 2e^-$$
$$\underline{PbO_2 + 4H^+ + SO_4^{2-} + 2e^- \rightarrow PbSO_4 + 2H_2O}$$
$$Pb + PbO_2 + 4H^+ + 2SO_4^{2-} \rightarrow 2PbSO_4 + 2H_2O$$

To get the molecular equation, combine the H^+ and the SO_4^{2-} to give H_2SO_4 (sulfuric acid). The liquid in the lead–acid battery in an automobile is about 10 M sulfuric acid:
$$Pb + PbO_2 + 2H_2SO_4 \rightarrow 2PbSO_4 + 2H_2O$$

23. $$2 \times [H_2 + 2OH^- \rightarrow 2H_2O + 2e^-]$$

$$\underline{1 \times [4e^- + O_2 + 2H_2O \rightarrow 4OH^-]}$$

$$2H_2 + O_2 \rightarrow 2H_2O$$

24. $$5 \times [O_2 + 2H_2O + 4e^- \rightarrow 4OH^-]$$

$$\underline{1 \times [C_3H_8 + 20OH^- \rightarrow 3CO_2 + 14H_2O + 20e^-]}$$

$$C_3H_8 + 5O_2 \rightarrow 3CO_2 + 4H_2O$$

25. $$Cd + 2OH^- \rightarrow Cd(OH)_2 + 2e^-$$

$$\underline{2e^- + NiO_2 + 2H_2O \rightarrow Ni(OH)_2 + 2OH^-}$$

$$Cd + NiO_2 + 2H_2O \rightarrow Cd(OH)_2 + Ni(OH)_2$$

26. $$2 \times [Li \rightarrow Li^+ + e^-]$$

$$\underline{1 \times [I_2 + 2e^- \rightarrow 2I^-]}$$

$$2Li + I_2 \rightarrow 2Li^+ + 2I^-$$

27.

$$Na^{+-}C{\equiv}N:$$

There are $+3 - 1 = +2$ charges around the C atom. There are -3 charges around the N atom. The oxidation number of C is $+2$, and the oxidation number of N is -3.

CHAPTER 19

1. $[H^+] = 2.51 \times 10^{-7}$ M

2. log $10,000,000 = 7$ and log $100,000,000 = 8$

3. log $0.00001 = -5$ and log $0.000001 = -6$

4. log $962 = 2.9832$ and log $7.76 \times 10^5 = 5.8899$

5. log $0.087 = -1.0605$ and log $3.24 \times 10^{-6} = -5.4896$

6. You get an ERROR message. Logs of negative numbers don't exist.

7. antilog $4 = 10^4 = 10,000$

8. antilog $0.86 = 10^{0.86} = 7.24$ and antilog $4.66 = 10^{4.66} = 45,709$

9. antilog $(-0.67) = 10^{-0.67} = 0.214$ and antilog $(-2.33) = 10^{-2.33} = 0.00468$

10. antilog $7.3 = 10^{7.3} = 1.995 \times 10^7$ and antilog $(-7.3) = 10^{-7.3} = 5.01 \times 10^{-8}$

11. **8.91, EE, +/−, 8, log, +/−**   pH = 7.05

12. pH = 11.80

13. pH = 2.40

14. $[H^+] = 3.16 \times 10^{-7}$

15. $[H^+] = 3.98 \times 10^{-8}$

16. a. 8   c. −8
    b. 0

17. a. 2.515   d. 23.991
    b. 3.917   e. 13.813
    c. 3.678   f. 9,623

18. a. −0.873   d. −9.265
    b. −2.331   e. −17.036
    c. −4.206   f. −22.996

19. a. 1×10^5 c. 1×10^{-5}
    b. 1

20. a. 1660
    b. 3.80×10^7
    c. 9.77×10^{10}

    d. 5.75×10^{13}
    e. 2.70
    f. 2.00

21. a. 0.500
    b. 2.95×10^{-5}
    c. 0.987

    d. 3.55×10^{-12}
    e. 4.27×10^{-23}
    f. 0.0776

22. pH = 4.10

23. pH = 3.50

24. $[H^+] = 7.94 \times 10^{-6}$ M

25. $[H^+] = 1.26 \times 10^{-7}$ M

INDEX

Page numbers in **boldface** indicate photographs.